AF846086

Pour Nicolas Chemanoff,

Ce résultat de nos recherches sur les politiques publiques ... de recherche !

Avec toute mon amitié

R. Justan

RESEARCH AND INNOVATION POLICIES IN THE NEW GLOBAL ECONOMY

New Horizons in the Economics of Innovation

General Editor: Christopher Freeman, *Emeritus Professor of Science Policy, SPRU – Science and Technology Policy Research, University of Sussex, UK*

Technical innovation is vital to the competitive performance of firms and of nations and for the sustained growth of the world economy. The economics of innovation is an area that has expanded dramatically in recent years and this major series, edited by one of the most distinguished scholars in the field, contributes to the debate and advances in research in this most important area.

The main emphasis is on the development and application of new ideas. The series provides a forum for original research in technology, innovation systems and management, industrial organization, technological collaboration, knowledge and innovation, research and development, evolutionary theory and industrial strategy. International in its approach, the series includes some of the best theoretical and empirical work from both well-established researchers and the new generation of scholars.

Titles in the series include:

Knowledge, Spillovers and Economic Growth
Regional Growth Differentials across Europe
M.C.J. Caniëls

Successful Innovation
Towards a New Theory for the Management of Small and Medium-Sized Enterprises
Jan Cobbenhagen

Industrial Diversification and Innovation
An International Study of the Aerospace Industry
François Texier

Trade Specialisation, Technology and Economic Growth
Theory and Evidence from Advanced Countries
Keld Laursen

Firm Size, Innovation and Market Structure
The Evolution of Industry, Concentration and Instability
Mariana Mazzucato

Knowledge Flows in National Systems of Innovation
A Comparative Analysis of Sociotechnical Constituencies in Europe and Latin America
Edited by Roberto E. López-Martínez and Andrea Piccaluga

The Anatomy and Evolution of Industries
Technological Change and Industrial Dynamics
Orietta Marsili

Managing New Product Development and Innovation
A Microeconomic Toolbox
Hariolf Grupp and Shlomo Maital

Technological Learning and Competitive Performance
Paulo N. Figueiredo

Research and Innovation Policies in the New Global Economy
An International Comparative Analysis
Edited by Philippe Larédo and Philippe Mustar

RESEARCH AND INNOVATION POLICIES IN THE NEW GLOBAL ECONOMY

An International Comparative Analysis

Edited by

PHILIPPE LARÉDO

and

PHILIPPE MUSTAR

CENTRE DE SOCIOLOGIE DE L'INNOVATION
ÉCOLE NATIONALE SUPÉRIEURE DES MINES DE PARIS, FRANCE

NEW HORIZONS IN THE ECONOMICS OF INNOVATION

Edward Elgar
Cheltenham, UK • Northampton MA, USA

Published by
Edward Elgar Publishing Limited
Glensanda House
Montpellier Parade
Cheltenham
Glos GL50 1 UA
UK

Edward Elgar Publishing, Inc.
136 West Street
Suite 202
Northampton
Massachusetts 01060
USA

A catalogue record for this book
is available from the British Library

Library of Congress Cataloguing in Publication data

Research and innovation policies in the new global economy: an international comparative analysis / edited by Philippe Larédo and Philippe Mustar.
p. cm. – (New horizons in the economics of innovation)
Includes bibliographical references and index.
ISBN 1-84064-784-1
1. Technological innovations–Government policy–Case studies. 2. Technology and state–Case studies. I. Larédo, P. (Philippe) II. Mustar, Philippe. III. Series.

HC79.T4 R465 2001
338'.064–dc21

2001033075

ISBN 1-84064-784-1

Printed and bound in Great Britain by MPG Books Ltd, Bodmin, Cornwall

CONTENTS

CONTRIBUTORS

BARRY BOZEMAN is Professor of Public Policy, Georgia Tech, and director of the Research Value Mapping Program. He specializes in R&D evaluation and technology transfer and development.

PARASKEVAS CARACOSTAS, economist, is Advisor to the Director-General, Research Directorate General at the European Commission. He recently published, with Ugur Muldur, *Society, the Endless Frontier*.

SUNG CHUL CHUNG is Senior Research Fellow at Science and Technology Policy Institute (STEPI), Seoul, Korea. As an economist, he has been engaged in S&T policy research for 15 years. His current research interests are in technology and international trade and Chaebol's innovation behavior.

JAMES S. DIETZ is Senior Research Associate, Research Value Mapping Program, Georgia Tech. His research interests include scientific and technical human capital theory and research evaluation. He previously served as a Program Officer at the National Science Foundation.

LUKE GEORGHIOU is Professor of Science and Technology Policy and Management at the University of Manchester and Director of PREST, an institute of the University. His research interests are in the fields of science technology and innovation policy, evaluation and foresight.

PHILIPPE LARÉDO is *Directeur de recherche* at the *Centre de Sociologie de l'Innovation* of the *École Nationale Supérieure des Mines de Paris*. He specializes in science, technology and innovation policies, and in the organisation and dynamics of research collectives.

BAREND VAN DER MEULEN is Senior Researcher at the Centre for Studies of Science, Technology and Society. He has published on science policy, engineering sciences, and Europeanisation of research systems. In his studies, recent advances in science and technology studies and in policy sciences are used to further understanding the dynamics of research systems.

FRIEDER MEYER-KRAHMER is the director of the Fraunhofer-Institute for Systems and Innovation Research (ISI), Karlsruhe, and Professor of Innovation Economics at the University Louis Pasteur, Strasbourg. His research subjects are technology and innovation policy, evaluation of governmental R&D programmes, barriers to innovation in industry, innovation in small and medium-sized firms, innovation and regional development.

DAVID C. MOWERY is Milton W. Terrill Professor of Business at the Walter A. Haas School of Business at the University of California, Berkeley, and Director of the Haas School's Ph.D. program. His research deals with the economics of technological innovation and with the effects of public policies on innovation.

UGUR MULDUR, economist, is Head of Unit, "Competitiveness, economic analysis, indicators", Research Directorate General at the European Commission. He recently published, with Paraskevas Caracostas, *Society, the Endless Frontier*.

EMILIO MUÑOZ is Research Professor in Biology and Biomedicine at the Spanish Research Council (CSIC), *Unidad de Políticas Comparadas (Ciencia, Tecnología y Sociedad)*. He served different managerial positions in the Spanish R&D Administration and has been member of the CREST Committee and Chairman of the COST Programme.

PHILIPPE MUSTAR is Professor of Socio-economics of innovation at the *École Nationale Supérieure des Mines de Paris* (*Centre de Sociologie de l'Innovation*). His research subjects are science, technology and innovation policy, high growth SMEs and university start-ups.

ERKKI ORMALA is Director, Technology Policy at the Nokia Corporation. Between1987 and 1999 he was the Secretary of the Science and Technology Policy Council of Finland. Between 1996 and 1999, he was the Chairman of the OECD Working Group for Technology and Innovation Policy.

ARIE RIP, Professor of Philosophy of Science and Technology, University of Twente, the Netherlands. Starting as a chemist, Arie Rip switched to science, technology and society studies. He has widely published in the sociology of science, science policy studies, technology dynamics and technology assessment.

YUKIO SATO is Executive Director at the Japan Atomic Energy Research Institute. Previously he was Director-General of the National Institute of Science and Technology Policy, Science and Technology Agency (Japanese Government).

ALBERTO SILVANI, Research Director at the Institute for Studies on Scientific Research and Documentation, National Research Council of Italy, and Professor of Management of Innovation at University of Cassino. He acts as Advisor to the Ministry of University and Scientific and Technological Research in the field of science and technology policy, technology transfer and regional development.

GIORGIO SIRILLI, Research Director at the Institute for Studies on Scientific Research and Documentation, National Research Council of Italy, and Professor of Economics at "Tor Vergata" University in Rome. He has extensively published in the fields of science and technology policy and indicators, the economics of technical change, R&D evaluation, organisation and management of innovation.

ACKNOWLEDGEMENTS

This book would not have been possible without the role of the *Commissariat Général du Plan* in its initiation. Special recognition is given to Brigitte Rémy (the head of the Studies and Research Service), and the *Commissariat*'s working group on benchmarking the French situation, presided over by Jean-Jacques Duby with the two editors of this book as *rapporteurs*. The results of this seminar have been integrated in the 1999 official report on research and innovation in France. The contributors to this book especially wish to thank the President and the participants in the working group* for the pertinence of their questions and their critiques of the preliminary versions of the texts.

Following this stage a two-year programme was financed by the *Centre de Sociologie de l'Innovation* of the *École des Mines de Paris*, which enabled the editors and the authors of this book to adopt a common framework, develop their analyses and prepare the chapters. We would like to express our gratitude to all for enabling this collective work.

We would also like to thank Dymphna Evans and Bob Pickens for their help in finalizing this project; Martine Mène, who has laid out this book, for her patience; and Professor Jean Pavlevski for his advice. A significant effort to harmonise the varied and often quite diverse usage of the English language is required when bringing together authors of 11 nationalities. Our colleague Mark Boden, of PREST, willingly accepted this role. Mark has also translated Chapters 1, 13 and 14, which were originally written in French. This book owes him much and we thank him deeply for his contribution.

It was our initial intent to include one chapter on Israel. Our colleague Ehud Zuscovitch had willingly accepted to undertake it. Illness forbade it. We take the opportunity of these acknowledgements to recall his unique professional and human qualities.

Finally, we would particularly like to thank our colleagues and friends, the authors of the chapters of this book, not only for the numerous fruitful discussions to which it has given rise, but also for their contributions to the construction of a shared framework that we hope will yield yet richer comparisons.

The Editors.

* The group was composed of Rémi Barré, Philippe Bourgeois, Paraskevas Caracostas, Michel Combarnous, Alain Costes, Guy Crespy, Dominique Deberdt, Jean-Jacques Duby, Robert Ducluzeau, Laurence Esterle, Dominique Guellec, Emmanuel Jolivet, Jean-Paul Karsenty, Jean-Claude Laprie, Patrick Lavarde, Patrick Llerena, Marie-Catherine Lecocq, Joseph Mariani, Jean Quessada, Brigitte Rémy, Michel Ronis, Jean-Luc Rossignol, Christian Sayettat and Jacques Sevin.

CHAPTER 1

GENERAL INTRODUCTION: A FOCUS ON RESEARCH AND INNOVATION POLICIES

PHILIPPE LARÉDO AND PHILIPPE MUSTAR

The analysis of science and technology policies is now an exercise of some long standing, having been initiated by the OECD at the beginning of the 1960s. Reading these first analyses today, one can see the extent to which they anticipated the conceptual developments of the 1980s and, in particular, the pioneering works of Freeman and Lundvall on national systems of innovation (Freeman, 1987 and Lundvall, 1988). These authors have rekindled interest in international comparisons, a primary result of which was the collection edited by Richard Nelson in 1993 (Nelson, 1993). Since then, the 1990s have been characterised by contradictory changes: the emergence of Europe is offset by the break-up of the Soviet Union and the pronounced "Balkanisation" of the former Eastern bloc countries; the end of the Cold War (and thus of the military imperative which underpinned a significant proportion of the public intervention relating to R&D) is a concomitant of the increased involvement of Western countries in a growing number of local conflicts (from Kuwait to Kosovo); and the "globalisation" of firms and markets (which requires previous notions of "multi-nationalisation" to be reconsidered) is associated with growing "sub-national" public interventions, particularly those related to innovation, which are becoming stronger almost everywhere.

The authors of this book have therefore considered it both opportune and necessary to re-examine the analyses conducted a decade ago. They have chosen to adopt Nelson's method, favouring national approaches (later we will explain the selection of countries made – and, in particular, the inclusion of a specific chapter on the European Union). Consequently, it has been necessary to establish a methodology and an experimental approach, which clearly identifies and emphasises the points of comparison, and ensures that the reader will come across them in each chapter. Finally, it goes without saying that such a work had to take into account the numerous lessons accumulated on systems of innovation, which brings to mind the title of the collection edited by Edquist (Edquist, 1997). These lessons have led us, as this Introduction will explain in greater detail, to a reorientation, which is described by the very title of this book, *Research and Innovation Policies*.

FROM NATIONAL SYSTEMS TO RESEARCH AND INNOVATION POLICIES

The shared thesis conveyed by the notion of a national system of innovation is that each country can be characterised by a particular body of sectoral specialisations, of rules and routines, of institutional organisations, and also of achievements. Works referring to what has now become a concept of the economics of innovation, emphasise the learning process and the institutional, social and political factors. In so doing, they are akin to a somewhat forgotten earlier development, proposed by a French economist, J. Weiller, who in 1949 put forward the concept of "national structural preferences", that is to say "preferences regarding the utilisation and construction of the productive forces within each country". Built on the basis of work on international exchanges and the "problems of international economics" (the title of his manual published at the same time), this concept aimed to explain the persistent disparities between countries, less in relation to international exchanges or growth, than to the choices of political economy implemented and state intervention. Contrary to Schumpeter, for whom "all economic argument almost completely loses its interest, in the context of the nation" (Schumpeter, 1962, p. 110), the nation for him is an "operational concept" of economics. In a world which is increasingly transnational, it should be understood as a "zone of structural preference and balance of power" (Weiller and Desroussilles, 1974, p. 167). This notion, formulated in this way, assumes its full meaning when "the most express reservations towards the idea of a worldwide generalisation of a certain type of progress", that is to say "of a certain type of economic organisation" (ibid., pp. 179-180), are put forward.

Two dimensions are thus emphasised, "operational" and "doctrinal", which are involved in the adoption of the term "national system of innovation".

No sooner had the concept of a national system of innovation been put forward than debates began to proliferate on the exact meaning of this combination of three constituent concepts. By insisting on the notion of system, authors emphasise the role of "agents, policies and institutions supporting the process of technical advance in an economy" (Bozeman, Chapter 3). They underline, subsequently, the importance of the processes leading to current organisational arrangements. At the core of their analysis are the history and the trajectories, the interactions between actors, and the roles of institutions and organisations. It is thus not surprising to see the nation, taken in the sense proposed by Weiller, occupying a central position, even if the "transnational" dynamics, which have been in operation for a long time, continue to reinforce the importance of the interactions between systems in their respective evolutionary developments.

But what system is under discussion? The analysts propose two versions of it: "Scholars looking at the 'narrow' NIS focus on organizations and institutions involved directly in the processes of scientific and technological exploration. The 'broad' NIS concept includes all economic, political, and other social institutions affecting learning, searching, and exploring activities" (Bozeman, Chapter 3). Nelson stresses the importance of the second approach, bearing in mind that specialists in innovation tend to "play down the existence of active coherent industrial policies" and of a "well-structured and thought through general policy" (Nelson, 1993, p. 515). From the fifteen national studies carried out, he sees an imperative emerging: the existence of "strong firms", in other words, firms that are capable, whatever their size, of being competitive in global markets. While it is clear that they must make the "the bulk of the effort" themselves, the studies carried out did "indicate strongly that aspects of the national background in which firms operate matter greatly" (ibid., p. 511). It is therefore appropriate to concentrate attention on this "infrastructure" which is important for the competitiveness of firms and also largely falls within the competence of public policies. This choice has been explicitly made in this book.

It is then necessary to define the relevant dimensions for such an analysis. Nelson distinguishes three principal components. He gives priority to the "package of fiscal, monetary and trade policies" which make exporting attractive or difficult. He goes on to underline the importance of the education system and, more particularly, of "the responsiveness of university systems to the training needs of industry". Finally, he questions the actual role of "dedicated government policies". All the country studies carried out share the characteristic of having "significant pockets of government overview, funding and protection", which include defence, space, electricity and telecommunications. He underlines that all the countries share a common investment – clearly of varying degrees of importance – in basic research. Lastly, he cites the impact of the Japanese model in the development of policies directly dedicated to civil technologies: "The rise of Japan as a model has enhanced the belief that an explicit national technology policy can be effective; indeed it is now widely argued that a nation will fall progressively behind if it does not have an explicit technology policy" (ibid., p. 17). It drives him to "first acknowledge the wide range of policies targeted at technological advance" and then to conclude one page later that "because they are so diverse, I cannot see any strong generalizations that can be drawn" (ibid., p. 512). Such a conclusion is the reverse side of the approach adopted by this comparative study. The "broad" approach of the systems of innovation considered has limited the capacities for analysis of the diversity of public interventions in favour of innovation. We have thus chosen to concentrate this book on the "institutions and policies that affect the creation, development, commercialization, and adoption of new technologies within an economy. As such, a national innovation system includes not just the institutions performing R&D and the level and sources of funding for such R&D, but policies – such as antitrust policy, intellectual property rights, and regulatory policy – that affect technology development, the training of scientists and engineers, and technology adoption" (Mowery, Chapter 2). The title adopted – *Research and Innovation Policies* – reflects this choice.

THE COUNTRIES CONSIDERED

What policies should be considered? Three principles and one constraint explain the selection made. First, the constraint: to focus attention on research and innovation policies (which should be clearly located in the overall national landscape) paradoxically leads to longer analyses – consequently, there are two chapters dealing with the United States. To keep the book to a reasonable length (500 pages even so) required that the number of countries under consideration be limited to around ten. The three principles guiding the choice of countries can be summarised as: continuity, diversity of sizes and trajectories, and the importance accorded to the European Union. The objective of continuity is to allow the reader comparisons over time, with the books previously published on the science policies of Western countries. This explains our first group: the United States, Japan, Germany, the United Kingdom, Italy and France. The diversity of sizes and natural resources are two aspects underlined by Nelson and Rosenberg (p. 19). We have brought them back into focus with the choice of Finland and the Netherlands. The maintenance of a certain equilibrium between established industrialised countries and newly industrialised ones is also a dimension emphasised by Nelson. He sees it as a way in which to assure, in the analysis, a significant presence of those innovations, which economists would describe as incremental. This led us to add Korea, and also Spain, the rapid growth of which is a concomitant of its entry in the larger context of the Common Market, and subsequently the European Union.

With seven out of the ten countries included amongst its 15 members, the European Union, which is our third choice, is intentionally over-represented. This stems from a hypothesis of the editors of this book, although it would be improper to consider it as shared by all the contributing authors. This hypothesis can be expressed in a few simple words: the European Union is on the way to becoming a system of innovation, or to employ the terminology proposed by Jean Weiller, a veritable zone of structural preference and balance of power. This is why we were anxious to include a specific chapter on its research and innovation policy. It is up to the readers to form their own opinion, comparing it to the national studies of the seven member states.

THE APPROACH USED AND THE CONSTRUCTION OF THE BOOK

As with Nelson's book, the comparative element is largely implicit. It is up to the reader to take responsibility for carrying out his or her own analysis while reading the chapters. The editors will allow themselves, by way of conclusion, to make a preliminary suggestion, which is both limited to a few main characteristics and quite obviously provisional.

For this comparative analysis to be possible, it was necessary both to identify a series of points, that the national analyses should tackle and to devise an experimental protocol, which enables the feasibility of a comparative analysis to be tested.

This dual approach has been rendered possible by the project of the *Commissariat général du plan*, to study, at the request of the French Prime Minister, the challenges facing research and innovation in France. In the broader context of this work (cf. the global report by R. Majoie, Majoie, 2000), we proposed to the Commissariat to carry out a benchmarking exercise on the French situation. The procedure adopted was as follows. A working group was convened under the presidency of J.J. Duby with the two editors acting as rapporteurs. It was composed of about 20 representatives of a range of French institutions (universities, public organisations, technopoles, firms, technical centres, the government). A number of experts – the authors of this book – came to present the developments in progress in their respective countries.[1] Each presentation and the ensuing discussions generally lasted a good half-day. These exchanges had a dual effect. On the one hand they effectively allowed the working group to progressively identify five core issues for France (cf. the report made by J.J. Duby: Duby, 2000) and for each of them to carry out an actual comparative analysis, thus reinforcing the approach adopted by the editors. On the other hand, they brought out numerous aspects, which were not dealt with or were dealt with only briefly by the experts, or were not made sufficiently explicit for a reader who is not *au fait* with the current national situation. In so doing, they initiated a genuine joint effort between the editors and each author, conducted through numerous electronic exchanges and visits by the editors. This book is thus the outcome of a project which spread out over almost three years: initiated in mid-1998, the hearings took place in early 1999 and the exchanges involved in finalising the different chapters took two years.

SIX CENTRAL QUESTIONS

What consequently is the series of points that the authors were systematically asked to consider? There are six.

All the works on national systems of innovation give a central place to the education system, more particularly to the universities, to their development and to the relations that they maintain with the economic world. This constitutes an important part of all the national analyses carried out. Both the growing importance of universities within national systems, the exponential

---

1. It goes without saying that there was no presentation and debate on the French situation (Chapter 13). This was the same for the chapter on European policy, but it should be noted that P. Caracostas was a member of the working group and participated in all the debates.

development and the diversity of their links with firms, and the extraordinary variety of the institutional arrangements, which govern these developments, can be seen.

Among the policies specifically dedicated to technological progress, Nelson underlines the place occupied in a number of countries by defence and by large civil programmes, which concern space, telecommunications and energy. This clearly has a quite specific resonance for the French editors of this book, with France being described in the majority of analyses as an extreme case of voluntarist policy structured around this approach. In most countries, with the possible exception of the United States, one can see that there have been particularly important reorientations in this regard. These include the very significant decrease in expenditure on military research, and also, with the exception of space, the considerable withdrawal of public authorities from this type of intervention. They also emphasise the growing importance of the European dimension. In effect, it is on the European level, notably via the European Space Agency, that national investments are decided and organised. Similarly, when it comes to the changes observed in telecommunications, electricity, the postal system or transport, it is impossible to avoid the question of the role of the European Union and of the common rules that it has progressively adopted with regard to "public services".

The third aspect brought out by Nelson concerns "explicit national technology policies". It seemed necessary to us to distinguish two realities in this formulation. On the one hand, there are the types of incentives for research and industrial innovation. It can be seen that numerous countries have active policies to support the innovation efforts of firms, particularly SMEs, whether these interventions are direct (project support) or indirect (fiscal deductions or public procurement policies such as SBIR in the United States). On the other hand, these interventions often take the form of "priority programmes" which seek, in domains such as biotechnology or information and communication technologies (ICT), to mobilise all of the public and private actors involved. For Nelson "a strong case can be made that such technology and industry-oriented public programs have made a big difference in many fields" (p. 513). They should therefore have witnessed their importance grow or at least remain constant. This has quite clearly not been the case. The national analyses, in effect, bring out their limited importance (cf. the marginal significance of the Advanced Technology Program in the United States) or the very appreciable decrease in the place they occupy in national policies. For European countries, this decrease is due less to their disappearance than to their progressive transfer away from the national level (but which is often not expressed as such). These latter now constitute the core of European Union interventions. Also evident is a second source of investigation of the emergence of a European system, all the more relevant given the recent reorientation of the Framework Programme and its shift in focus to societal problems, resulting in a significant change in perspective.

In all countries, universities are an essential component of the basic research landscape, which, in all countries, is largely financed by the public authorities. However, they do not act alone. All countries, led by the United States, have also gradually established numerous "government" or "national" laboratories. Certain of these, such as the CNRS in France or the *Max Planck Gesellschaft* in Germany have been created to carry out fundamental research. However, for the majority, the missions for which they have been created were "targeted". Various works[1] have shown that, with the help of time, it is no longer possible to take this description at face value and that it is necessary to question the often significant changes experienced. We have asked the authors to describe the changes in this second component, which, together with the universities, constitute public sector research. We have also asked them to examine the composition of the whole of public sector research, its recent changes, and also the importance and the modalities of its public financing. The studies confirm that everywhere the scales tilt increasingly in favour of the universities, but also that everywhere public organisations remain in existence, even in the United Kingdom where only some of them have been privatised. However, governments everywhere have sought to bring about changes in their laboratories, particularly to promote or renew their links with the economy and with society. Consequently they constitute, despite an ongoing decrease in their relative importance, an enduring component of national systems.

In the introduction, we have underlined at what point the national–global dichotomy seems to us to be simplistic. It has led us to provide the reader with a chapter specifically on Europe. However, it has also led us to examine the sub-national dimension and the involvement of regions or states in the support of firms' innovation capacities. The analyses show a growing role of the regions, both because in those places, where regional or state involvement has existed for a long time, such as in the United States or Germany, they have increasingly seized on these issues, and because, in other

1. See amongst others the works by Crow and Bozeman in the US and their recapitulation in Crow and Bozeman, 1998, or Boden et al., 1998 for UK or Joly and Mangematin, 1996 for France.

countries, such as France or Spain, their recent creation has been accompanied by rapid and significant activity. The construction of a favourable background is no longer only the business of national policies but of a whole set of public interventions, which, in Europe, mix regions, nations and the European Union in an increasingly intimate manner.

Consequently, it has been important to resume Nelson's line of enquiry, on the global coherence of public interventions. A particular emphasis has thus been put on public governance, on the mechanisms and processes which prepare and follow through public interventions, on the modalities of co-ordination between public policies on the national level, but also between public authorities. This has brought out great differences in the role of "advisory bodies" from one country to another, the growing place of foresight activities in the very large majority of the countries studied, and the accent which is increasingly often placed on evaluation as a central means of retroactive effect on the conduct of activities and the formulation of public interventions. In the majority of countries, the co-ordination of interventions is currently a central issue which has led, in the case of Japan, to a deep reform of the whole of the public structures associated with higher education, research and innovation.

STRUCTURE OF THE BOOK

Moving on to the national chapters, each chapter is, of course, introduced by a presentation of the overall landscape and a setting in historical perspective, without which it is difficult to position research and innovation policies as well as the evolutions that they have witnessed. It finishes with a "prospective" reflection by the authors who highlight the challenges which will confront the country in the next five to ten years.

The first two chapters are devoted to the United States. David Mowery, and Barry Bozeman and James Dietz have shared this task. David Mowery puts forward a perspective on the American innovation system, before focusing his attention more closely on the overall organisation of federal intervention, university research and on aspects connected with intellectual property issues relating to the knowledge produced by public sector research. Barry Bozeman and James Dietz go back over the three paradigms (market failure, mission, co-operative technology) which have organised American public policy before turning their attention to defence research, national laboratories, federal support for technology and state intervention.

In many aspects, Japan in the 1990s has experienced a unique dynamic in its research and innovation policy. The accent has been placed more on research than on innovation, with a doubling of the public budget over five years. Yukio Sato presents the far-reaching reform in the public sector which has been initiated. This is leading, on the one hand, to a restructuring of ministerial structures and advisory bodies and, on the other hand, to a reorganisation of the public research system with autonomisation of public laboratories and exposure of the universities to the economic world.

Sung Chul Chung emphasises the extraordinary growth in Korean investment in research and innovation, largely carried out by firms. He underlines the initial role, then the progressive marginalisation of public research institutes, paralleling the continuous growth of the involvement of universities in research. Quite evidently, one cannot fail to question the future of this situation in the wake of the crisis, which, besides the question of industrial involvement, paradoxically brings out the weakness in the co-ordination of public interventions.

The reader enters next into the eight European chapters. We have chosen to start with the chapter on the innovation and research policy of the European Union. Paraskevas Caracostas and Ugur Muldur put into perspective the challenges facing Europe as a whole. They underline the extent to which the investments in research and innovation are inferior in comparison with the United States and Japan. They bring out the "European paradox", that is to say the discrepancy which exists between the role of Europe in global scientific production and its place in the production of patented inventions. They introduce us into the dynamic of the European construction and into the importance that innovation has progressively taken within it. They analyse the growing financial involvement of the Framework Programme and its impact on national systems, but they also show the role of other instruments, both regulatory and financial, with the Structural Funds, which support the efforts of the less-developed regions of Europe. Finally, they highlight the strong reorientations of the Fifth Framework Programme, which covers the transition to the third millennium, and the priorities of which, as much as the modalities of adoption and implementation, appear to them to promote renewed forms of co-operation between member countries and the Commission.

The analysis undertaken of the strengths and weaknesses of the German system by Frieder Meyer-Krahmer enables the reader to better understand how the federal and state systems complement each other, and why the accent is so strongly placed on technology on the federal level. It shows the multiplicity of modalities of public intervention in a system where two-thirds of research and innovation effort have, for a long time, been the preserve of firms. The central points of debate concern the flexibility of the system, that is to say, on the capacity of German firms to position themselves in emerging sectors. The author also considers that one of the challenges concerns the atomisation of university research. This poses the question of the restructuring of public research in its entirety. He underlines the change of context, which sees questions relating to public problems, such as the environment, growing in importance. Finally, it seems to him that the major challenge resides in the definition and implementation of technology policies and their co-ordination with the different spatial scales.

There is a marked contrast with the presentation of the United Kingdom provided by Luke Georghiou. In this case, the full extent of the centralisation of a national system articulated around its processes of foresight and evaluation is considered. The combined movements of privatisation and of budgetary restrictions have provided an impetus for a convergence of all the public research actors in a "contractual research space". Finally, the question is posed of the means of public intervention with regard to innovation, while the gap between public research capacities and industrial capacities for innovation is considered as the major problem of this country.

Barend van der Meulen and Arie Rip present, with the case of the Netherlands, a radically different system of operation. They oppose "steering" from above which in their view characterises the United Kingdom with the specific institutional competences developed in the Netherlands to define policy priorities. This capacity for aggregation rests on a "densely populated intermediary layer" between the government and research institutions, which, lacking financial responsibilities, promotes public debate and mutual adjustments. This is, it seems to them, the principal force, which allows flexibility and adaptability in the European and global environments, which directly confronts Dutch firms.

Erkki Ormala shows us the voluntarist approach of Finland, which, since the end of the 1980s uses the concept of national system of innovation to discuss, devise and implement its research and innovation policy. The reorientations in the public governance system, with the role and operation of the Science Policy Council, the reorganisation of research organisations, the evolution of the mode of operation of the agency supporting technology and innovation accompany the model repositioning of the Finnish economy.

This description could, in part, be that of one large region, such as a German *Land* or the Spanish Basque Country and its exemplary policy for the promotion of firms' innovation capacities. The growing role of the Spanish regions, reinforced by the progressive transfer to them of the responsibility for universities, is clearly brought out by Emilio Muñoz. He underlines the recent features of research and innovation policies, the difficulty of government co-ordination, despite the existence of national plans, and the disparity that currently exists between those sectors that are major providers of jobs and the competences of public sector research. In this discrepancy, he sees one of the major challenges confronting the country.

Such analyses could be carried out on Italy, where the dynamism of its industrial districts has played a central role in the considerations of analysts of innovation. However, Alberto Silvani and Georgio Sirilli have chosen to bring to light the profound changes in the conduct of research and innovation policies. These range from a reshuffle of governmental organisation, via the development of new instruments to support university research projects and the follow-up of university activities (evaluation), and focus on "simple to implement" procedures in supporting industrial research and innovation.

These changes characterise a strong transformation in the approach of research and innovation by public authorities. The analysis that we propose brings out the reverse transformation that France has undergone in the 1990s, which has moved it significantly away from the Colbertist model of a powerful and centralising state described by Chesnais in Nelson's book. The near disappearance of large programmes and the growing role of European Union programmes in information and communication technologies, the strong and continuous decrease in military research, the rapid increase in the research capacities of universities and their move closer towards the CNRS, and the involvement of the regions in support for innovation in SMEs are some of the traits which explain this repositioning. It is reflected in a double movement: a reorganisation of the ministerial machinery and a centring of research and innovation policies on public sector research. New challenges follow from this, largely underlined by the strategic exercise recently conducted by the *Commissariat général du Plan*, linked to the "good formatting" of public sector research, as well as to the capacity to provide impetus and co-ordination of multiple public participants.

This brief journey underlines a series of convergences that, in particular, concern the growing role of the universities, governments' disengagement from industry with the virtual end of large programmes and the less central role of defence policies. Are these moves likely to reduce the interest of approaches concerned with systems of innovation? The response is far from being obvious, since these changes in national policies also relate to the appearance of and the growing role of other public powers. These latter, which encircle governments (at least in Europe) seem destined for a bright future. More than national or federal governments, the regions and the states seem to be closer to actors and their problems and specific needs (in this respect the country-regions of Europe, such as Finland, are particularly rich in lessons). When the debate moves to the global level, it is striking to see that the *de facto* delegation to the European Union (whatever mechanisms are adopted) does not concern only the global organisation of markets or societal responsibilities (such as climate change), it also concerns, through the Framework Programme, preparation for the future. We will tackle these points later by way of conclusion.

REFERENCES

Boden, R., P. Gummett, D. Cox and K. Barker, 1998, Men in White Coats ... Men in Grey Suits, *Accounting, Auditing and Accountability Journal* 11, 3, 1998, pp. 267-291.

Crow, M. and B. Bozeman, 1998, *Limited by Design: R&D Laboratories in the U.S. National Innovation System*, New York, Columbia University Press.

Duby, J.J., 2000, La politique française d'innovation face à la concurrence internationale. Cinq meilleures pratiques étrangères pour la recherche française, in J. Majoie, op. cit.

Edquist, C. (ed.), 1997, Systems of Innovation: Technologies, Institutions and Organizations, London: Pinter Publishers.

Freeman, C., 1987, *Technology Policy and Economic Performance: Lessons from Japan*, London: Pinter Publishers.

Gibbons, M. et al., 1994, *The New Production of Knowledge*, London: Sage.

Joly, P.B. and V. Mangematin, 1996, Profile of Public Laboratories, Industrial Partnerships and Organisation of R&D: the Dynamics of Industrial Relationships in a Large Research Organisation, *Research Policy*, 25, 6.

Lundvall, B.A., 1988, Innovation as an Interactive Process: From User-Producer Interaction to the National System of Innovation, in G. Dosi, C. Freeman, R. Nelson, G. Silverberg and L. Soete (eds), *Technical Change and Economic Theory*, London: Pinter Publishers.

Majoie, J., 2000, *Recherche et innovation: la France dans la compétition mondiale*, Rapport au CGP, Paris: La Documentation française.

Nelson, R. (ed.), 1993, *National Innovation Systems, A Comparative Analysis*, New York: Oxford University Press.

Schumpeter, J., 1962, *Esquisse d'une histoire de la pensée économique*, Paris: Dalloz.

Weiller, J., 1946 (tome 1) and 1950 (tome 2), *Problèmes d'économie internationale*, Paris: PUF.

Weiller, J., 1949, Les préférences nationales de structure et le déséquilibre structurel, *Revue d'économie politique*, pp. 414ff.

Weiller, J. and G. D.-Desroussilles, 1974, *Les cadres sociaux de la pensée économique*, Paris: PUF.

CHAPTER 2

THE UNITED STATES NATIONAL INNOVATION SYSTEM AFTER THE COLD WAR

DAVID C. MOWERY

1. INTRODUCTION

Recent changes in government policy, industrial R&D strategies, and international competition have produced significant structural change in the US and other OECD members' national innovation systems. In this chapter, I summarize and survey recent trends in the structure of the US national innovation system. Although much of the post-1980 structural change in US public R&D spending closely resembles trends in other large OECD economies, the United States has pursued an intellectual property rights policy since the early 1980s that goes beyond those of other economies in favoring the owners of intellectual property. This policy, which in many respects represents a large-scale application of the "linear model", has affected many elements of the US national innovation system in both the public and private sectors.

I devote particular attention to the evolving structure of collaboration between US research universities and industrial firms in the United States, emphasizing two points: (1) much of the increased collaboration since 1980 that has attracted the attention of policymakers, critics and university administrators represents a revival of long-standing relationships, rather than something unprecedented in US history; and (2) this renewed collaboration, which has potential costs as well as benefits, relies as much on the unusual structure of the US university system and other federal policies as it does on the Bayh–Dole Act of 1980. The development of similarly widespread collaborative relationships between universities and industry in other industrial economies may well require more extensive structural change than the modest policy initiatives embodied in the Bayh–Dole Act.

Although the US national innovation system has been affected by numerous changes in federal policy since 1980, these changes have been incremental rather than sweeping. The end of the Cold War, which sharply reduced the importance of a central rationale for the public R&D system that emerged in the United States during the 1945-1995 period, has produced only modest pressure for more significant institutional or policy change. The national security motivation that underpinned much of the postwar federal R&D investment has been largely replaced with a less well-defined economic rationale. But the basic institutional structure for US R&D policy, which continues to combine large public investments with little central oversight of resource allocation patterns within either the Executive or Congressional branches of government, has changed very little. The endorsement of the current system in the 1998 "Ehlers Report"[1] suggests that political pressure

1. In February 1997, the leader of the House of Representatives, Congressman Newt Gingrich, asked the House Committee on Science to develop a long-range science and technology policy, arguing that "The United States has been operating under a model developed by Vannevar Bush in his 1945 report to the President entitled *Science: The Endless Frontier*. It continues to operate under that model with little change. This approach served us very well during the Cold War, because Bush's science policy was predicated upon serving the military needs of our nation, ensuring national pride in our scientific and technological accomplishments, and developing a strong scientific, technological, and manufacturing enterprise that would serve us well not only in peace but also would be essential for this country in both the Cold War and potential hot wars. With the collapse of the Soviet Union, and the de facto end of the Cold War, the Vannevar Bush approach is no longer valid".
Congressman Vernon Ehlers, a Ph.D. physicist from Michigan, was selected to lead the review of US science policy, and issued a report, "Unlocking Our Future: Toward a New National Science Policy" (commonly referred to as the Ehlers Report) in September 1998. Having been billed initially as "another Bush Report", laying the foundations for a new, 21st century US science policy, the report contained little criticism of current policies, institutions, and resource allocation trends in US science and technology policy. As such, the report serves as an important endorsement by the Republican Party of current policy, and suggests that in the absence of major national security or economic crises, these policies would continue under a Republican President.

for such change remains weak. Moreover, despite the prominence of the economic rationale for these investments in political debate, investments by the federal government in programs and projects that are primarily designed to strengthen the technological capabilities of US industry remain small, accounting for no more than 1% of federal R&D spending.

An economic downturn could change this situation, but political, ideological, and legislative restrictions on federal spending are so constraining that any programmatic response (for example, the SEMATECH R&D consortium of the early 1990s) is likely to be modest in scope and brief in duration.

2. THE ORIGINS AND STRUCTURE OF THE POSTWAR "US NATIONAL INNOVATION SYSTEM"

The concept of a "national innovation system" is discussed and defined by Freeman (1987) and Nelson (1993), among others. A "national innovation system" refers to the collection of institutions and policies that affect the creation, development, commercialization and adoption of new technologies within an economy.

As such, the US national innovation system includes not just the institutions performing R&D and the level and sources of funding for such R&D, but policies, such as antitrust policy, intellectual property rights and regulatory policy, that affect technology development, the training of scientists and engineers and technology adoption.

But the structure of a nation's innovation system is the result of complex historical processes of institutional development, in addition to public policy. Moreover, the performance of these systems depends in part on the actions and decisions of private enterprises, and these decisions can reinforce or offset the effects of public policies. No assessment of the US or other nations' innovation systems therefore is complete without some discussion of both their public and private sector components.

The foundations of the structure of the postwar US "national innovation system" were largely put in place during 1945-1950, as demobilization for peace was replaced by Cold War rearmament. The federal government assumed a role as a financial supporter that dwarfed its pre-1940 presence, as the federal share of national R&D spending rose from roughly 20% in 1939 to more than 50% by 1962. Federal spending supported R&D activity in industry and universities, rather than being concentrated in federal government laboratories. As of 1980 (a date chosen arbitrarily as representative of the apogee of the Cold War US innovation system), 12.2% of R&D performed was located in the public sector, 13.2% took place in universities, and 71.1% was located in industry.

During this period, US science and technology "policy" was the product of loosely coordinated and often inconsistent decisions made in diverse policy areas (including procurement policy) designed to further the missions of individual federal agencies. These policies were motivated more by national security concerns than by any comprehensive economic strategy. The key elements of this decentralized system arose during and after World War II. Its loose coordination and weak central oversight are attributable in part to the failure of Congress and the Executive branch to agree on the structure and powers of a federal "science agency" to replace the wartime Committee on Medical Research and the Office of Scientific Research and Development, as recommended by Vannevar Bush in his famous 1945 report, *Science: The Endless Frontier.*

Indeed, the federal agency originally charged by Bush (under a different name, the National Research Foundation) with responsibility for supporting non-defense basic research, the National Science Foundation (NSF), was established only in 1950. By that time, the major mission agencies (for example, the military services, the Atomic Energy Commission and the National Institutes of Health) had begun ambitious intramural and extramural R&D programs.[1] The NSF budget remains dwarfed by the research budgets of these other agencies.

No single agency, office or committee in either the Executive or the Congressional branch of the federal government determines the allocation, costs and benefits of the entire federal R&D budget (including military R&D) on a regular basis. In Congress, jurisdiction is divided among more than 25 authorizing and appropriations committees. Recent administrations have attempted to coordinate R&D priorities among the many

1. "In the absence of a National Science Foundation between 1945 and 1950, no comparable institutional sponsor of fundamental physical research for civilian needs had existed, nor had any comparable institutional mechanism to plan, if not to enforce, a research program in the physical sciences better balanced between civilian and military purposes ... In retrospect, the delay in the establishment of the National Science Foundation was critically important in the evolution of postwar policy for research and development, not least because it cost the nation a program balanced between civilian and military patronage and purpose" (Kevles, 1977, pp. 359-360).

Executive agencies (for example the Bush Administration's Federal Coordinating Councils for Science, Engineering, and Technology; the Clinton Administration's National Science and Technology Council and President's Council of Advisors on Science and Technology), but decision-making on R&D spending priorities and program development remains decentralized.

Postwar US civilian technology policy consisted largely of federal government funding for scientific research, especially basic research in biomedical science. Federal funding of development or applied research was minimal in areas other than national defense. During the late 1950s, defense-related spending accounted for as much as 80% of federal R&D outlays. Federal and state governments invested little in programs designed to assist firms in adopting technology, which may have contributed to the relatively slow adoption by US manufacturing firms of advanced manufacturing technologies (Mowery and Rosenberg, 1993; Edquist and Jacobsson, 1988). US agriculture was an important exception to this characterization, with its elaborate network of federal and state-funded extension agents (see Evenson, 1982).

This dramatic transformation in the scope and sources of funding for R&D aided the emergence of the United States as a world leader in basic scientific research. The rise to world prominence of US scientific research was associated with several other distinctive features of the US national innovation system. Defense-related R&D and procurement programs provided a powerful impetus to the development and commercialization of new civilian technologies in commercial aerospace, semiconductors, computers and computer software (such defense "spillovers" proved to be far less beneficial for civilian firms in nuclear power).

In almost all of these industries other than commercial aerospace, new firms played a prominent and, in some cases, dominant role in the commercialization of important technological advances. The important role of these new firms appears to be unique among the major postwar industrial economies, and reflected the unusual conjunction of defense procurement policies with US antitrust policy and a domestic financial system that funded new enterprises in high-technology industries (Mowery, 1997a, provides additional details).

3. PUBLIC AND PRIVATE R&D SPENDING, 1980-1996

3.1 – *An evolving structure*

The most dramatic shift in spending trends within the US R&D system during the past 15 years is the decline in R&D spending by the federal government. Having grown at an average rate of 6% per year in real terms during 1980-1985, inflation-adjusted federal R&D spending declined by roughly 1.5% during 1985-1990, and by 7.8% during 1991-1996.[1] The other major source of R&D spending within the US system is industry, which accounted for 63% of total R&D spending in 1996. Industry-financed R&D scarcely grew at all in real terms during the early 1990s. This trend was reversed in 1993, however, and National Science Foundation data reveal that real industry-funded R&D spending grew at an average annual rate of nearly 10% during 1993-1995 (National Science Foundation 1998).

These shifting growth trends in industry and government-funded R&D have produced wide swings in the growth in overall US R&D spending since 1980. Total national R&D spending grew by nearly 7% annually in constant dollar terms during 1980-1985, but during 1985-1993, the average annual rate of growth in total constant dollar R&D spending declined to 1%. More recently, however, total US R&D spending has grown more rapidly, driven by growth in industrially financed R&D. During 1993-1996, total national R&D spending grew by more than 10% in constant dollars.

Declines in federal R&D spending are largely due to reductions in defense-related R&D spending. Defense-related federal R&D spending increased from 50% of federal R&D spending in 1980 to almost 70% by 1986, a level from which it has declined once again to approximately 52% of total federal

---

1. R&D spending from "Other non-federal sources" (R&D funded by state and local governments, as well as universities and colleges) grew by 2% in real terms during 1994-1995. The National Science Foundation data currently available from the forthcoming *National Patterns of R&D Resources: 1998* provide only estimated levels of R&D spending for 1996 and 1997, and these are less reliable, particularly for industry-funded R&D investment, than the actual spending levels reported (with a lag) by the NSF. In addition, revisions in NSF data collection procedures mean that the data on industry-funded R&D before and after 1991 are not strictly comparable with one another, especially for the disaggregated components of R&D spending and for individual industries (see National Science Foundation, 1996). Accordingly, our discussion of spending trends covers only the period through 1995, and we confine the analysis of trends in the components of industry-funded R&D investment to the 1991-1995 period.

R&D spending.[1] Political pressure for higher defense spending has become more intense since 1999, and further reductions in the defense share of the federal R&D budget therefore appear unlikely. Another important trend in federal R&D spending during the 1980-1998 period is expanded federal funding for biomedical research (most of which is allocated to the National Institutes of Health), which has grown at an average annual rate of roughly 8% in real terms since 1975.[2]

These shifts in federal R&D spending priorities have affected the allocation of public R&D funds among scientific and engineering disciplines. Historically, the Defense Department, NASA and the Department of Energy were the largest sources of federal research support in electrical, mechanical and materials engineering, as well as physics, chemistry and aeronautical engineering. Decreases in these federal agencies' budgets for basic research and for development have translated into lower academic and industrial research funding for these disciplines. During the FY1993-1997 period, for example, constant dollar federal obligations for academic research in chemical engineering declined by 12.7%. Research funding for academic electrical engineering declined by 31.9% during the same period, and research funding for academic mechanical engineering declined by 40.9% (all figures taken from McGeary and Merrill, 1999). Federal research obligations for academic physics research declined by 21.8% in constant dollars during this period, and federal funding for academic geological research declined by 60%. The growth in biomedical research funding is also revealed in these data, as constant dollar federal obligations for biological sciences research increased

1. The economic consequences of this sharp reduction in defense-related R&D spending are difficult to project. Technological "spillovers" from defense to civilian applications of this spending now are less significant than was true of the 1950s and early 1960s, as the requirements for military and civilian applications in such technologies as aerospace and electronics have diverged. In addition, a considerable portion of federal defense-related R&D spending was directed to applied research, such as weapons testing, that generated few civilian economic benefits. Nevertheless, the enormous defense-related R&D budget contained a substantial basic research component, and defense-related R&D accounted for a considerable share of federally funded research in US universities in such areas as electronics. Reductions in spending in these areas could have negative consequences for civilian innovative performance.
2. Rapid growth in federal biomedical research funding has been more than matched by industrially funded expenditures. The share of total US biomedical R&D financed from federal sources has declined from more than 60% in 1980 to less than 50% in 1996.

by 16.1% during FY1993-1997, and federal funding for "other life sciences" grew by 74.3%.

The outlook for growth in overall federal R&D spending is uncertain. Legislative actions by the Senate and House of Representatives increased the federal R&D budget for both fiscal 1998 and 1999. In the case of research funding for the National Institutes of Health, legislators raised spending well above the levels requested by the White House. In the near term, however, projected budgetary surpluses may result in further increases in federal R&D spending, especially in biomedical research (the National Institutes of Health fiscal 1999 budget was increased by more than 14% over its fiscal 1998 level). Longer-term trends, however, are less favorable for civilian R&D spending growth, since spending growth in so-called "civilian discretionary" categories (which includes civilian R&D) is tightly constrained by spending ceilings passed in 1997.[1] Even if these budgetary restrictions are eliminated by legislative action, growth in entitlement spending for the elderly and healthcare will constrain growth in federal R&D spending. Federal R&D spending has declined from 38% of total national R&D spending in 1991 to 32% in 1996, and is not likely to increase significantly above this share in the foreseeable future.

Another important shift in the profile of total US R&D spending during the 1990s is the reduction in the share of "research" within overall "R&D" spending. During 1991-1996, spending on basic research from all sources in the United States grew by less than 1% per year, or a total increase of less than 5%. Industry-funded basic research grew by only $100 million during this period, from almost $7.4 billion in 1991 to less than $7.5 billion in 1996 (in 1992 dollars) – real federal spending on basic research increased by almost 5%, from $15.5 to almost $16.3 billion in constant dollars. Industry-funded investments in applied research grew by 4.9% during this period, while federal spending on applied research declined by nearly 14%.[2]

---

1. The large increases in R&D spending originally proposed by the Clinton Administration for fiscal 1999 relied on optimistic assumptions about funds from significant increases in cigarette taxes, and the even larger increases in FY1999 R&D spending embodied in the October 1998 spending legislation avoided Congressional spending restrictions only by making generous spending allowances for "emergencies".
2. Interestingly, despite the increasing share of basic research within the federal R&D budget (in fiscal 1998, 21.8% of total federal R&D spending, up from 14.2% in fiscal 1980 and 17.1% in fiscal 1990), the relative decline in federal R&D spending within the overall R&D budget means that basic research spending is likely to shrink within national R&D spending.

In other words, the upturn in real R&D spending during 1991-1996 that has resulted from more rapid growth in industry-funded R&D investment is largely attributable to increased spending by US industry on development, rather than research. Industry-funded real spending on "development" grew by more than 24% during 1991-1996, from $65 billion to almost $81 billion (federal development spending declined by more than 11% during this period, reflecting cutbacks in defense-related R&D spending). In 1991, basic and applied research spending together accounted for 40.7% of total US R&D spending from all sources; by 1996, these categories accounted for 38.1% of total R&D spending (National Science Board, 1996, 1998).

If they persist, these post-1990 spending trends could alter the profile of US R&D spending. The reduction in the federal government's share of overall R&D means that increased federal R&D spending will do less to offset the effects of any future reductions in the rate of growth in industry-financed R&D spending on overall US R&D spending levels. Since industry-funded R&D investment tends to move pro-cyclically, future trends in total US R&D spending are likely to be more sensitive to the domestic business cycle. In addition, the reduction in the federal government share of total R&D spending and the apparent shift in the profile of industry-funded R&D spending to favor development more heavily than "upstream" research activities (basic and applied research) could shorten the time horizon of overall US R&D investment. Finally, the recent changes in federal R&D spending priorities will have long-lasting effects on graduate training and advanced degree production in US universities, raising the fraction of biomedical disciplines in the output of advanced degree holders and reducing the share of engineers.

3.2 – Comparing structural change in the US R&D system with that in other OECD economies, 1981-1993

Data from the OECD suggest that the structural change observed during the past 20 years in the US R&D system parallels trends in other OECD economies. Table 2.1 contains data on trends during 1971-1993 in the distribution of R&D performance and funding among government, academia and industry in the OECD member states with the largest R&D budgets. These indicators suggest that the post-1981 restructuring of the sources of funding in the US R&D system is proceeding along lines that closely resemble those of the UK, France, Germany and Japan. The share of public funding of R&D has declined and industry R&D funding has grown; although the

Japanese government made a public commitment in 1996 to significantly increase public R&D spending, slow economic growth may constrain growth in public R&D spending (see Normile, 2 February 1997). The sharpest decline in public sector R&D funding among these five nations during 1981-1993 occurred in the United Kingdom, where the share of national R&D spending funded by public sources dropped by roughly one-third. In both the United States and France, public funding declined by approximately 10% of R&D spending, a decline in the public share of roughly one-fifth. The data for Germany and Japan reveal smaller declines in the public share through 1993.

TABLE 2.1 – Structural change in the five largest OECD R&D systems, 1971-1993

	Source of R&D finance (percentage)							
	Industry				Government			
	1971	1981	1991	1993	1971	1981	1991	1993
USA	39.3	48.8	57.5	58.7	58.5	49.3	40.5	39.2
Japan	64.8	67.7	77.4	73.4	26.5	24.9	16.4	19.6
France	36.7	40.9	42.5	46.2	58.7	53.4	48.8	44.3
Germany	52.2	57.9	61.7	60.2	46.5	40.7	35.8	37.0
UK	43.5	42.0	50.4	52.1	48.8	48.1	34.2	32.3

	Shares of total R&D performance (percentage)							
	Industry				Government			
	1971	1981	1991	1993	1971	1981	1991	1993
USA	65.9	70.3	72.8	71.2	15.5	12.1	9.9	10.2
Japan	64.7	66.0	75.4	71.1	13.8	12.0	8.1	10.0
France	56.2	58.9	61.5	61.7	26.9	23.6	22.7	21.2
Germany	63.7	70.2	69.3	66.9	14.2	13.7	13.9	14.8
UK	62.8	63.0	65.6	65.9	25.8	20.6	14.2	13.8

Note: The total to 100% is made of other national sources.

The shifts during 1981-1993 among universities, industry, and government in the performance of R&D within these five economies are less significant (again, with the exception of the United Kingdom, where a number of public research laboratories have been privatized). The data for the United States reveal very small increases (a shift of less than 1% in the share of each) in the share of R&D performed by industry and universities, and a slightly larger decline (of nearly 2%) in the share of R&D performed in government laboratories. The share of publicly performed R&D in both Japan and France declined by comparable or slightly larger amounts, while the German data (which include the effects of unification) rise modestly. The United Kingdom data, however, reveal a sharp decline of nearly 7% in the share of R&D performed in government facilities.

If there is an "outlier" in these measures of structural change since the early 1980s, it is the United Kingdom, rather than the United States. In general, the shifts in funding sources in the US and these other economies are slightly larger than the shifts in performance. Indeed, the contrasting magnitude of the trends in funding, as opposed to performance, reflects the difficulties of undertaking radical structural changes in national R&D systems of the sort illustrated by the United Kingdom. The high political costs of closing or privatizing large public research establishments vastly outweigh those associated with the gradual shrinkage of such facilities through incremental shifts in the shares of public and private R&D spending.

US policy appears to have moved further in extending formal intellectual property protection to the results of publicly funded research than has occurred in other large OECD economies (see below for further discussion). Indeed, the overall growth in the use of formal intellectual property protection in the United States since the early 1980s contrasts with trends in other large OECD economies. Patent applications from US inventors in the United States have soared since the mid-1980s,[1] and patenting by US universities has grown even more dramatically (see below). Patent applications by domestic inventors in Germany, France, Japan and the United Kingdom display no comparably sharp surge in growth from the mid-1980s (Kortum and Lerner, 1999).

4. OTHER CHANGES IN THE UNITED STATES SINCE 1980

Despite their Administrations' rhetorical rejection of intervention in the market, significant technology policy initiatives were undertaken under Reagan

1. Kortum and Lerner (1999) find a significant surge in patenting by domestic inventors in the United States, noting that "until the mid 1980's, [patent] applications fluctuated within a band of 40 to 80 000 per year, but in 1995 US inventors applied for over 120 000 patents on their inventions" (p. 1). Kortum and Lerner's data reveal a steep increase in patents applied for and granted to domestic inventors in the United States since 1980.

and Bush.[1] Faced with more intense foreign competition from firms that had improved their abilities to exploit scientific advances made in US and other foreign laboratories, federal policymakers began to experiment with programs that sought to strengthen civilian technological capabilities by subsidizing and promoting joint research, encouraging collaboration between US universities and industry in technology development, and supporting collaboration between US industry and the federal laboratories.

Programs such as the National Center for Manufacturing Sciences (NCMS), the semiconductor research consortium SEMATECH, the Advanced Technology Program (ATP) of the Department of Commerce, and even the National Science Foundation's Engineering Research Centers all represented significant departures from the postwar structure of federal science and technology policy. The SEMATECH and NCMS programs relied on expanded funding by the Defense Department for civilian technology development in so-called "dual-use" technologies, based on the premise that the economic benefits of "spin-offs" from military to civilian technology applications had declined.[2] These initiatives expanded the federal role in supporting civilian technology development, especially in specific high-technology sectors that were believed to be important to both civilian economic competitiveness and national security. A small but growing share of the defense R&D budget flowed to "dual-use" technology development projects, under the approving oversight of Congress.

Other US initiatives in technology policy during the Reagan and Bush Administrations reduced antitrust restrictions on collaboration in research and improved enforcement of intellectual property protection. Justice Department guidelines and review procedures for mergers were relaxed somewhat, and major federal antitrust suits against high-technology firms were dropped or settled in the early 1980s. The Reagan Administration also supported the 1984 National Cooperative Research Act, which reduced the antitrust penalties for collaboration among firms in pre-commercial research. The NCRA has been credited with facilitating the early growth and operation of the Microelectronics and Computer Technology Corporation, a research consortium involving US computer and electronics firms. In 1993, the NCRA was extended to cover joint production ventures.

---

1. The discussion in this section draws on Ham and Mowery (1995, 1997).
2. Federal funding for the SEMATECH consortium ended in 1996.

Shifts in US policy toward intellectual property rights began with the 1982 legislation that established the Court of Appeals for the Federal Circuit, which strengthened the protection granted to patent holders.[1] In addition, the US Supreme Court upheld the validity of a broad patent in the new industry of biotechnology in its 1980 *Diamond v. Chakrabarty* decision, facilitating the patenting and licensing of inventions in this sector. The US government also pursued stronger international protection for intellectual property rights in the Uruguay Round trade negotiations and in other bilateral venues. The faith in intellectual property rights as a critical policy tool in improving US competitiveness was exemplified in two other statutes of the 1980s. The Federal Technology Transfer Act of 1986 and amendments passed in 1989 authorized federal laboratories to conduct cooperative research and development agreements (CRADAs)[2] with private firms.

The Bayh–Dole Patent and Trademark Amendments Act of 1980 (sponsored by a leading Democratic and a leading Republican senator) permitted performers of federally funded research to file for patents on the results of such research and to grant licenses for these patents, including exclusive licenses, to other parties. The Act facilitated university patenting and licensing, discussed below in greater detail, in at least two ways. First, it replaced a complex web of agreements that had been negotiated between individual universities and federal agencies with a uniform policy. Second, the Act's provisions represented a strong Congressional expression of support for the negotiation of exclusive licenses between universities and industrial firms for the results of federally funded research.

The Bayh–Dole Act responded to a belief by policymakers (based on little or no evidence) that stronger protection for the results of publicly funded R&D would accelerate their commercialization and the realization of these

1. According to Katz and Ordover (1990), at least 14 Congressional bills passed during the 1980s focused on strengthening domestic and international protection for intellectual property rights, and the Court of Appeals for the Federal Circuit created in 1982 has upheld patent rights in roughly 80% of the cases argued before it, a considerable increase from the pre-1982 rate of 30% for the Federal bench.
2. A CRADA specifies terms under which a private organization provides personnel, equipment or financing for R&D activities that are consistent with a specific laboratory's broader mission. Most CRADAs include provisions that cover the sharing of intellectual property rights to any technologies resulting from the project.

economic benefits by US taxpayers.[1] These new policies of support for patenting the results of public R&D programs also appeared to promise increased economic returns for little additional investment of public funds, an attractive feature for policymakers dealing with severe fiscal constraints.

The significance of this new US policy towards the protection of the results of public R&D programs has been noted in work by Eisenberg (1996), Nelson and Romer (1997) and David and Foray (1996). Among other things, it upends the classic market failure analysis of Arrow (1962) and Nelson (1959). Both of these scholars, especially Arrow, argued that the "non-excludable" characteristics of the results of fundamental research meant that pricing these results at a sufficiently high level to reward their discoverer was socially inefficient. In addition, Arrow noted that the market for inventions itself was prone to failure, because of the paradox of information, uncertainty and other problems.[2] The Bayh–Dole Act and its various modifications adopt the opposite view: unless the discoverer can establish property rights to these results, mechanisms for their distribution and transfer will fail, preventing the commercialization of these results and the realization of a return to investments of public funds.

Rather than emphasizing public funding and relatively liberal disclosure and dissemination, the Bayh–Dole Act assumes that restrictions on dissemination of the results of many R&D projects will enhance economic efficiency by supporting their commercialization. In many respects, the Bayh–Dole Act is the ultimate expression of faith in the "linear model" of innovation – if basic research results can be purchased by would-be developers,

1. The "evidence" of unsuccessful commercialization of federally owned patents consisted mainly of numerous citations to the tiny percentage of the 28 000-30 000 patents owned by the Federal Government that had been commercialized. Advocates of the Bayh–Dole Act overlooked the fact that title to most of these patents, which resulted from federal defense contracts, had been ceded to the Federal Government by private contractors who had not invoked their rights under the policies then prevalent in the Defense Department to retain title to the patents. In other words, as Eisenberg (1996) points out, the statistical data cited in support of the Bayh–Dole Act suffered from a serious selection bias.
2. Indeed, these recent policy initiatives rely heavily on the efficiency of markets for licensing intellectual property, despite the extensive scholarly evidence on the transactional and other problems associated with such markets. Inefficiencies or failures in these markets arise from a number of sources, including the Arrow paradox of information, the high levels of uncertainty over the quality of the technology offered for license and over the behavior of the licensee, and the small numbers of buyers and sellers in markets for specific technologies (Caves, Crookell and Killing, 1983; Williamson, 1979, 1985; Mowery, 1983, 1988).

commercial innovation will be accelerated. However, as Nelson and Romer or David and Foray point out, the effects of raising the private returns to invention are likely to differ from those of R&D subsidies, the Arrow recommendation.[1] In an illustration of the difficulties of radical shifts in domestic policy, as well as the strength of the bipartisan support for the incremental changes made thus far, the Clinton Administration has not dramatically shifted these elements of US technology policy. During its first term, the Clinton Administration increased support for civilian technology adoption by expanding the Advanced Technology Program (ATP), which provides matching funds for firms and consortia for the development of "pre-commercial" technologies. Funding for the Manufacturing Extension Partnership (MEP) program, which is administered by the National Institute for Standards and Technology (NIST) of the Commerce Department, increased under the Clinton Administration. The Clinton Administration also expanded initiatives in the "dual-use" technology area.[2]

Although they assuredly had little precedent in postwar US R&D policy, none of these technology policy initiatives consumed significant budgetary resources; together they accounted for no more than 2%-3% of federal R&D spending in the first Clinton Administration. Moreover, a combination of factors, including the election of a Republican Congress in 1994 and the significantly improved performance of US firms in industries such as semiconductors (see Macher et al., 1998), led to reductions in funding for even these limited initiatives (federal funding for SEMATECH, for example, ceased in 1996). The Advanced Technology Program (ATP) in the Commerce Department continues to operate, but at a much more modest scale than originally planned. Interestingly, with the exception of the ATP, these technology programs have received minimal formal evaluation.

---

1. The internal tensions, not to say contradictions, in this policy posture can be illustrated by reference to the Advanced Technology Program of the US Department of Commerce. ATP was created in 1989 to fund "high-risk, high-spillover" R&D by private firms, providing matching funds to projects that are scrutinized on the basis of their technical and economic merit. Having set out to provide public funds to support the creation of spillovers, however, the architects and managers of this program also insist that protection of the intellectual property created in the program is essential for the commercial applications of ATP-funded research results to be achieved. For further discussion, see Yager and Schmidt (1997).
2. Illustrating the elastic character of current definitions of "dual-use" R&D spending, the February 1995 estimate omitted "dual-use S&T investments" in the Defense Nuclear Agency, the Ballistic Missile Defense Organization and the three uniformed services that had been included in earlier tabulations, presumably by redefining them as "military-only" R&D programs (US Department of Defense, 1995, p. 10).

5. STRUCTURAL CHANGE IN THE US INNOVATION SYSTEM

The changes in federal policy, along with change in the competitive and technological environment of many US firms, appear to be creating a new "division of labor" in the US innovation system. The major elements of structural change, many of which affect industrial R&D, include the following:

- Increased reliance by US firms on sources of R&D outside their organizational boundaries, through such mechanisms as consortia, collaboration with US universities and federal laboratories, and strategic alliances with other US and foreign firms.
- Expanded performance of offshore R&D by US firms and increased performance by non-US firms of industrial R&D within the United States.
- Increased reliance by US universities on US and foreign industrial firms for research funding and expanded efforts by US universities to patent and license the results of research performed within them.

Space does not permit a detailed descriptive analysis of these trends. But along with the reduced presence of federal funding within the US innovation system, at least some of these elements of change are likely to result in a US innovation system that more closely resembles those in other industrial economies.

5.1 – "Externalization" of industrial R&D

Beginning in the 1980s, a combination of severe competitive pressure and the perception of disappointing returns from their rapidly expanding investments in internal R&D contributed to the decision by many US firms to "externalize" a portion of their R&D operations. Pioneers of industrial R&D, including General Electric, AT&T and Du Pont, reduced their central corporate R&D laboratories and a number of alternative arrangements appeared, many of which relied on collaboration between US firms and other R&D performers.

R&D collaboration was also given an impetus by the shifts in federal policy during the 1980s and 1990s from neglect and/or hostility to encouragement and occasionally, financial support. This policy shift, along with the desire of large US firms to rely more heavily on external sources of R&D, has spawned a diverse array of collaborative arrangements. Among the most important of these are industry-led consortia; collaborations between industrial firms and US universities; and collaborations between federal laboratories and industrial firms. US firms formed more than 450 "research joint ventures", as reported by US firms under the terms of the National Cooperative Research Act, during 1985-1994 (Link, 1995); Link's survey found that the majority of these research joint ventures focused on process R&D. US firms have expanded their domestic alliance activities as well, and the formation of "intranational" strategic alliances involving US firms has outstripped the formation of international alliances.

In addition to research joint ventures and consortia, US firms entered into numerous "strategic alliances" with foreign firms during the 1980-1994 period. The majority of the international strategic alliances in the National Science Foundation's database link US and Western European firms (National Science Board, 1996). Alliances between US and Japanese firms were also widespread. Both intranational and international alliances involving US firms appear to be most numerous in biotechnology and information technology. In contrast to the bulk of domestic research consortia, a large proportion of US firms' alliances with foreign firms focused on joint development, manufacture or marketing of products. In addition to the cost-sharing and technology-access motives that also underpinned the formation of many domestic research joint ventures, the international alliances of US firms were motivated by concerns over access to foreign markets (Mowery, 1988).

5.2 – "Internationalization" of the US R&D system

A second important characteristic of the structural change in the US innovation system during the 1980s was the progressive "internationalization" of the US industrial R&D system, as US firms expanded offshore R&D operations and as foreign firms expanded their R&D activities within the United States. Table 2.2 reproduces data from the 1996 and 1998 editions of *Science and Engineering Indicators* (National Science Board, 1996; 1998) on trends during 1980-1995 in US outward R&D investment, measured as a share of industry-financed R&D spending overall and in each of 11 broad sectors. One of the most interesting points to emerge from this table is the modest growth in the share of total US industry-financed R&D that is invested in foreign-sited R&D. Rather than a consistent increase, this share experienced a significant decline during 1980-1985, a steady increase through 1990, and an increase from 1990-1995. Across the entire time 1980-1995 period, the share of industry-financed R&D devoted to foreign R&D grew by slightly more than 2%.

The table also underscores the intersectoral differences in these trends: electrical equipment, petroleum extraction and refining, pharmaceuticals and non-electrical machinery display declines during this period in the shares of their R&D spending allocated to foreign activities, while that of scientific instruments increases slightly. The chemicals industry and non-manufacturing industry (for which the time series coverage is especially imperfect) both display significant increases in the share of their R&D spending devoted to foreign sites.

The share of industrial R&D performed within the United States that was financed from foreign sources grew substantially during this period, from 3.4% in 1980 to more than 11% in 1995 (National Science Board, 1998). Despite this growth, as of 1994 foreign sources financed a smaller share of industrial R&D performed within the United States than is true of Canada, the United Kingdom or France. Foreign firms also participated in the formation of research joint ventures with US firms. According to Link (1995), 32% of the research joint ventures filing under the terms of the National Cooperative Research Act during 1985-1994 listed foreign firms among their members. Finally, a number of foreign firms operating R&D facilities in the United States pursued collaboration with US universities. According to Florida (1997), more than 50% of the Japanese R&D laboratories in the

TABLE 2.2 – Share of company-financed R&D performed abroad by US firms and their subsidiaries, 1980-1995

Industry	1980	1985	1990	1995
Food	na	6.6	3.3	9.1
Chemicals	9.8	9.2	9.9	22.9
Pharmaceuticals	20.3	11.5	21.8	25.1
Petroleum	10.1	2.1	3.3	4.3
Stone, etc.	na	na	11.0	7.0
Primary metals	1.9	na	3.6	4.5
Fabricated metals	na	2.7	12.9	11.8
Machinery	11.4	6.4	10.7	5.2
Electrical machinery	8.3	6.4	8.3	5.1
Transport equipment	14.7	8.5	14.4	na
Scientific instruments	7.6	3.7	9.7	11.6
Non manufacturing ind.	0.7	0.4	2.5	8.0
Total	10.4	6.4	9.7	12.0

Source: National Science Board (1998).

United States, more than 80% of the US-sited French R&D laboratories, and almost 75% of German corporate R&D laboratories in the United States were involved in such collaborative agreements.

5.3 – A changing role for university research?

The historic involvement of publicly funded universities in the United States with agricultural research, much of which was applied in character, and the involvement of these universities with the agricultural users of this research, are well-known aspects of US economic history. But throughout this century, the decentralized structure of US higher education and the dependence of public and private universities on local sources of funding also meant that in a broad array of non-agricultural fields, ranging from engineering to physics and chemistry, collaborative research relationships between university faculty and industry were common (Rosenberg and Nelson, 1994).[1]

Thus it is a fallacy to think of US university research as traditionally basic and conducted with no attention to practical objectives. Many important advances in applications have emerged from academic research in the United States, and much industry–university collaboration historically has focused on engineering and applied sciences. Through much of the 1900-1940 period, US universities, especially public universities, pursued extensive research collaboration with industry. Indeed, the academic discipline of chemical engineering was largely developed through such collaboration between US petroleum and chemicals firms and M.I.T. and the University of Illinois (Rosenberg, 1998).

In view of the applied character of a good deal of their research, it is not surprising that a number of US research universities were active in patenting and licensing faculty inventions through much of the twentieth century. Beginning in 1926, the University of California required all employees to report patentable inventions to the university administration. Other universities, such as M.I.T. and the University of Wisconsin, also developed administrative units to help patent and license inventions resulting from research.

1. Indeed, the Research Corporation, which served for many years as a leading "broker" and licensor of university inventions for many US universities, was founded in 1912 by Frederick Cottrell, a University of California Berkeley professor, to commercialize his electrostatic antipollution innovations. The Corporation's licensing activities now are managed by an independent organization, Research Corporation Technologies, founded in 1987 (for additional discussion of the history of the Research Corporation, see Mowery and Sampat, 1999.)

During the pre-1940 period, most of this patenting and licensing activity was conducted directly with industrial sponsors of academic research and through organizations such as the Research Corporation.

World War II transformed the role of US universities as research performers and the sources of their research funding. Universities' share of total US R&D performance grew from 7.4% in 1960 to 14.5% in 1997, and universities accounted for 59% of the basic research performed within the United States in 1997 (National Science Foundation, 1998). Much university research nevertheless retained an applied character, reflecting the importance of research support from such federal "mission agencies" as the Defense Department.

The share of industry funding declined within the greatly expanded research budgets of postwar US research universities through the 1950s and 1960s. Beginning in the 1970s, the share of industry funding within academic research began to grow. In 1970, federal funds accounted for 70.5% of university-performed research and industrial support accounted for 2.6%. By 1997, the share of federal funds in supporting academic research had been reduced to 59.6%, and industry's contribution had tripled to 7.1% (National Science Board, 1998). Most of the increase in industry funding occurred during the 1980s; since 1990 this industry share has remained roughly constant.

The increased importance of industry in funding university research is reflected in growth in the number of research institutes at US universities seeking to support research on issues of direct interest to industry. By 1992, more than 1050 of these had been established, and data from Cohen, Florida, and Goe (1994) indicate that 57% of all such institutes in existence as of 1992 were established during the 1980s. Nearly 45% of these institutes involve one to five firms as members, and more than 46% of them rely on government funds for support in addition to (or in some cases, in lieu of) support from industry.

This renewal and strengthening of links between university and industrial research accelerated during the 1980s, and was especially visible in expanded university licensing of patented inventions. Having grown by approximately one-third between 1969 and 1974, the number of patents issued to US universities and colleges more than doubled between 1979 and 1984, more than doubled again between 1984 and 1989, and doubled yet again between 1989 and 1997 (Table 2.3). Trajtenberg, Henderson, and Jaffe (1994) noted that the share of all US patents accounted for by universities grew from less than 1% in 1975 to almost 2.5% in 1990. Moreover, the ratio of patents to R&D

spending within universities almost doubled during 1975-1990 (from 57 patents per $1 billion in constant dollar R&D spending in 1975 to 96 in 1990), while the same indicator for all US patenting displayed a sharp decline (decreasing from 780 in 1975 to 429 in 1990). In other words, universities increased their patenting per R&D dollar during a period in which overall patenting per R&D dollar was declining.

In tandem with increased patenting, US universities expanded their efforts to license these patents. The Association of University Technology Managers (AUTM) reported that the number of universities with technology licensing and transfer offices increased from 25 in 1980 to 200 in 1990, and licensing revenues of the AUTM universities increased from $222 million in fiscal 1991 to $698 million in fiscal 1997 (Association of University Technology Managers, 1998).

TABLE 2.3 – Utility patents issued to US universities and colleges

Year	Number of US patents
1969	188
1974	249
1979	264
1984	551
1989	1 228
1994	1 780
1997	2 436

Note: Year of issue. Source: US Patent and Trademark Office (1998).

This transformation in the role of the research university within the US innovation system is widely attributed to the Bayh–Dole Act of 1980. The Act appears to have been an important factor, but it is likely that US universities (even those not active in patenting and licensing before 1980) would have expanded patenting and licensing even in the absence of the Bayh–Dole Act. Other factors that influenced the growth of university patenting and licensing include changes in US intellectual property rights policy, along with the continuing growth in federal biomedical research funding and the important scientific advances achieved in this field.

Although US universities have expanded their patenting activities since 1981, the data in Table 2.3 (as well as similar data reproduced in Henderson et al., 1998) suggest that this growth represented an acceleration of a trend that is discernible in the pre-1980 period. Moreover, this increased patenting was dominated by growth in biomedical patents: biomedical patents issued to US universities increased by 123% during the 1969-1979 period, while non-biomedical patents grew by only 22%. Part of this growth in overall university patenting during the 1970s was accounted for by universities not previously active in patenting: the number of US universities with at least one patent increased from 36 in 1970 to 86 in 1980. The post-1980 growth in university patenting and licensing thus appears to be an acceleration of an established trend.

A more detailed analysis of the effects of Bayh–Dole on university patenting and licensing is contained in Mowery et al. (2001), which examines data on the patenting and licensing activities of three academic institutions that have been the leading recipients of licensing and royalty income for much of the 1990s: Columbia University, the University of California, and Stanford University.[1] As their dominant position in the rankings of licensing and royalty income suggests, these three research universities are among the most important practitioners of the new approach to university technology transfer. Moreover, two of the three universities (Stanford and the University of California) were active in technology licensing well before the passage of the Bayh–Dole Act, but Columbia University was not. Comparison of the Stanford and University of California. data "before and after" Bayh–Dole, as well as comparison of the data from all three universities for the 1980s and 1990s, supports some preliminary findings about the effects of the Bayh–Dole Act on university technology transfer.

At both the University of California and Stanford University, the number of annual faculty invention disclosures and (of greater importance) the share of biomedical technologies within these annual invention disclosures, increased before 1980 (Mowery et al., 2001). Inasmuch as the licensing activities and income of both of these universities during the 1980s and early 1990s were dominated by biomedical technologies, we conclude that the direct effects of Bayh–Dole on these universities' licensing activities were modest. Other evidence suggests that Columbia University, with very little involvement in patenting and licensing of faculty inventions before 1980, was also

1. According to the annual report of the Association of University Technology Managers (AUTM), FY1993 gross licensing revenues for the University of California system, Stanford University, and Columbia University respectively amounted to $45.4 million, $31.2 million, and $21.1 million; the fourth and fifth leading academic institutions in this ranking, the University of Wisconsin-Madison and the University of Washington, respectively earned $15.8 million and $14.8 million (Association of University Technology Managers, 1994, p. 8).

beginning to re-evaluate its policies and had filed for several key patents before the passage of the Bayh–Dole Act.

Although this evidence indicates that the Bayh–Dole Act had surprisingly modest effects on the activities of universities long active in patenting and licensing, the effects of Bayh–Dole on the marketing efforts of US universities appear to have been considerable. The passage of this law hastened or caused the entry by many universities (such as Columbia) into patenting and licensing activities that they formerly avoided as a matter of policy. Our evidence also suggests that even at universities long active in patenting and licensing of faculty inventions, administrators intensified their efforts to gain access to and/or market these inventions. Despite their very different histories of patenting and licensing before and after 1980, the data in Table 2.4 reveal remarkable similarities in the patenting and licensing activities of these three universities by the end of the first decade of the Bayh–Dole Act.

Another issue of considerable interest to observers of the growth in US universities' patenting and licensing activities concerns the effects, if any, of increased patenting and licensing on the mix of applied and basic research within academia. Specifically, have the increased financial incentives for patenting and licensing of their inventions by universities caused faculty to pursue more applied research topics? This question is very difficult to answer, but most evidence suggests that the "research culture" of US universities thus far has not been broadly transformed. As Mowery et al. (2001) point out, the upsurge in patenting and licensing remains confined to a relatively narrow portion of most universities' faculty, mainly in the biomedical sciences and a few areas of engineering.

A closely related issue concerns change since Bayh–Dole in the "importance" of US universities' patented inventions. Henderson et al. (1998) have concluded that US university patents issued after 1980 are less "general" (less heavily cited across a broad range of technical fields) and less "important" (less heavily cited overall), relative to industrial patents, than pre-1980 university patents. On its face, this evidence may suggest some changes in the internal "research culture" of US universities in the wake of Bayh–Dole, revealed in a shift of faculty research toward more applied topics that yield patents of lower generality and importance.

A further analysis of this issue (Mowery and Ziedonis, 1999), however, indicates little change in the characteristics of the patents issued to the University of California and Stanford after 1980. Although limited to only two leading research universities, this finding suggests that the results in Henderson et al. (1998) may reflect the entry into patenting and licensing of universities with little experience in these activities, rather than a change in the characteristics of the patenting of the US universities that had more substantial pre-1980 patent portfolios.

Mowery and Ziedonis (1999) also analyzed patenting before and after 1980 on a larger sample of US university patents that distinguishes between institutions with substantial experience in patenting before 1980 ("incumbents"), universities with fewer than ten patents issued during 1975-1980 ("low-intensity incumbents"), and universities that received no patents before 1980 ("entrants"). The patents issued after 1980 to the "low-intensity incumbents" and the "entrants" are of significantly lower importance and

TABLE 2.4 – Selected data on University of California, Stanford University and Columbia University licensing income, FY1970-1995

Fiscal year	1970	1975	1980	1985	1990	1995
University of California						
Gross income	1 140	1 471	2 114	3 914	13 240	58 556
Gross income from top 5 earners	900	1 075	1 083	1 855	7 230	38 666
Share of gross income from top 5 earners (%)	79%	73%	51%	47%	55%	66%
Share of income of top 5 earners associated with biomedical inventions (%)	34%	19%	54%	40%	91%	100%
Share of income of top 5 earners associated with agricultural inventions (%)	57%	70%	46%	60%	9%	0%
Stanford University		FY76				
Gross income	180	843	1 084	4 891	14 757	35 833
Gross income from top 5 earners		579	938	3 331	11 203	30 285
Share of gross income from top 5 earners (%)		69%	86%	69%	76%	85%
Share of income of top 5 earners associated with biomedical inventions (%)		87%	40%	64%	84%	97%
Columbia University						
Gross income				542	6 903	31 970
Gross income from top 5 earners				536	6 367	29 936
Share of gross income from top 5 earners (%)				99%	92%	94%
Share of income of top 5 earners associated with biomedical inventions (%)				81%	87%	91%

Note: In thousands of dollars 1992.

generality than those issued to the "incumbents". These two groups account for a growing share of all patents issued to US universities after 1980 (their share increased from 8.4% in 1976 to 31.7% in 1992). We therefore conclude that the entry of less experienced universities into patenting, rather than change in the patenting activities of universities that have long been active in this area, is an important factor in the changing characteristics of the post-1980 patent portfolios of US universities. As these less experienced patenters and licensors learn more about the costs and returns to these activities, their patenting strategies should become more selective and the differences between their patent portfolios and those of more experienced academic patenters may decline.

In summary, the data compiled by leading US research universities on their rapidly expanding post-1980 patenting and licensing activities suggests that the Bayh–Dole Act was but one of several factors underpinning the transformation in universities within the US innovation system. The growing importance of biomedical research, much of which relied on federal support that expanded significantly during the 1970s, appears to have been at least as important as Bayh–Dole in explaining increased university patenting and licensing after 1980. Other factors, especially the shifts in US government policy toward intellectual property rights enforcement in general, and especially toward intellectual property rights in biotechnology, also encouraged the growth of university patenting in the biomedical and other areas. Growth in federal funding for biomedical research if anything accelerated during the 1990s, and expanded biomedical research funding should provide an additional impetus to growth in university patenting and licensing. The shift during the 1980s in the role of the research university within the US national innovation system seems likely to be an enduring one.

Is Bayh–Dole a useful model for other nations' university systems?

A number of other industrial-economy governments, such as Japan and Australia, have passed legislation with provisions that resemble those of the Bayh–Dole Act, in an effort to strengthen the contributions of their universities to industrial innovation. A number of other industrial-economy governments are considering similar policies. But the data discussed above suggest that the effects of the Bayh–Dole Act on US universities' technology transfer activities are easily overstated. Although Bayh–Dole drew additional universities into technology licensing activities, the effects of this law probably would have been far more modest in the absence of large-scale federal funding for biomedical research. In addition, US universities would have found it far more difficult or impossible to patent and license the results of biomedical research in the absence of supporting initiatives in intellectual property rights.

Another reason for caution in emulating the Bayh–Dole Act stems from the earlier discussion of the historical role of US universities in the US national innovation system. As I noted in that discussion, US universities have long maintained relatively close links with industrial research, although the strength of these links has waxed and waned over the course of this century. But the factors that gave rise to these links in the 1920s are equally relevant to an assessment of the effects of the Bayh–Dole Act in the 1990s. The US university system is virtually unique among those of the OECD in combining (1) large scale; (2) a dominant role in the performance of publicly funded basic research; and (3) an extremely decentralized structure, characterized by fierce competition among universities for research funds, faculty, students and distinction. In the absence of these supporting "structural" factors (especially the third), legislation similar to the Bayh–Dole Act would have less effect on university–industry links and technology transfer activities. Similar legislation may be necessary, but it is by no means sufficient, in other national university systems.

In addition to skepticism about the effectiveness of similar legislation on other national university systems, the inherent desirability of such policies deserves careful consideration and debate. Although university patenting need not inevitably limit technology diffusion, restrictive licensing terms (for example, exclusive licenses that cover a broad array of possible fields of use) could reduce the speed and coverage of "knowledge distribution" within the US national innovation system and within the international innovation system (David and Foray, 1995).

Bayh–Dole and related initiatives that seek to strengthen the contributions of research universities to economic growth are based on a narrow view of the channels through which universities and faculty interact with industry and affect the innovation process. Not only publishing, but conferences, consulting, graduate education and service on scientific advisory boards all operate to transfer academic research results to industry and vice versa. The incentives created by Bayh–Dole may lead to one channel being favored at the expense of others. Emphasis on patenting, rather than publishing, the results of research may well reduce the volume of information flowing through conventional channels of dissemination in favor of market-based transactions.

6. THE GOVERNMENT PERFORMANCE AND RESULTS ACT

In 1993, Congress passed the Government Performance and Results Act (GPRA), which sought to bring strategic planning practices widely employed in the private sector into wider use within government agencies. The GPRA mandated that all federal agencies must: (1) draft a strategic plan that lays out agency goals over a five-year planning horizon; (2) develop annual performance plans that link these broad goals to agency activities; and (3) provide Congress with annual reports assessing agency performance in meeting their strategic goals. Implementation of the GPRA has been gradual, since Congressional oversight committees insisted on evaluating agencies' draft strategic plans and draft annual performance reports. Nevertheless, by the end of 1997, most agencies had developed statements of long-term goals and frameworks for evaluating their annual progress toward achieving these goals.

Federal agencies most heavily involved in supporting R&D expressed considerable concern over the GPRA, because of its heavy emphasis on quantifiable goals and performance measures. Since R&D, especially basic research, is an inherently uncertain process, quantifying outcomes and measuring progress toward agency goals is very difficult. Moreover, meeting the (alleged) requirements of the Act for quantifiable targets and indicators might lead senior managers in many agencies to emphasize short-term, incremental research strategies and programs. In other words, the GPRA's insistence on measurement of goals and performance would cause major R&D-funding agencies to reduce their commitments to long-term, fundamental research. In addition, measurement of the "quality" of research is difficult, and therefore likely to receive insufficient emphasis in agencies' GPRA reports.

Are these fears justified? Have R&D agencies emphasized incremental, short-term goals in their reports, and has this emphasis affected the types of research they support? At present, these questions cannot be answered with more than anecdotal evidence. As Vogelsang (1998) points out, the GPRA explicitly allows for non-quantitative implementation designs, and several agencies have adopted such an approach. On the other hand, Cozzens (1998) suggests that a number of R&D funding agencies have in fact concentrated on short-term goals and goals that emphasize process, rather than outcomes. No credible evidence has yet been developed, however, that suggests that these plans or other agency reactions to GPRA have affected agencies' allocation of resources to short- or long-term research.

The GPRA implementation process is in its early stages. Thus far, this process has highlighted the limited ability of most US R&D funding agencies to develop or conduct evaluations of their performance. There is little indication that techniques from the "foresight" exercises employed elsewhere in the OECD are being adapted to the demands of the GPRA. Indeed, the almost complete absence of "foresight" planning exercises in the federal government's R&D programs is a striking contrast with the situation in most OECD governments. Over time, the full implementation of the GPRA could affect other institutional actors in the publicly funded sector of the US R&D system. For example, universities may be asked to compile indicators of the performance of individuals or institutional research performance for use by their major sources of federal funding. At present, however, any such "ripple effects" are in their earliest stages and scarcely perceptible.

7. CONCLUSION

Virtually all of the central components of the innovation system that emerged in the postwar US economy now are undergoing change. The future US innovation system is likely to be characterized by:

- lower levels of overall federal R&D funding;
- proportionately lower levels of investment in research, as opposed to development;
- a higher level of internationalization, both in terms of US R&D investment in foreign economies and in terms of higher levels of non-US R&D investment within the domestic US economy;
- more stringent domestic and international protection of intellectual property rights;
- higher levels of inter-firm collaboration, university–industry collaboration, and collaboration between US and foreign firms in R&D;
- greater efforts by US universities to seek to protect and license the results of publicly and privately funded research.

The implications of these changes for the performance of the US innovation system and for the role of this system within the global science and technology system are unclear. The effects of some of these changes, such as the efforts by US universities to protect and market the results of their research, may be modest. But these changes, along with those affecting defense-related procurement, the role of universities, and intellectual property rights and antitrust policies, could reduce the importance of new firms in the commercialization of new technologies and in the creation of new industries. The effects of these changes are mediated and possibly offset, however, by the abundance of venture capital for the foundation of new technology-intensive firms within the United States, as well as the relatively modest entry barriers in segments of such rapidly growing "new industries" as computer software and multimedia.

What may be most striking about the US R&D system at the end of the twentieth century is the remarkably limited extent of any change in the structures and institutions through which science and technology policy are formulated. As I pointed out earlier, US science and technology policy in the 1990s, no less than in the 1950s, is really the outcome of decisions made independently in various federal agencies and Congressional committees, with remarkably little attention to a consistent set of overall national goals. Although the end of the Cold War is widely acknowledged to have replaced a "national security" rationale for federal R&D policy with an "economic" rationale, the influence of even this original rationale was always limited, and its alleged replacement has yet to be made operational.

In a complex federal political system such as that of the United States, the very slippery nature of this new "economic rationale" may serve as a justification for the use of the federal R&D budget by Members of Congress and Senators as a new "pork barrel", a source of public largesse for district- and state-level economic and political benefit. Indeed, some evidence suggests that redistributive politics are playing a more prominent role in R&D funding decisions. During its 1999 session, Congressional intervention in the allocation of research funds among US universities reached unprecedented levels, affecting nearly $1 billion in spending (Brainard and Cordes, 1999).

Structural change in the US innovation system is not occurring in isolation from change in the structure of other industrial economies' innovation systems. Recent trends in the mix of public and private R&D spending in the United States parallel those observed in many other OECD economies. Indeed, one of the defining characteristics of such structural change in the United States is increased links with non-US firms and government-supported programs. Structural change in the US and foreign nations' innovation systems may well result in some "convergence" in structure, as competition policy in the European Union becomes more stringent and venture capital becomes more abundant in the EU and elsewhere. In the long run, these tendencies toward convergence may reduce some of the "structural frictions" in trade and technology policy that have pervaded recent bilateral and multilateral negotiations (Ostry, 1989; Mowery and Rosenberg, 1989). But the pace of such "convergence" will be slow and frictions will continue to arise, creating a crowded policy agenda for the foreseeable future.

ACKNOWLEDGEMENT

An earlier version of this chapter was presented at a workshop at the Commissariat General du Plan, and comments from the participants in the workshop, as well as the editors of this volume, are appreciated. Research for this chapter was supported by the Alfred P. Sloan Foundation, the Andrew Mellon Foundation, the California Policy Seminar, and the US–Japan Industry and Technology Management Training Program. Michael Barnes, Rose Marie Ham and Arvids Ziedonis provided invaluable assistance with data collection and analysis.

REFERENCES

Arrow, K.J., 1962, Economic Welfare and the Allocation of Resources for Invention, in R.R. Nelson, ed., *The Rate and Direction of Inventive Activity*, Princeton, NJ: Princeton University Press for the NBER.

Association of University Technology Managers, 1994, *The AUTM Licensing Survey: Executive Summary and Selected Data, Fiscal Years 1993, 1992, and 1991*, Norwalk, CT: Association of University Technology Managers.

Brainard, J. and C. Cordes, 1999, Pork-Barrel Spending on Academe Reaches a Record $797 Million, *Chronicle of Higher Education*, 23 July, p. A44.

Cohen, W., R. Florida and R. Goe, 1994, *University–Industry Research Centers in the United States*, technical report, Center for Economic Development, Carnegie-Mellon University.

Cohen, W., R. Florida, L. Randazzese and J. Walsh, 1997, Industry and the Academy: Uneasy Partners in the Cause of Technological Advance, in R. Noll (ed.), *Challenge to the Research University*, Washington, DC: Brookings Institution.

Cozzens, S.E., 1998, Results and Responsibility: Science, Society, and GPRA, presented at *the AAAS Colloquium on Science and Technology Policy*, Washington, DC, 29 April-1 May.

David, P.A., D.C. Mowery and W.E. Steinmueller, 1992, Analyzing the Economic Payoffs from Basic Research, *Economics of Innovation and New Technology*.

David, P.A. and D. Foray, 1995, Accessing and Expanding the Science and Technology Knowledge Base, *STI Review #16*, 13-68.

Edquist, C. and S. Jacobsson, 1988, *Flexible Automation*, Oxford: Blackwell.

Eisenberg, R., 1996, Public Research and Private Development: Patents and Technology Transfer in Government-Sponsored Research, *Virginia Law Review* 82, 1663-1727.

Ergas, H., 1987, "Does Technology Policy Matter?", in B. Guile and H. Brooks (eds), *Technology and Global Industry*, Washington, DC: National Academy Press.

Evenson, R.E., 1982, Agriculture, in R. Nelson (ed.), *Government and Technical*

Progress: A Cross-Industry Analysis, New York: Pergamon.

Florida, R., 1997, The Globalization of R&D: Results of a Survey of Foreign-Affiliated R&D Laboratories in the United States, *Research Policy* 26, 85-103.

Freeman, C., 1987, *Technology Policy and Economic Performance: Lessons from Japan*, London: Pinter.

Ham, R.M. and D.C. Mowery, 1995, Enduring Dilemmas in US Technology Policy and the Clinton Administration, *California Management Review*.

Ham, R.M. and D.C. Mowery, 1997, The US Response to Globalization: Searching for the Keys Under the Lamp Post, in J. Dunning, ed., *Governments, Globalization, and Competitiveness*, Oxford: Oxford University Press.

Ham, R.M., G. Linden and M. Appleyard, 1998, The Evolving Role of Semiconductor Consortia in the US and Japan, *California Management Review* 41, 137-163.

Henderson, R., A.B. Jaffe and M. Trajtenberg, 1994, Numbers Up, Quality Down? Trends in University Patenting, 1965-1992, presented at the CEPR conference on "University Goals, Institutional Mechanisms, and the 'Industrial Transferability' of Research", Stanford University, 18-20 March.

Henderson, R., A.B. Jaffe and M. Trajtenberg, 1998, Universities as a Source of Commercial Technology: A Detailed Analysis of University Patenting, 1965-1988, *Review of Economics & Statistics*, 119-127.

Katz, M.L. and J.A. Ordover, 1990, R&D Competition and Cooperation, *Brookings Papers on Economic Activity: Microeconomics 1990*, 137-192.

Kortum, S. and J. Lerner, 1999, What is Behind the Recent Surge in Patenting?, *Research Policy* 28, 1-28.

Langlois, R. and D.C. Mowery, 1996, The Federal Role in the Development of the American Computer Software Industry: An Assessment, in D.C. Mowery (ed.), *The International Computer Software Industry*, New York: Oxford University Press.

Lundvall, B.-A., 1992, *National Systems of Innovation: Towards a Theory of Innovation and Interactive Learning*, London: Pinter.

Malakoff, D., 1998, Legislators Get Creative with New Crop of Earmarks, *Science* 281, 9 April, p. 1436.

Malakoff, D., 1998, House Report Takes Middle Ground, *Science* 282, 10 February, p. 23.

McGeary, M. and S. Merrill, 1999, Recent Trends in Federal Spending on Scientific and Engineering Research: Impacts on Research Fields and Graduate Training, Appendix A in National Research Council Board on Science, Technology, and Economic Policy, *Securing America's Industrial Strength*, Washington, DC: National Academy Press.

Mowery, D.C. (ed.), 1988, *International Collaborative Ventures in US Manufacturing*, Cambridge, MA: Ballinger Publishers.

Mowery, D.C., 1994, *Science and Technology Policy in Interdependent Economies*, Dordrecht: Kluwer.

Mowery, D.C., 1989, Collaborative Ventures Between US and Foreign Manufacturing Firms, *Research Policy* 18, 19-32.

Mowery, D.C., 1997a, US Postwar Technology Policy and the Creation of New Industries, in *Creativity, Innovation, and Job Creation*, Paris: OECD.

Mowery, D.C., 1997b, Technological Innovation in a Multipolar System: Analysis and Implications for US Policy, prepared for the study group on The Globalization of Industrial R&D, Council on Foreign Relations, New York.

Mowery, D.C. and N. Rosenberg, 1989, *Technology and the Pursuit of Economic Growth*, New York: Cambridge University Press.

Mowery, D.C. and N. Rosenberg, 1993, The US National Innovation System, in R. Nelson (ed.), *National Innovation Systems: A Comparative Analysis*, New York: Oxford University Press.

Mowery, D.C. and A.A. Ziedonis, 1998, Market Failure or Market Magic? Structural Change in the US National Innovation System, *STI Review 22*.

Mowery, D.C. and B. Sampat, 1999, Patenting and Licensing of University Inventions: Lessons from the History of Research Corporation, presented at the conference on R&D and Economic Growth in the 20th Century, U.C. Berkeley, 30 March.

Mowery, D.C. and A.A. Ziedonis, 1999, The Characteristics of US University Patents Before and After the Bayh–Dole Act: Analyzing Data from Entrants and Incumbents, presented at the National Bureau of Economic Research, 19 July.

Mowery, D.C., R. Nelson, B. Sampat and A.A. Ziedonis, 2001, The Effects of the Bayh–Dole Act on US University Research and Technology Transfer: An Analysis of Data from Columbia University, the University of California, and Stanford University, *Research Policy* 30, 99-119.

National Science Board, 1996, *Science and Engineering Indicators: 1996*, Washington, DC: US Government Printing Office.

National Science Board, 1998, *Science and Engineering Indicators: 1998*, Washington, DC: US Government Printing Office.

National Science Foundation, 1994, *National Patterns of R&D Resources: 1994*, Washington, DC: US Government Printing Office.

National Science Foundation, 1996, *National Patterns of R&D Resources: 1996*, Washington, DC: US Government Printing Office.

National Science Foundation, 1998, *National Patterns of R&D Resources: 1998*, Washington, DC: US Government Printing Office.

Nelson, R., 1959, The Simple Economics of Basic Scientific Research, *Journal of Political Economy*, 297-306.

Nelson, R. (ed.), 1993, *National Systems of Innovation*, New York: Oxford University Press.

Nelson, R. and P. Romer, 1997, Science, Economic Growth, and Public Policy, in B. Smith and C. Barfield (eds), *Technology, R&D, and the Economy*, Washington, DC: Brookings Institution.

OECD, 1996, *Technology and Industrial Performance*, Paris: OECD.

Ostry, S., 1990, *The Political Economy of Policy Making: Trade and Innovation Policies in the Triad*, New York: Council on Foreign Relations.

Trajtenberg, M., R. Henderson and A. Jaffe, 1994, *University Versus Corporate Patents: A Window on the Basicness of Inventions*, CEPR Working Paper 372, Stanford University.

US Council of Economic Advisers, 1994, *Economic Report of the President, 1994*, Washington, DC: US Government Printing Office.

US Patent and Trademark Office, 1998, *University Patent Report*; processed.

Vogelsang, A.J., 1998, Performance of R&D, Accountability of R&D, and the Government Performance and Results Act, presented *at the AAAS Colloquium on Science and Technology Policy*, Washington, DC, 29 April-1 May.

Yager, L. and R. Schmidt, 1997, *The Advanced Technology Program: A Case Study in Federal Technology Policy*, Washington, DC: American Enterprise Institute.

CHAPTER 3

RESEARCH POLICY TRENDS IN THE UNITED STATES: CIVILIAN TECHNOLOGY PROGRAMS, DEFENSE TECHNOLOGY AND THE DEPLOYMENT OF THE NATIONAL LABORATORIES

BARRY BOZEMAN AND JAMES S. DIETZ

1. INTRODUCTION

Having developed according to fragmented plans and, sometimes, no plan at all, the US national innovation system is inordinately complex. Most research policymakers and performers spend little time conceptualizing the US National Innovation System (NIS), trying instead to develop a fuller understanding of their own subsystem. Many of these subsystems – such as university research providers, national laboratories, federal agencies and state government technology development programs – are themselves quite varied and complex. As a result, empirical knowledge of the US NIS is limited.

With over 15 000 new research and development (R&D) laboratories having been designed and built since 1950 and with the growth of national R&D expenditures from under $1 billion in 1940 to more than $220 billion in 1998 (National Science Foundation, 1999), simply keeping track of the elements of the system proves a formidable task. Generally, R&D institutions and R&D policies have grown without consideration of system level issues, coordination issues, phase-in or phase-out issues or systematic consistent review.

In the US, the term "national innovation system" is one used often by science and technology policy analysts and rarely by others. As we use the term here, it refers to the intricate network of agents, policies and institutions supporting the process of technical advance in an economy. Following Lundvall (1992), we distinguish between a "narrow" sense and a "broad" sense of NIS. Scholars looking at the "narrow" NIS focus on organizations and institutions involved directly in the processes of scientific and technological exploration. The "broad" NIS concept includes all economic, political and other social institutions affecting learning, searching and exploring activities. Thus the narrow definition would include a nation's patent system, its universities and its research laboratories, while the broad definition also includes many other subsystems and processes, such as a nation's financial system, its monetary policies, norms of competition and the internal organization of private firms in the nation.

The essence of the NIS concept is the focus on nation-specific factors affecting innovation. Underlying the concept of the NIS is the argument that there are important differences, at the national level, in policies and institutions affecting the rate, direction and character of technical advance in a nation's firms and industries. Perhaps surprisingly, the idea that nation-specific factors are vital to innovation has, until recently, received little attention from scholars studying technical change and economic growth. In particular, most orthodox economic analyses have generally ignored the role of any institutional diversity in shaping patterns of innovation. This no doubt primarily reflects the difficulty of modelling institutions within the neoclassical economic paradigm, but probably also reflects the assumptions that economists have long held about technology (i.e., that "knowledge" is a public good, freely available to all). Though the "new growth theory" has been more accepting of institutions than the neoclassical growth theory which preceded it, the treatments of institutions in these models is not very rich and is often done in an *ad hoc* way (Nelson 1996).

As the final piece of our "National Comparative Research and Development Project", we attempted to provide a perspective on the US NIS (Crow and Bozeman, 1998). Our approach proceeded not from the assumptions and constraints of neoclassical economics but, instead, on an organization theory-based "environmental context technology". In this and other contexts (Bozeman, 1994; Bozeman, Papadakis and Coker, 1995; Bozeman and Crow, 1990) we have undertaken broad, encompassing approaches to analysis of the US NIS. Here our focus is more limited; we examine (per our commission) three interrelated components of the NIS:

- civilian technology programs at both the national and state levels that are designed to promote economic growth;
- defense R&D with an emphasis on dual-use technologies and conversion of defense R&D from military to civilian applications;
- the Federal Government's national laboratories.

These "components" are best viewed as a mix of public policies, legal initiatives and the institutions most responsible for implementing them. In each case, we provide a brief history of the respective NIS component and then focus on major current trends. We conclude with a brief discussion of the role of the component in the NIS and the likely future evolution. Before examining the three components, it is useful to provide a theoretical backdrop that may help understand US science and technology policy "philosophies". Most aspects of US science and technology policy are fundamentally influenced by their constituent ideological underpinnings, ones which are in some ways unique to the US NIS.

2. THE CONTEXT OF US RESEARCH POLICY: THREE COMPETING POLICY PARADIGMS[1]

US research policy has been dominated by three policy paradigms, each in ascendance at different points, depending on the relative strength of political forces and upon economic conditions in the US and the rest of the world. From its earliest days, the US has wrestled with the difficult question of how far the government should go in regulating or influencing individual and commercial behavior. The belief that government should not intrude or meddle unduly in private affairs, is sewn throughout the fabric of US history and imprinted on the American psyche. As a sponsor of R&D, market failure has historically been the cornerstone rationale for US science and technology policy. As a performer of R&D, the chief orientation for government lies within its missions – government should perform R&D in the service of well-specified missions for which there is a national interest not easily served by the private sector. Among the important elements of the mission philosophy are defense- and national security-related R&D, but also energy production and conservation, medicine and public health, space and agriculture. More recently, as a promoter of economic development, government plays

1. This discussion draws from Bozeman (1994) and Crow and Bozeman (1998).

TABLE 3.1 – Three competing R&D policy models

	Market	Mission	Cooperative
Core Assumptions	1. Markets are most efficient allocator of information and technology. 2. Government role is limited to market failures such as extensive externalities, high transaction costs and information distorsions. Small mission domain chiefly in defense. 3. Innovation flows from and to private sector, minimal government role.	1. The government role should be closely tied to authorized prog-rammatic mission of agencies 2. Government R&D is limited to missions of agencies, but not confined to defense. 3. Government should not compete with private sector in innovation and technology. But a government role in connection with traditional activities of line agencies.	1. Markets are not always the most efficient route to innovation and economic growth. 2. Global economy requires more centralized planning and broader support for civilian technology development. 3. Government can play a role in developing technology, especially precompetitive technology, for use in the private sector.
Peak influence	Highly influential during all periods.	1945-1965, 1992-present.	1992-1994.
Policy examples	Deregulation; contraction of government role; R&D tax credits; capital gains tax roll back. Little or no need for federal laboratories except in defense support.	Creation of energy R&D, agricultural laboratories, and other such broad mission frameworks.	Expansion of federal laboratory roles in technology transfer and cooperative research; manufacturing extension policies.
Theoretical roots	Neoclassical economics.	Traditional liberal governance with broad definition of government role.	Industrial policy theory.

a cooperative role by performing research and developing technology to be consumed by industry or by brokering cooperative R&D among firms where government plays no direct role in the conduct thereof. Table 3.1 provides a summary of the assumptions of each of the respective policy philosophies.

The market failure paradigm

The market failure paradigm for science and technology policy is based on familiar premises: it is the free market that most efficiently allocates goods and services and, left to its own devices, an unfettered market will lead to optimal rates of science production, technical change and economic growth. The market failure approach recognizes that there may be a role for government in science and technology affairs, but only when clear externalities are present (i.e., benefits cannot be captured in the market), when transactions costs are prohibitively high, or when information is unavailable or distorted so that market signals are unclear. Public policy in general in the US (i.e., not just science and technology policy), is strongly influenced by the market failure paradigm. In its simplest form, the market failure paradigm assumes that there is a significant need for new scientific and technical resources for economic productivity, and that the private sector will, through the competitive workings of the market, sense that need and respond in an economically efficient manner. Despite recent experiments with an expanded government role in science and technology-based economic development, the market failure paradigm in science and technology policy is alive and well.

For example, such comparatively recent initiatives as the R&D tax credit provisions of the Economic Recovery Tax Act of 1981 assume industry is the progenitor of innovation and the role of government is chiefly limited to staying out of the way (Bucy, 1985; Bozeman and Link, 1984; Bozeman and Link, 1983; Eisner, 1985; Landau and Hanney, 1981). Similarly, efforts to roll back early Clinton Administration initiatives such as the Advanced Technology Program of the National Institute for Standards and Technology (NIST) and the Technology Reinvestment Program of the Department of Defense (DOD) are framed within the assumptions and rhetoric of the market failure paradigm.

In fact, in early 1997, the Republican majority of the 105th Congress issued a declaration to end "corporate welfare". Using the market failure argument, they targeted for cuts a number of federal programs designed to stimulate new technology development in critical industries, including all programs at NIST and DOD directly linked to civilian technology development (US Congress, 1997).

Historically, the market failure paradigm has been dominant in science and technology policy, excepting defense/space-related R&D, and R&D of direct relevance to agency missions. Until the 1980s there was much validity to the stereotype that government R&D was for missions, industry R&D for technology innovation and development, and universities for basic knowledge and people production.

The mission paradigm

The earliest government involvement in science and technology policy has been within the framework of the mission paradigm. The mission paradigm assumes that the Federal Government's role in science and technology should flow directly from the legitimated missions of agencies and should not extend beyond those missions in pursuit of more generalized technology development, innovation or competitiveness goals. As such, the view is not radically different from the market failure paradigm, except that the mission paradigm encourages the performance of non-defense missions in a wide array of policy domains.

The roots of the mission paradigm can be traced to early government R&D roles in national defense, public health and, to some extent, agriculture (which also has cooperative technology paradigm elements). But recently there have been several reminders of the significance of mission paradigm thinking. The Secretary of Energy Advisory Board's report (1994) on the future of the Department of Energy national laboratories, widely known as the Galvin Commission report, emphasized that many laboratories had strayed from their traditional energy- and defense-related missions and that any future work in technology development and commercialization should flow directly from those missions. In many ways, this was a sharp reaction against policy initiatives and program expansion under the cooperative technology paradigm. At the same time, the counter-reaction to dual-use technology initiatives has much the same tone: defense money for defense research.

The cooperative technology paradigm

During the period of economic uncertainty in the 1980s and the perceived crisis in US competitiveness (MIT Commission on Industrial Productivity, 1989; National Academy of Sciences, 1978; National Governors' Association, 1987; President's Commission on Industrial Competitiveness, 1985; Council on Competitiveness, 1993, 1996a, 1996b) many core assumptions were re-examined, including bedrock faith in the private sector as the source of virtually all significant innovation. As other nations, especially Japan, forged ahead with government support of technology development, the market failure and mission paradigms seemed less compelling (Bozeman and Crow, 1990; Crow and Bozeman, 1987a; 1987b).

During the 1980s, a number of science and technology policy initiatives challenged the preeminence of the market failure paradigm. As used here, the "cooperative technology paradigm" is an umbrella term for a set of values emphasizing cooperation among sectors – industry, government, and university – and cooperation among rival firms in the development of pre-competitive technologies and "infratechnologies" (Link and Tassey, 1987). Kash and Rycroft (Kash and Rycroft, 1994; Kash, 1989; Rycroft and Kash, 1994) have provided strong arguments for cooperative technology policy approaches.

The 1980s and early 1990s was a period in which the dominant market failure paradigm received its strongest challenge. At first, challenges to market failure thinking evoked a "permissive" tone: changing patent policy to expand the use of government technology (Patent and Trademark Laws Amendment, 1980), relaxing antitrust laws, and altering guidelines for disposition of government-owned intellectual property (Bagur and Guissinger, 1987; Gillespie, 1988). The later brand of cooperative technology philosophy and its attendant policy revealed a more "promotional" dimension: inducing cooperative research and development (Link and Bauer, 1989; National Cooperative Research Act of 1984), establishing research consortia and multisector centers (Smilor and Gibson, 1991), and promoting the conversion of defense technology to civilian applications.

Finally, the Clinton–Gore technology plan of then Presidential and Vice Presidential candidates Bill Clinton and Al Gore (Clinton–Gore, 1992) outlined a national plan for movement toward enhanced cooperation, national science and technology planning, national technology projects, and defense conversion. Generally, this was the strongest cooperative technology development policy statement ever provided. However, as the new Administration took office, science and technology policy took a back seat to health care and to putting out political fires. The ideas in the Clinton–Gore plan were never fully implemented.

The cooperative technology development policies that have attracted the most attention are those pertaining to domestic technology transfer, especially the use of federal laboratories as partners in the commercialization of technology (Herrmann, 1983; Rahm, Bozeman and Crow, 1988; US General Accounting Office, 1989). Previously aloof from commercial concerns, indeed prohibited by law from developing technology specifically for private vendors, the legislation of the 1980s gradually changed the mission, tenor and climate of the federal laboratories (Bozeman, 1994) and, to some extent, the companies interacting with the laboratories (Roessner and Bean, 1991). The intellectual property dictum "if it belongs to everyone, it belongs to no one" began to take hold as the government laboratories increasingly moved from a sole focus on public domain research to a mandated role as a technology development partner to industry.

Recently, the pendulum seems to have swung away from the cooperative technology paradigm. While there are still many elements of the policy model evident in federal policies and activities, the enthusiasm of the first two years of the Clinton Administration barely survived the election of the 1994, and the 1996 and 1998 victories of the Republican-majority Congresses ensured the decline of cooperative technology. The Clinton Administration then pressed its cooperative technology paradigm more gingerly and Congress continued to oppose an expanded role for the Federal Government.

3. FEDERAL LABORATORIES IN THE US NIS

Federal laboratories have in recent times served as a political football in US science and technology policy, partly because this set of research institutions is amenable, at least in principal, to direct political control. Thus, as the political climate changes, the federal laboratories are among the first to attract the attention of political change agents, such as new Presidential Administrations or Congresses with a new political majority. Another reason for the political vulnerability of the federal laboratories is that some of the most notable laboratories have recently been "adrift", searching for new missions with the end of the Cold War and, earlier, the reduced prominence of nuclear technology.[1]

The US federal laboratories, most famous for their role in the Manhattan Project, represent much more than a post-Cold War holding company. In the first place, only a handful of the 726 federal laboratories are weapons laboratories. Federal laboratories focus on such diverse topics as medical and biotechnology research, environmental research, standards and testing, agricultural research, energy and materials sciences and space science.

In the second place, the resources available through the federal laboratory system are enormous by most any standard. There are 58 Nobel laureates working in the national laboratories (Crow and Bozeman, 1998), more Nobel laureates than can be found in all but a few nations. More than

1. In the interests of space constraints, our discussion of the federal laboratory system is brief. However, the reader may consult Crow and Bozeman (1998) for a book-length treatment of federal laboratories and their role in the US NIS. This section draws in part from Crow and Bozeman (1998).

$20 billion per year are spent on R&D in the federal laboratories, representing more than one-third of all federal R&D money expended. Federal laboratories employ about 60 000 scientists and engineers, a significant fraction of the US scientific and technical human capital. In addition to producing tens of thousands of scientific and technical publications and papers each year, federal laboratory personnel file nearly 1000 patent applications each year (Crow and Bozeman, 1998).

The range of functions performed by federal laboratories is remarkable. The core functions of such "superlabs" as Sandia National Laboratory and the Naval Research Lab are familiar, but the superlabs have a wide range of functions that go well beyond their core technical mission. Indeed, this mission accretion has been the source of considerable criticism (Secretary of Energy Advisory Board, 1995; White House Science Council, 1983).

Federal laboratories have been at the center of many of the most significant cooperative technology policy initiatives, including technology transfer (Stevenson–Wydler Act of 1980; National Cooperative Research Act of 1984; Technology Transfer Act of 1985). This effort has not always been a huge success, at least not when measured by crude quantitative indicators. A detailed review of the technology transfer outputs of federal laboratories in 1994-1995 indicated that the total number of new licensed technologies was only about 300 per year with very little economic return (less than $10 million annual) to the laboratories themselves (Federal Laboratory Commission, 1996; US Congress, 1996). But the effort continues and, indeed, is redoubled. In 1999, Secretary of Energy Bill Richardson announced new technology transfer initiatives and reiterated the Department's commitment to technology transfer.

Despite the importance of technology transfer and of cooperative research and development agreements (CRADAs), the federal laboratories' true significance in the US NIS comes from their core scientific and technology development work, especially as these pertain to missions in national defense, public health medicine, and energy and environment. Indeed, the chief conclusion of a panel reviewing the DOE national laboratories was that the laboratories should stay closer to their historic and statutory missions. It seems unlikely, however, that this advice will be taken, not with the current environment for federal laboratories, an environment that entails high uncertainty and competition.

4. THE ROLE OF THE CIVILIAN TECHNOLOGY PROGRAMS IN THE US NIS

4.1 – Federal civilian technology programs

The Advanced Technology Program (ATP)

Operated by the National Institute for Standards and Technology (NIST) of the Department of Commerce, the ATP[1] has been the flagship cooperative technology program of the Clinton Administration due to its relative size and prominence. Because of its intent to use government as a stimulus for private sector development, the program has remained controversial, especially with a Republican-dominated US Congress grounded strongly in the market failure paradigm.

The goal of ATP is to support the development of risky, pre-competitive technologies through cost-sharing research with industry. These technologies are intended to create opportunities for new, world-class products, services and industrial processes that benefit not just the organizations participating in the project, but other companies and entire industries as well. The ATP program has two components: single company projects and joint ventures (http://www.atp.nist.gov, 1999).

The ATP was created during the Bush Administration under the Omnibus Trade and Competitiveness Act of 1988 (P.L. 100-418) and modified by the American Technology Preeminence Act of 1991 (P.L. 102-245). According to the legislation, the ATP is charged with assisting US businesses in creating and applying the generic technology and research results necessary to commercialize significant new scientific discoveries and technologies rapidly, and to refine manufacturing technologies. The initial funding for ATP stipulated that funds were to be used for research, not product development. By 1999, ATP had funded a total of 431 projects, involving more than 1000 participants (not including subcontractors and informal partners and collaborators). The projects have entailed approximately $2.8 billion of research. This figure includes both ATP funds and industry funds, with industry funds comprising slightly more than 50%. Most ATP projects are three to five years in duration.

The ATP awards are made to individual companies and to joint research ventures (comprising two or more companies, often in combination with

1. The Clinton Administration requested $239 million for ATP in FY2000 which is an increase of 21% over the FY1999 budget ($203.5 million).

universities and research laboratories). The projects are proposed by businesses and carried out with company researchers and other researchers from partnering organizations. Projects generally require cost sharing by industry. According to ATP promotional material, ATP benefits are distributed widely among companies of all sizes: "To date, over half (53%) of the ATP awards have gone to individual small businesses or to joint ventures led by a small business. Large firms can work with the ATP, especially in joint ventures, to develop critical, high-risk technologies that would be difficult for any one company to justify because, for example, the benefits spread across the industry as a whole. Universities and non-profit independent research organizations also play significant roles as participants in ATP projects" (http://www.atp.nist.gov, 1999, p. 3).

ATP does not itself utilize the project results, thus taking a quite different posture to that of agencies sponsoring R&D under a mission paradigm. The objective, instead, is for ATP-funded results to have an impact in the market-place. The ATP seeks a position in the middle ground between basic research and product development, with the goal of stimulating commercial activity. As Ruegg (1997, p. 2) notes, "ATP relies on the presence of expected private returns to induce companies to be willing to cost-share the research with the ATP and subsequently to carry out commercial development of the new technology with private capital, but it selects projects for awards for which it thinks the potential social rate of return (the return to the nation) far exceeds the private rate of return on investment (the return to the award-recipient innovators)".

The ATP provides awards on the basis of competitive peer review. The review process is performed by panels (Bozeman, 1994) of diverse experts, including people with technical, business and economic expertise. They are charged with assessing the technical and economic merit of applicant proposals.[1]

---

1. Selection criteria include: (1) the potential of the research to contribute significantly to the scientific and technical knowledge base of the industry or industries; (2) the potential of the technology to generate broad-based benefits to the nation and the difference ATP funding would make in that respect; (3) the proposed pathways to market and to broader diffusion; (4) the commitment of the proposers to follow through with future development of products and processes if they are successful in overcoming the challenging technical barriers addressed in their ATP research project; (5) the organizational structure of the project, and its management plan; and the proposers' experience, qualifications and resources (Ruegg, 1997).

TABLE 3.2 – ATP award statistics (1990-1998)

Number of proposals submitted	3 585
Number of awards	431
Number of participants awarded	1 010
Dollar value of commitments	$2.8 billion
ATP funds committed	$1.4 billion
Industry cost sharing	$1.4 billion

Source: National Institutes of Standards, Advanced Technology Program
Note on commitments: Cost sharing does not necessarily involve a direct commitment of money but, instead, can entail personnel and equipment use donations (among other qualifying cost sharing categories).

Between 1990 and 1997, ATP conducted 30 competitions. As seen in Table 3.2 (adapted from Ruegg, 1997), more than 3500 proposals have been submitted and 431 awards have been made. The award range for General Competition Projects was between $480 000 and $20 million and for Focused Competition Projects between $520 000 and $31 million. Over 50% of ATP projects are led by small businesses and over 125 universities are involved in 400 projects (www.atp.nist.gov).

Despite the unpopularity of the ATP program with adherents of the market failure paradigm, the many evaluations of the program have been quite positive. For the most part, the major evaluations of ATP have occurred under the auspices of the ATP Economic Assessment Office and have been charged with identifying the economic benefits of the program. In Table 3.3, we present a brief review of the major evaluation studies and their results.

The Manufacturing Extension Partnership (MEP)

The goal of the Manufacturing Extension Partnership[1] is to assist small- and medium-sized manufacturing enterprises to adopt advanced technologies that are appropriate for their needs and will make them more economically competitive in national and global markets. It does this primarily in two ways: by funding regionally-based manufacturing centers and by providing matching funds to states that create their own extension services. Due to this latter charge, MEP is often characterized as the federal cooperative technology program most closely tied to the states (Coburn, 1995).

---

1. For FY2000, the Clinton Administration is requesting $99.8 million for the MEP, which represents approximately a 7% reduction from its FY1999 level of $106.8 million.

TABLE 3.3 – Summary of ATP evaluation findings

This overview examines original source material but also draws from Ruegg (1997).

SILBER, 1996

- Relied on telephone interviews to assess the progress of selected early ATP projects.
- Found that 70% of respondents reported that they probably would not have developed the technology at all without the ATP award. Of those who said that they probably would have proceeded without the ATP, over 90% reported that their level of effort, goals, and rate of progress would have been substantially lower without the ATP award.
- Of course, surveys based on respondents reports of alternative non-occurring scenarios do not provide much confidence with respect to rater validity.

LAIDLAW, 1997

- Surveyed participants in 28 projects funded in 1991, studying research cycle-time reduction, especially acceleration of commercialization and increased project benefits.
- 96% of the companies reported that their participation in the ATP reduced their research cycle time by a median value of 50% or three years.
- 86% of the companies expected the reductions in research time to flow through to later stages in the technology development and commercialization cycle.
- Many of the companies gave ballpark estimates of the value of reducing research time by just one year, which ranged from one million dollars to "billions". Again, the substantive findings are more encouraging than is the apparent validity of the indicators.

CONSAD, 1997 AND VONORTAS, 1997

- ATP has also sponsored evaluation studies using macroeconomic models.
- Consad (1997) used a regional econometric model to examine the economic impact of an ATP-funded project to improve control of dimensional variation of stamped metal parts in the production of automobile bodies and other discrete-parts manufacturing. The study projected increases in Gross Domestic Product of at least $3 billion and employment gains of several hundred thousand jobs.
- An extensive comparison of joint ventures funded by the ATP with non-ATP joint ventures was undertaken. Preliminary findings show that the two sets of joint ventures appear complementary. Membership in the ATP-funded joint ventures appears to involve more small and medium-sized enterprises than membership in the non-ATP ventures. Moreover, ATP-funded joint ventures have much lower participation by foreign-owned companies and a higher participation of universities than the non-ATP ventures.

MARTIN, ET AL., 1998 AND FAUCETT, 1997

- A set of case studies examined ATP-funded tissue engineering technologies.
- Using an approach to modeling social and private returns similar to the one used by Mansfield (1996) the study uses non-market valuation methods to value the benefits of new medical treatments developed under the ATP-funded project.
- The researchers determined that the expected social return on ATP public investments in the tissue engineering technologies is about $34 billion in net present value, while the expected total social return on public and private investment is estimated at about $112 billion. These figures were derived using calculations of consumers' value for improved health and longevity (see Viscusi, 1995).
- Another set of case studies by Faucett assessed the economic significance of ATP projects aimed at improving the performance of material surfaces under harsh conditions.

MEP was created in 1993 by uniting two previously independent NIST programs (the State Technology Extension Program, STEP) and the Manufacturing Technology Centers Program (MTC) originally authorized by the Omnibus Trade and Competitiveness Act of 1988 (Coburn, 1995). Each of the Partnerships is sponsored by a not-for-profit or government organization, typically an economic development entity of the state or a university or technical college (www.nist.gov). Partnership funding can come from a number of sources in addition to NIST, including other federal agencies, state and local governments, universities and private fees (US General Accounting Office, 1996b). Partnerships usually have a number of centers located throughout the state or region that work directly with firms requesting services. Centers draw expertise from their own staff as well as from independent consultants.

Partnerships commonly provide a wide range of services, including technical consulting and factory site-reviews, seminars and direct hands-on training, technology demonstrations and assistance in selecting equipment or software (Coburn, 1995). They also provide referrals for assistance in management, finance, and marketing via links to other established programs. Although Partnerships do not conduct their own research, they may draw upon research conducted within the state or elsewhere. Partnerships are funded as "cooperative agreements" between NIST and the awarded institution, where NIST supplies no more than 50% of the total funding for the first three years, with support decreasing to 33% by year six (Coburn, 1995).

A 1996 US General Accounting Office (GAO) study of the program found that relatively small companies (those with 1994 sales less than $1 million), and relatively new companies (those started since 1985), were most likely to report that their overall business performance has improved due to assistance of a MEP center (US General Accounting Office, 1996b). Companies reported they were satisfied with the expertise of staff (88%) and with the timely response (83%) of the MEP center with which they worked (US General Accounting Office, 1996b). Companies also reported that program assistance had positively affected their use of technology (63%), the quality of their product (61%) and the productivity of their employees (56%) (US General Accounting Office, 1995).

The US Census Bureau surveyed over 4400 MEP customers whose projects were completed during 1997. These companies reported a $236 million boost in revenues, while saving $31 million in inventory and over $24 million in labor and materials. They also reported having invested close to $200 million in modernization (www.nist.gov).

Small Business Innovation Research (SBIR) Program

The Small Business Innovation Research (SBIR) program was created by the National Science Foundation in 1977. In 1982, Congress passed the Small Business Innovation Development Act mandating that all federal agencies with annual extramural research budgets in excess of $100 million implement SBIR programs. Such initiatives are now operated by 11 federal agencies.[1]

The SBIR program was created to increase the participation of small businesses in federal R&D programs, to assist firms financially in innovation and product development, and to increase the overall competitiveness of US industry (Coburn, 1995). The State Science and Technology Institute, which monitors state level R&D policy, regards the SBIR program as the first successful, large-scale cooperative technology program (Coburn, 1995).

While it exists (legislatively) to help agencies meet their missions, the program has been a powerful source of research funding for often investment-poor small businesses. Under the program, each agency is required to award 2.5% of its extramural R&D budget to small business R&D projects. Each agency administers its own version of the SBIR program with its own set of requirements, while adhering to guidelines laid down by Congress and the Small Business Administration (SBA). The SBA, which is an independent agency of the US government, is charged by Congress to coordinate SBIR programs across the agencies. And although SBA does not operate its own SBIR program, it is the entity responsible for establishing broad policy guidelines, monitoring the agencies and reporting to Congress.

To be eligible to compete for funds, a small business must be a US-owned and independently operated for-profit company with 500 or fewer employees. In addition, the principal researcher on the proposed research project must be an employee of the company (www.sba.gov/SBIR/sbir.html). Agencies make SBIR awards based on small business competence, the innovative nature of the proposed research, technical merit and future market potential. There are three award phases of the program:

- Phase I "proof-of-concept" grants of up to $100 000 for approximately six months to support exploration of the technical merit or feasibility of an idea or technology.

---

1. These include the Departments of Agriculture, Commerce, Defense, Education, Energy, Health and Human Services, and Transportation, as well as the Environmental Protection Agency, the National Aeronautics and Space Administration, the Nuclear Regulatory Commission and the National Science Foundation.

- Phase II "pre-prototype" awards of up to $750 000 for two years for furthering the results of Phase I. During this time, the R&D work is performed and the developer evaluates its commercial potential. Only Phase I award winners are invited to submit Phase II proposals.
- Phase III is the "commercialization" period during which the Phase II innovation moves from the laboratory into the marketplace. No SBIR support is offered; the business must find private or other non-SBIR funds to support commercialization (www.sba.gov/SBIR/sbir.html).

Federal agencies provide Phase I and II funding which, in principle, permits the small business to compete for and acquire capital to produce, market and sell the product (Phase III). The rationale, then, for the program has been accepting of the market failure paradigm, although in 1992 Congress required agencies to select more commercially oriented proposals and to double the SBIR set aside from 1.25% to 2.5% each year. A 1998, GAO study found that about 50% of the SBIR companies had sales of products or services related to their research or received additional developmental funding after receiving SBIR awards (US General Accounting Office, 1998). The program remains popular in business communities and state and federal political circles.

Most states have small business development programs that offer a range of services to assist companies participating in the federal program. Assistance varies from educational seminars and databases, funding for proposal preparation, and grants or loans to assist companies during the Phase III commercialization efforts (see Table 3.4 for a summary of SBIR awards by state).

In 1992, during the height of the cooperative technology paradigm ascendancy, Congress amended the SBIR Act to authorize the Small Business Technology Transfer (STTR) Program, which expands the scope of the SBIR program to include non-profit organizations and academic research institutions unable to participate directly in SBIR. The act directs (five) federal agencies[1] with extramural research budgets in excess of $1 billion to reserve 0.15% for STTR operations.

The major goal of the STTR program is to transfer technology from research institutions to small businesses for the purposes of commercial product development. The research institutions (typically universities) are the originators of the technology and have usually established some assessment

1. These include the Departments of Defense, Energy, and Health and Human Services; the National Aeronautics and Space Administration and the National Science Foundation.

TABLE 3.4 – SBIR awards by state and total dollars, 1997

State	Phase 1 Awards	Phase 1 Dollars	Phase 2 Awards	Phase 2 Dollars	Total Awards	Total Dollars
California	693	58 757	326	192 505	1 019	251 262
Massachusetts	508	43 007	202	123 124	710	166 131
Virginia	199	16 071	80	52 676	279	68 747
Maryland	184	15 537	55	34 125	239	49 662
New York	125	10 331	67	37 229	192	47 560
Colorado	146	11 993	55	33 193	201	45 186
Pennsylvania	114	9 494	56	32 895	170	42 389
Texas	135	11 264	41	25 696	176	36 960
Ohio	121	10 369	41	25 834	162	36 203
New Jersey	97	8 002	40	24 982	137	32 984
Connecticut	79	6 385	39	24 498	118	30 883
Alabama	56	4 739	36	22 759	92	27 498
Florida	76	6 390	35	20 454	111	26 845
Michigan	71	6 082	31	18 507	102	24 589
Washington	85	7 039	29	16 778	114	23 817
Arizona	76	5 982	29	14 931	105	20 913
New Hampshire	39	3 233	24	14 860	63	18 093
New Mexico	58	4 505	24	13 440	82	17 945
Oregon	43	3 703	24	14 018	67	17 721
Minnesota	54	4 440	20	12 893	74	17 333
North Carolina	36	3 233	20	12 606	56	15 839
Illinois	53	4 593	14	9 341	67	13 934
Georgia	36	2 864	12	8 538	48	11 402
Utah	30	2 583	15	8 113	45	10 696
Wisconsin	32	2 800	12	7 683	44	10 483
Tennessee	30	2 449	12	5 990	42	8 439
Indiana	14	1 158	11	6 143	25	7 301
Delaware	15	1 120	9	4 424	24	5 544
Dist. Columbia	8	649	6	3 856	14	4 505
Kentucky	8	655	3	3 352	11	4 007
Missouri	15	1 260	5	2 323	20	3 583
Rhode Island	1	6	5	3 340	6	3 346
Oklahoma	9	655	3	1 993	12	2 648
Hawaii	11	901	3	1 533	14	2 434
Nevada	9	705	3	1 615	12	2 320
Kansas	6	569	3	1 699	9	2 268
Wyoming	3	204	4	1 944	7	2 148
Vermont	14	1 133	1	750	15	1 883
Maine	3	190	2	1 389	5	1 579

Note: Only states with awards over $1.5 million are listed.

of commercial feasibility of the product. The role of the small business is then to develop the technology into a commercial product. For this reason STTR projects tend to be more development-oriented than SBIR projects although the overall three-phase award process is similar. According to the GAO, early prospects for the program are promising (US General Accounting Office, 1996c) although political prospects are uncertain as the program remains on a authorization due to expire in 2001.

4.2 – The role of the US States in civilian technology

The role of state governments in technology development is substantial yet, in some ways, supplementary to the role of the Federal Government. Most state government programs are at heart economic development programs that recognize a role for technology (and to a lesser extent, science) as a means toward this end (Melkers and Cozzens, 1997a; 1997b). In fact, many such programs were created in response to the economic downturns of the mid- to late 1980s or in recognition of the need to diversify state economies. As such, state programs tend to adhere somewhat closely to the cooperative technology paradigm in objective, although state legislatures tend to couch such policies and programs in the language of "pragmatism" (Melkers and Roessner, 1997; Melkers and Cozzens, 1997b).

All states have established some kind of state-level science and technology coordinating entity, although variation in organization structure across states is enormous. Even within states, there seems to be large variance in organizational centralization or coordination (www.ssti.org). Some states house the technology policy-making function within a state department of commerce or economic development (for example, Office of Advanced Technology of the Texas Department of Commerce, Massachusetts Economic Development Department, the Office of Strategic Technology within the California Trade and Commerce Agency, the Michigan Economic Development Authority). Others have created an explicit science and technology agency or center (for example, Tennessee's Science and Technology Office, the Utah Office of Technology Development, the Washington Technology Center, Oklahoma Center for Advancement of Science and Technology). Still others have created an independent foundation, nonprofit, or public corporation (for example, the Utah Technology Finance Corporation, Minnesota Technology Inc., the Indiana Business Modernization and Technology Corporation, the Alaska Science and Technology Foundation, the New York State Science and Technology Foundation). For states with large investments in technology development, Georgia has, perhaps, the most decentralized system. It includes an Economic Development Institute headquartered on the campus of the Georgia Institute of Technology, along with the "Georgia Research Alliance" founded by the business community as a statewide non-profit entity, which includes representation from the university system and state government.

Collectively, according to the State Science and Technology Institute, as recently as 1995 states were allocating over $3 billion[1] annually to R&D efforts (see Table 3.5) (National Science Foundation, 1999b).

During the 1990s, states put increasing emphasis on science and technology within their economic development plans. According to the State

TABLE 3.5 – States with the largest R&D and R&D plant expenditures, 1995

	Millions of dollars		Dollars per capita	Share of total state spending	%
Texas	326	Kansas	32.73	Nebraska	1.14
California	274	Hawaii	29.83	Kansas	1.13
New York	267	Nebraska	28.45	Georgia	1.01
Florida	256	North Dakota	26.75	North Dakota	0.97
Pennsylvania	205	Georgia	25.31	South Dakota	0.94
Georgia	182	South Dakota	21.63	Texas	0.83
Ohio	114	Montana	21.00	Montana	0.78
Illinois	112	Florida	18.06	Arizona	0.73
North Carolina	105	Iowa	17.33	Florida	0.66
Kansas	84	Arizona	17.24	Pennsylvania	0.65
All States	3 009	All States	11.48	All States	0.42

Sources: National Science Foundation. Battelle Memorial Institute and State Science and Technology Institute, Survey of State R&D Expenditures: Fiscal Year 1995. Columbus, 1998.

1. State science and technology programs are typically funded from general operating revenues, lottery proceeds, revenue bonds and specifically designated tax funds. State-funded academic R&D expenditures reported in both the NSF academic surveys and Battelle/SSTI state survey comprise only funds directly targeted to academic R&D activities. The state totals do not include undirected general-purpose state government appropriations that are used – at the universities' discretion – for research or to cover the indirect costs for conducting R&D. The reported funding sources for those types of R&D expenditures are the universities themselves, not their state governments. Consequently, the actual contribution of state governments to academic R&D – even with the adjustments made in this report – is understated, particularly for public institutions (NSF, 1999b).

Science and Technology Institute (1996), more than a dozen states adopted stand-alone science and technology strategic plans in the early 1990s.

State efforts tend to support the creation of technology-based companies and the modernization of process technologies within the existing industrial base (Congressional Research Service, 1998). State programs commonly focus on (www.ssti.org/states.htm):

- Developing the R&D capacity of the states' colleges and universities, and, importantly, their role in working with the private sector through efforts such as technology licensing, consulting services, science parks and incubators, or joint industry-university centers (Coburn, 1995; NSF, 1999b; Melkers and Cozzens, 1997b).
- Providing financial support to entrepreneurs and small businesses through programs similar to, or that complement, SBIR.
- Providing or facilitating technical assistance to industry on new technology processes and products via extension services or other mechanisms, and/or operating awareness and outreach seminars for industry on available state, federal or local programs.
- Providing matching funds required for state-level programs of the federal government such as MEP, the Experimental Program to Stimulate Competitive Research (EPSCoR),[1] and the State Industry-University Cooperative Research Centers (SIUCRC).[2]
- Providing tax incentives for R&D. According to the State Science and Technology Institute, in 1996, 35 states offered some kind of R&D tax incentive program for industry – many of which are modelled after the federal R&D tax credit.
- Operating a venture capital, start-up assistance fund for high-tech companies.

1. The Experimental Program to Stimulate Competitive Research (EPSCoR) was established by the National Science Foundation (NSF) in 1978 under Congressional mandate to avoid "undue concentration" of federal research funding. The program supports academic R&D in 18 states and Puerto Rico that have historically been among the most underdeveloped in terms of their scientific and technical capability. Since that time eight other federal agencies have adopted their own EPSCoR programs, including the Department of Commerce's Experimental Program to Stimulate Competitive Technology (EPSCoT). EPSCoR requires states to match the NSF contribution (www.ehr.nsf.gov/ehr/epscor).
2. The SIUCRCs are NSF-funded, university-based research centers that receive base funding of an equal amount from NSF and the state government. Industry contributed support must equal the NSF or state contribution. Each center has a core engineering research program that supports generic, fundamental research relevant to industrial needs within the state (www.eng.nsf.gov).

Overall, state R&D support increased at an annual rate of 3.3% (adjusted for inflation) between 1965 and 1995 compared with 2.5% per year at the federal level (National Science Foundation, 1999b). Continued expansion of the role of the states seems likely at this juncture. States seemed to have thus far avoided many of the ideological trappings that continue to contribute to policy instability at the federal level. A 1992 Carnegie Commission report recognized the increasing importance of the states and recommends that the federal government regard the states as "full partners" in science and technology policy-making (Carnegie, 1992).

5. DEFENSE R&D, CONVERSION AND DUAL-USE TECHNOLOGY

The rebuttal to those insisting that the US has no strategic technology policy is: "of course the US has a strategic technology policy, it is to be found in the Department of Defense". Defense procurement and R&D has played a major role in advancing a wide variety of technologies including, among others, computers and related semiconductor technologies, communication and telecommunication, materials science, aerospace and photonics. The billions of dollars the Defense Department, especially the Defense Advanced Projects Research Administration (DARPA)[1] invests in dual-use and generic technologies may not be the most efficient means of producing technological innovation, but the enormous scale of defense R&D ensures technology development.

Despite this, the R&D budget of DOD has fallen dramatically in recent years, and for the first time in modern history, the Administration is requesting more for civilian ($39.8 billion) than for defense R&D ($38.5 billion) for fiscal year 2000.[2] Since 1993, the civilian R&D budget has increased 15% in real terms while defense R&D has dropped by 20% (Congressional Research Service, 1999). Since 1990, defense R&D has decreased from 63% of the federal R&D budget to 50% (as proposed for 2000) (National Science Foundation, 1999a).

---

1. Aside from DARPA whose budget in FY1999 was approximately $1.9 billion, the Defense Department armed services (Army, Navy, Air Force) operate their own R&D programs and laboratories. The Ballistic Missile Defense program (BMD) is the largest single R&D project (FY1999 $3.8 billion) in the Department (operated out of the Office of the Secretary).
2. The defense R&D budget is projected to remain flat until 2005.

Since the collapse of the Soviet Union and the concomitant focus on defense downsizing and defense conversion, many have argued that the traditional technology innovation and development role placed by defense laboratories should be channelled more directly into civilian technology concerns. This view has, in recent times, proved controversial. While dual-use technology and defense-related technology enjoyed bipartisan support during the early 1990s, they became much more controversial when Republican Party leaders began to rethink the appropriateness of a DOD commercial technology role in 1995 (Morrison, 1995).

It is important to understand the context of early 1990s thinking about defense conversion and dual-use technology. While the US, Germany, and Japan were each spending between 2.5% and 3.0% of GDP on R&D, the enormous defense spending in the US made its R&D investment not comparable. This point cannot be overemphasized when comparing US R&D expenditures with such nations as Japan and Germany. Excluding defense R&D, the US was spending less than 2% of GDP on R&D in 1992 (National Science Foundation, 1992). Moreover, there had been relatively little fluctuation in defense R&D as a proportion of total federal R&D in the more than 20 years between 1970 and 1992. Thus, the prospect of a sharp decline in overall defense spending and an expected attendant reduction of defense R&D were, in the early 1990s, viewed in some quarters as a cause for alarm. To bring the importance of defense R&D into perspective, we can consider that in 1992, total spending for the National Science Foundation (the agency responsible for funding general science) was less than 6% of the more than $38 billion devoted to defense R&D. Indeed, the $11.6 billion devoted to intramural research at such DOD laboratories as the Naval Research Laboratory or the Air Force Arnold Engineering Center far surpassed the NSF's relatively meager $2.1 billion 1992 budget. Moreover, not only were the DOD laboratories affected by the early 1990s reduction in defense R&D, many of the Department of Energy multi-program laboratories, among the largest laboratories in the world, were strongly affected. In the ten years before the defense downturn of the early 1990s, the budgets for the DOE weapons laboratories (including Los Alamos, Sandia, and Lawrence Livermore) had risen 58% (compared to a 15% rise for all DOE laboratories). By 1992, each of the weapons laboratories received more than $1 billion in R&D funding, more than four times the amount allocated to such well known and accomplished national laboratories as Lawrence Berkeley, Oak Ridge, and Argonne (US DOE, 1992). With the defense downturn, the weapons laboratories steeled themselves for sharp cuts in personnel and budgets.

Not only is the issue of defense spending important to the overall US R&D effort, there are enormous stakes for particular industries and particular regions. Thus, the state of Connecticut, to take an example of one state whose early 1990s economy was particularly dependent on defense spending, faced the prospect of a *de facto* economic depression with the sharp downturn in defense spending (Prokesch, 1993). Prospects in California were similarly bleak.

The US Congress was particularly wary about the meaning of possible reductions in defense spending for the overall R&D effort in the US. One response was a series of what could only be viewed as economic relief legislative packages. These included, most prominently, a $1.7 billion defense conversion package passed in 1992, which included, among other provisions, a $97 million Defense Dual-use Assistance Extension Program that provided cost-shared grants to R&D centers sponsored by federal, state or local governments that offered commercial firms assistance in developing, producing and manufacturing for civilian use technology developed in defense R&D.

OTA and defense conversion

In 1993, the US Office of Technology Assessment (OTA) was asked by the US Congress to review issues pertaining to the closeness of the link between military research and civilian needs. The US Congress was interested in the overall issue of how to best merge military R&D objectives with the possible secondary benefits realized in the civilian sector. The assumption was that investments in military R&D, which at the time totalled more than 55% of the entire federal R&D investment, might, through careful planning, be parlayed into specific civilian applications for industry. As a result of this general thrust a number of new programs were developed. Although shortened in 1995 by a 104th Congress hostile to civilian objectives driving military R&D, this effort did involve an attempt to rethink the design of defense R&D laboratories.

The OTA study considered three quite divergent possibilities, including shrinking the DOE laboratory system, reinforcing the laboratories' focus on nuclear and energy technologies, and assigning new civilian missions to the weapons laboratories. Each alternative had strong political advocates. The latter alternative, new civilian missions, received the greatest attention.

With respect to the first approach, there has not been any actual policy statement urging the large-scale closure of DOE/DOD laboratories.

The extreme opponents of the laboratory system have made some public statements and voiced opinions in news media, but no policy initiative has followed. Most policy statements have either urged that the laboratories stay close to mission or have urged that the laboratories be put to use in commercial enterprise. Specific OTA recommendations included those detailed below:

- Speed up Cooperative R&D Agreements (CRADAs): It was assumed that CRADAs are the means by which technology transfer can be best accomplished. (This assumption is not warranted according to research findings, as reported in Crow and Bozeman, 1998.) The OTA recommendation to facilitate CRADAs focused on numerical assessments of technology transfer derived from government laboratory R&D.
- Enhance local autonomy of laboratories: Again, the focus is on the CRADAs. The "autonomy" recommendation, by and large, aimed at increasing the ability of local laboratory directors to make independent decisions and authorizations for cooperative R&D.
- Create a laboratory rationalization commission: A commission would be established to determine which DOE & DOD laboratories remain as critical assets for the US R&D missions. Such a commission would consider the design and structure of all federal laboratories and not just the DOD or weapons-oriented DOE laboratories. Focusing such efforts only on DOD and DOE laboratories would fail to recognize that the government R&D laboratory community and the university community are highly interwoven and potentially very interdependent (Bozeman and Crow, 1990; Crow and Bozeman, 1987a).

The Galvin Commission

Better known than the OTA defense conversion study is another study occurring about the same time and having many of the same purposes. The best-known statement of the second approach, staying with traditional missions, is the Energy Advisory Board monograph, also known as the Galvin Commission.

The Secretary of Energy turned to Robert Galvin, the former Chairman of Motorola, and an industry-dominated group to consider the issue of mapping out some alternative futures for the DOE national laboratories. The report, which received broad exposure from the press and public, concluded that major changes were required. Specifically the report recommended that the government carry out of a number of steps intended to enhance performance and return on national investment. These steps include the following:

- develop new management systems based on quality principles;
- develop mission to emphasize the use of federal laboratory core competencies as a national resource in collaboration with the private sector and other federal agencies;
- focus laboratories on long-term R&D as a supplement to address the decline in US industrial investment in long term R&D;
- expand all laboratory missions to provide industrial, educational resource and environmental assistance to the region in which laboratories are located;
- develop more partnerships with industry.

The parts of the Galvin Report that seemed to receive the most attention were those pertaining to the need for most DOE laboratories to stay close to traditional missions. The actual wording of the report was much more guarded concerning the prospects of laboratories to greatly expand missions as commercial partners for industry. Indeed, even those aspects urging a commercial role for the laboratories argued that "the government-funded technology transfer/industrial competitiveness activities of the national laboratories should be focused on industries and areas of technology that contribute directly to the DOE's primary missions in national security, energy and environment" (p. 151). Thus, the Galvin Commission represented a mission paradigm approach to laboratory policy, even while recognizing some of the seemingly new realities of an expanded commercial role.

The Technology Reinvestment Program

The centerpiece of the move to use defense R&D to promote civilian technology development was the Technology Reinvestment Program. The Technology Reinvestment Program (TRP) was a competitive, cost-sharing grants program specifically aimed to help defense contractors develop commercial technology. The mission of the program went to the heart of dual-use technology, embracing the assumption that defense laboratories and defense contractors could, with the federal government acting as broker, stimulate the US economy and carve out an expanded civilian role for defense R&D. In 1994, the Defense Department awarded $202 million in grants to industry to stimulate technology development in areas of dual-use technology, including flat-panel displays, high speed computing and materials research. Later that year, Congress authorized $575 million for the Technology Reinvestment Program (TRP).

By February, 1995, the House Appropriations Committee, now under Republican control, was moving to rescind $425 million already approved for TRP. Republican leaders ridiculed TRP as a "slush fund that drains scarce resources from the military" (Vartabedian, 1995). This period marked the abrupt end to Congress's fascination with defense conversion, defense downsizing and the redesign of the weapons and defense laboratories as potential partners for industry. It is easy to dismiss the TRP as having been a casualty of the new Republican majority, although such an interpretation is a bit too facile. As one analyst (Cassata, 1995) noted, "the [TRP] programs are proving ready targets for both defense hawks and budget hawks". Many of those opposing TRP were simply looking for expenditures to cut and seemed to be expressing no views about R&D policy *per se*. In fiscal year 1993, the US Congress authorized $3.2 billion for defense conversion programs of all types and many viewed these programs as a prime target for budget-cutting (US General Accounting Office, 1994).

Some critics (Brandt, 1994; Mussington, 1995) of dual-use were not ideological opponents but skeptical about the viability of the approach. Some applauding the goal of integrating civilian and defense technology development (for example, Alic, 1994) questioned the ability of specific defense conversion policies to achieve this goal. Some critics (Kutler, 1993) suggested that if the goal is to redirect the defense industry to the production of civilian technology, this can be achieved more easily with private capital than with government grants programs. No new funds were authorized for TRP after fiscal year 1995.

Defense R&D in 1999

In early 1999, dual-use technology and defense conversion seem from another era. The Department of Defense downsizing seems to have halted; there are no new policies urging an expanded commercial role for defense and weapons laboratories. Nevertheless, DOD R&D still far surpasses all other federal R&D spending. The next closest is the Department of Health and Human Services with $12.2 billion R&D obligations in fiscal year 1998. Consider the pervasiveness of DOD in US R&D:

- DOD is the source of nearly three-quarters of all federal money spent on development.
- DOD is the source of three-quarters of the federal R&D money spent by industrial R&D performers.
- With just a small fraction of its overall budget, DOD manages to be

one of the three (along with HHS and NSF) largest supporters of university research.
- About half of all money spent at US federal laboratories is DOD funds.
- DOD provides 56% of all R&D funding for federally financed research and development centers (FFRDCs). (Office of Technology Assessment, 1995.)

The Defense Department continues to play much the same "shadow" industrial policy roles as it has played for decades. In fiscal year 1996, DOD began funding the basic research component of six strategic technology development programs: biomimetics ($10 million), nanoscience ($23.9 million), smart structures ($8.7 billion), broadband communications ($17.2 billion), intelligent systems ($18.5 million) and compact power sources ($9.5 million). Perhaps one of these government-subsidized R&D programs will yield the next great leap forward dual-use technologies, such as the Internet or personal computers. However, the issue remains: is this the most efficient approach to industrial policy?

6. CONCLUSIONS: POLICY PARADIGMS AND THE FUTURE OF US RESEARCH POLICY[1]

Since 1994, the market failure paradigm for R&D policy has made a strong comeback. The fate of the TRP is, in many ways, representative of the difficulties of implementing programs that fly in the face of market failure thinking for R&D policy. Generally speaking, a cooperative technology paradigm succeeds only with the combination of a Republican retreat and a perceived national economic crisis. The history of US research policy unfolds as the various policy paradigms wax and wane in importance. We expect that each of the three paradigms will remain important, but their relative strength will vary according to assumptions one makes about the political and economic future of the US

The market paradigm

The market paradigm suggests a modest role for federal laboratories: work within the context of missions not easily privatized. Depending on the

1. This section draws from Crow and Bozeman (1998).

stringency of one's view as to what is and is not easily privatized, the market paradigm ranges from a focus on that most public of public goods, national defense, to a variety of quasi-market public goods with strong externalities, such as public health. Moreover, the market paradigm is not very relevant for energy or agricultural missions for laboratories and is immune to arguments that energy and agriculture missions not performed in the private sector need performing in the public. There is not yet much evidence that today's proponents of the market paradigm have strongly embraced environmental research, it clearly has substantial externalities and there is no evidence that the market will or even could produce an optimal level of environmental research. If there is a shortfall of environmental research today it is probably due to the high threshold of indifference of many policymakers and R&D sponsors rather than to the ideological premises for environmental research. Unless pollution and hazardous waste problems are checked, or unless pollution credits and other such market solutions begin to have more of an effect than they have heretofore, the growth of environmental research could occur even within the framework of a market failure paradigm.

During the next two decades, a federal laboratory system dominated by market paradigm thinking might have the following characteristics: (1) a smaller but stable set of federal defense-supply laboratories (including the phasing out of the weapons laboratories' "peripheral" missions); (2) the end of the federal energy R&D laboratory focus; (3) some support for laboratories focusing on quasi-public goods such as environment and hazardous waste and public health-related R&D (but not medical research and technology, in general). In sum, if the market paradigm continues to dominate, the role of the federal laboratory system may shrink dramatically and cooperative research and research partnerships may be de-emphasized. The result could be a shift of resources to university R&D or a return to "Reaganomics" approaches focusing on industry research and tax credits.

The mission paradigm

There is nothing in current political trends to suggest an impending expansion of the three missions that have historically underpinned the largest portion of federal laboratory R&D: defense, energy and agriculture. Thus, reliance on a mission paradigm could result in a diminished federal laboratory system, accompanying diminished core missions. The likely exception is the defense and national security mission that requires enormous expenditures just to preserve the base investment. Further, the microscopic "peace dividend" and the sound and fury of the base-closing commission provide ample testimony to the resilience of the defense establishment.

Both university and industry benefit from the mission paradigm focus, especially as the size and scope of federal laboratories continues to diminish. Absent an upward trend, highly unlikely, in federal laboratory support, mission R&D will continue to be contracted out to industry and universities and each laboratory type will continue to benefit from the mission paradigm and its consequences.

Ascendance of the cooperative technology paradigm

The cooperative technology paradigm has always been in the policy background (even when it was in the rhetoric foreground). It is likely to continue to have only episodic influence. If the US economy began to experience sharp declines (well beyond those of the 1980s) *vis-à-vis* its trading partners, and if the US balance of technology goods payments was to become highly unfavorable, then policymakers might be drawn to cooperative technology as a dominant framework. The likelihood further increases if an even larger number of US competitors (including the European Union and a reviving Japan) implemented a strong set of highly effective cooperative technology policies.

Gauging the likely impact of cooperative technology development on industry and university laboratories poses problems. First, the impact would depend upon the readiness of non-government laboratories to cooperate. Current evidence suggests that industry laboratories approach cooperation with great caution. Providing incentives, including matching R&D funding, could well alter industry receptivity.

The safest prediction about the future of the US NIS is that it will continue to be highly varied, decentralized and fragmented. It is a system that both consumes and produces enormous resources and it is so integral to the US economy that its inefficiencies are less important than its ultimate effectiveness.

REFERENCES

Alic, J., 1994, The Dual-use of Technology: Concepts and Policies, *Technology in Society*, 16 (2), 155-172.

Bagur, J. and A. Guissinger, 1987, Technology Transfer Legislation: An Overview, *Journal of Technology Transfer*, 12, 51-63.

Bozeman, B., 1994, Evaluating Government Technology Transfer: Early Impacts of the 'Cooperative Technology Paradigm', *Policy Studies Journal*, 22, 322-327.

Bozeman, B. and A.N. Link, 1983, *Investments in Technology: Corporate Strategies and Public Policy Alternatives*, Greenwood Publishing Group, Inc., New York, NY: Praeger.

Bozeman, B. and A.N. Link, 1984, Tax Incentives for R&D: A Critical Evaluation, *Research Policy*, 13(1), 21-31.

Bozeman, B. and M. Crow, 1990, The Environments of US R&D Laboratories: Political and Market Influences, *Policy Sciences*, 23, 25-56.

Bozeman, B., M. Papadakis and K. Coker, 1995, *Industry Perspectives on Commercial Interactions with Federal Laboratories: Does the Cooperative Technology Paradigm Really Work?* Report to the National Science Foundation, Research on Science and Technology Program (January).

Bucy, J.F., 1985, Meeting the Competitive Challenge: The Case for R&D Tax Credits, *Issues in Science and Technology* (summer), 69-78.

Brandt, L., 1994, Defense Conversion and Dual-Use Technology: The Push toward Civil-Military Integration, *Policy Studies Journal*, 22 (2), 359-370.

Carnegie Commission on Science, Technology and Government, 1992, *Science, Technology, and the States in America's Third Century*, New York: Carnegie Commission.

Cassata, D., 1995, Swords-to-Plowshares Program Faces Uncertain Future, *Congressional Quarterly Weekly*, 53, 620-624.

Coburn, C. (ed.), 1995, *Partnerships: A Compendium of State and Federal Cooperative Technology Programs*, Battelle Memorial Institute. Columbus, OH: Battelle Press.

Congressional Research Service, 1998, *State Technology Development Strategies: the Role of High Tech Clusters*, (21 October) p. 11, 98-859 STM.

Congressional Research Service, 1999a, Research and Development Funding: Fiscal Year 2000, *CRS Issue Brief*, Library of Congress.

Council on Competitiveness, 1993, *Industry as a Customer of the Federal Laboratories*. Washington, DC: US Government Printing Office.

Council on Competitiveness, 1996a, *Competitiveness Index: A Ten-Year Strategic Assessment*, Washington, DC: US Government Printing Office.

Council on Competitiveness, 1996b, *Endless Frontier, Limited Resources: US R&D Policy for Competitiveness*, Washington, DC: US Government Printing Office.

CONSAD Research Corp., 1997, *Advanced Technology Program A Case Study: The Development of Advanced Technologies and Systems for Controlling Dimensional Variation in Automobile Body Manufacturing*, Gaithersburg, MD: NIST (March).

Crow, M. and B. Bozeman, 1987a, A New Typology for R&D Laboratories: Implications for Policy Analysis, *Journal of Policy Analysis and Management*, 6 (3), 328-341.

Crow, M. and B. Bozeman, 1987b, R&D Laboratories' Environmental Contexts: Are the Government Lab-Industrial Lab Stereotypes Still Valid? *Research Policy* 13, 329-355.

Crow, M. and B. Bozeman, 1998, *Limited by Design: R&D Laboratories in the US National Innovation System*, New York: Columbia University Press.

Eisner, R., 1985, R&D Tax Credits: A Flawed Tool, *Issues in Science and Technology*, (summer), 79-86.

Gillespie, G., 1988, Federal Laboratories: Economic Development and Intellectual Property Constraints, *Technology Transfer*, 20-26.

Herrmann, J., 1983, Redefining the Federal Government's Role in Technology Transfer, *Research Management*, 21-24.

Kash, D.E., 1989, *Perpetual Innovation: The New World of Competition*, New York: Basic Books.

Kash, D.E. and R.W. Rycroft, 1994, Technology Policy-Fitting Concept with Reality, *Technol. Forecast SOC*, 47 (1) (September), 35-48.

Kutler, J., 1993, Wall Street's Role in Military Conversion, *New York Times*, Sec. 3, p. 11, 12 December.

Laidlaw, F.J., 1997, *Acceleration of Technology Development by the Advanced Technology Program: The Experience of 28 Projects Funded in 1991*, Gaithersburg, MD: NIST (September).

Landau, R. and N.B. Hannay (eds), 1981, *Taxation, Technology, and US Economy*, New York: Pergamon.

Link, A.N., 1997, *Early Stage Impacts of the Printed Wiring Board*, Report to the Advanced Technology Program. www.atp.nist.gov/atp/pubs.htm.

Link, A.N. and L.L. Bauer, 1989, *Cooperative Research in US Manufacturing: Assessing Policy Initiatives and Corporate Strategies*, Lexington, MA: Lexington Books.

Link, A.N. and G. Tassey, 1987, *Strategies for Technology-Based Competition: Meeting the New Global Challenge*, Lexington, MA: Lexington Books.

Lundvall, B. (ed.), 1992, *National Systems of Innovation: Towards a Theory of Innovation and Interactive Learning*, London: Pinter Publishers.

Mansfield, E., 1996, *Estimating Social and Private Returns from Innovations Based on the Advanced Technology Program: Problems and Opportunities*, Gaithersburg, MD: NIST (January).

Martin, S.A. et al., 1997, *Framework for Estimating the National Economic Benefits of ATP Funding of Medical Technologies*, Research Triangle Institute, Gaithersburg, MD: NIST.

Melkers, J.E. and D. Roessner, 1997, Politics and the Political Setting as an Influence on Evaluation Activities: National Research and Technology Policy Programs in the United States and Canada, *Evaluation and Program Planning*, 20 (1).

Melkers, J.E. and S. Cozzens, 1997a, A Review of Performance Measurement in State Science and Technology-Based Economic Development Programs, *Proceedings: Technology Transfer Metric Summit,* Technology Transfer: How Do We Know What Works? A Working Conference on Measurement and Evaluation, Chicago: Technology Transfer Society.

Melkers, J.E. and S. Cozzens, 1997b, Use and Usefulness of Performance Measurement in State Science and Technology Programs, *Policy Studies Journal*, 25 (3).

Melkers, J.E. and S. Cozzens, 1997c, Developing and Transferring Technology in State S&T Programs: Assessing Performance, *Journal of Technology Transfer*, 22 (2) (Summer).

Morrison, D., 1995, Technology Tussle, *National Journal*, 27, 799-803.

Mussington, D., 1995, Defense Conversion and Industrial Competitiveness in the United States: A New Test for Industrial Policy, *Business and the Contemporary World*, 7 (2), 146-156.

National Academy of Sciences, 1978, *Technology, Trade, and the US Economy*, Washington, DC: National Academy Press.

National Governors' Association, 1987, *The Role of Science and Technology in Economic Competitiveness*, New York: National Governors' Association.

National Science Foundation, 1992, *National Patterns of R&D Resources: 1992*, Arlington, VA: National Science Foundation.

National Science Foundation, 1999a, R&D as a Percent of GDP is the Highest in Six Years, Data Brief, NSF 99-302, Arlington, VA: National Science Foundation.

National Science Foundation, 1999b, What is the State Government Role in the R&D Enterprise?, NSF 99-348, Arlington, VA: National Science Foundation.

National Science Foundation, 1999c, President's FY 2000 Budget Includes Reduced R&D Request; Nondefense R&D funding Catches up to Defense R&D, Data Brief, NSF 99-353, Arlington, VA: National Science Foundation.

Nelson, R., 1996, The Agenda for Growth Theory: A Different Point of View. Draft Manuscript.

Office of Technology Assessment, US Congress, 1995, *A History of the Department of Defense Federally Funded Research and Development Centers*, OTA-BP-ISS-157, Washington, DC: USGPO.

Powell, J., 1997, *Development, Commercialization, and Diffusion of Enabling Technologies: Progress Report for Projects Funded 1993-1995*, Gaithersburg, MD: NIST (December).

President's Commission on Industrial Competitiveness, 1985, *Global Competition: The New Reality*, Washington, DC: US Government Printing Office.

Prokesch, S., 1993, Companies Struggle to Adjust as US Cuts Military Budget, *The New York Times*, 10 February.

Rahm, D., B. Bozeman and M. Crow, 1988, Domestic Technology Transfer and Competitiveness: An Empirical Assessment of Roles of University and Government R&D Laboratories, *Public Administration Review*.

Roessner, J.D. and A. Bean, 1991, How Industry Interacts with Federal Laboratories, *Research Technology Management*, 34 (4), 22.

Ruegg, R., 1997, The Advanced Technology Program's Evaluation Plan and Progress, *Proceedings of the 7th International Forum on Technology Management*, Kyoto, Japan, (November).

Rycroft, R.W. and D.E. Kash, 1994, Complex Technology and Community-Implications for Policy and Social-science, *Research Policy*, 23 (6), 613-626.

Silber and Associates, 1996, *Survey of Advanced Technology Program 1990-1992 Awardees: Company Opinion about the ATP and Its Early Effects*, Gaithersburg, MD: NIST (January).

Smilor, R.W. and D.V. Gibson, 1991, Technology-Transfer in Multi Organizational Environments: The Case of R&D Consortia, *IEEE Transactions on Engineering Management*, 38 (1) (February), 3-13.

Solomon Associates, 1993, *The Advanced Technology Program; An Assessment of Short-Term Impacts: First Competition Participants*, Gaithersburg, MD: NIST (February).

State Science and Technology Institute, 1996, *State Funding for Cooperative Technology Programs*, June.

The MIT Commission on Industrial Productivity, 1989, *Working Papers of the MIT Commission on Industrial Productivity*, two-volume set.

US Congress, House Science Committee, HR 2429, Small Business Technology Transfer Program Reauthorization Act of 1997.

US Department of Energy, 1992, *DOE Multiprogram Laboratories: 1979-1988 A Decade of Change*, Washington, DC: USGPO.

US General Accounting Office, 1989, Technology Transfer: Implementation Status of the Federal Technology Transfer Act of 1986.

US General Accounting Office, 1994, Defense Conversion: Status of DOD Funding and Spending, Washington, DC: USGAO/NSIAD-94-218BR.

US General Accounting Office, 1995, Manufacturing Extension Programs: Manufacturers' Views of Services, Washington, DC: USGAO/GGD-95-216BR.

US General Accounting Office, 1996a, Performance Measurement: The Advanced Technology Program and Private Sector Funding, Washington, DC: USGAO/RCED-96-47, Code 307723.

US General Accounting Office, 1996b, Manufacturing Extension Programs: Manufacturers' Views about Delivery and Impact of Services, Washington, DC: USGAO/GD-96-75.

US General Accounting Office, 1996c, Federal Research: Interim Assessment of the Small Business Innovation Research and Technology Transfer Programs, Washington, DC: USGAO/T-RCED-96-93.

US General Accounting Office, 1998, Federal Research: Observations on the Small Business Innovation Research Program, Washington, DC: USGAO/T-RCED-98-218.

Vartabedian, R., 1995, Defense Conversion Program Appears Likely to Lost Funding, *Los Angeles Times*, Sec. D, p. 1.

Viscusi, W.K., 1995, Carcinogen Regulation-Risk Characteristics and the Synthetic Risk Bias, *American Economic Review*, 85 (2) (May), 50-54.

Vonortas, N.S., 1997, *The Effectiveness of Research Joint Ventures Sponsored by ATP*, Gaithersburg, MD: NIST (October).

CHAPTER 4

THE STRUCTURE AND PERSPECTIVE OF SCIENCE AND TECHNOLOGY POLICY IN JAPAN

YUKIO SATO

1. INTRODUCTION

At the dawn of the 21st century, Japan is embarking upon drastic and dynamic change in her entire socio-economic system.

The Japanese Government has been determined to carry out reforms in a number of major areas since former Prime Minister Ryutaro Hashimoto launched the "Six Reforms" initiative in 1996. These areas are: administrative reform; reform of the social security system; reform of financial systems; economic structural reform: fiscal structural reform and educational reform. Although some of the reform processes have been delayed or altered to some extent because of the recent severe setback in the economy, these reforms are necessary and inevitable in order for Japan to be a nation with vitality in the 21st century.

In June 1998, the Diet passed the Administrative Reform Basic Law, which stipulates, among other things, the reinforcement of Cabinet functions and the reorganisation of the current 22 Ministries and agencies into 13 consolidated entities starting in the year 2001. In January 1999, the Government adopted the outline of bills for administrative reform in line with the Administrative Reform Basic Law. The specific bills, which were passed by the Diet in July 1999 and became laws, include provisions for establishing a Cabinet Office and other new government Ministries, as well as amendments to the Cabinet Law and the National Government Organisation Law.

This administrative reform, together with other reforms, is bringing about revolutionary change in Japan, which could end up being comparable to the changes following the Meiji Restoration in 1868 and after the end of World War II in 1945. The last two revolutionary changes took place in order for Japan to catch up with Western countries and to adapt to Western systems after the failures in Japanese social and political systems caused by isolation. The last two major changes were rather simple in their objectives and straightforward in their execution. Both reformations put heavy emphasis on the introduction of technologies and socio-economic systems from Western countries.

The principal driving force to proceed with new Japanese reforms for the 21st century also arises from the necessity for adaptation to the world system. However, the current situation is quite different and more complex than either of the two previous reformations. The world system surrounding Japan has itself dramatically changed and the pace of change is accelerating. Moreover, the Japanese system has become interwoven with other elements in a globalised economic system to such an extent that changes in Japan may exert great influence over worldwide developments. This is exemplified by the recent Asian economic crisis and the merger and acquisition practices across the boundaries between nations and between business types in a highly competitive world, undergoing a revolution in information and communication technologies.

In addition to the change in the relationship between the Japanese and world systems, underlying conditions for the economy and society in Japan have been changing profoundly. Past practices such as lifetime employment and seniority-based wage systems are coming to an end. However, the most influential change is demographic. The birth rate in Japan has been falling dramatically in recent years. The total fertility rate dropped to 1.42 in 1995, falling substantially below the 2.08 threshold, the level required to maintain the population at its current level.

Because of the trend towards fewer children, the total Japanese population is predicted to reach a peak in the year 2007 and then to decrease for the first time since World War II. As a result of increasing Japanese life expectancy (already the longest in the world), and the declining birth rate, the ageing of the Japanese population is proceeding at a higher rate than in any other major industrialised country. In addition, the proportion of the population aged 65 or over, now 15%, will exceed 30% by 2050, while the working age population (15-64 years of age) which has increased every year up until 1995, is now expected to undergo a continuous decline.

2. RECENT DEVELOPMENTS IN THE FRAMEWORK FOR SCIENCE AND TECHNOLOGY POLICY

The Government adopted a new approach to science and technology policy in the 1990s, in recognition of the important role that science and technology should play in addressing the challenges facing contemporary Japan.

In 1995, the Science and Technology Basic Law was enacted to achieve a higher standard of science and technology to contribute to the development of the economy and society in Japan and to the improvement of the welfare of the nation, as well as to contribute to global progress in science and technology and to the sustainable development of human society. Stating that the nation is responsible for formulating and implementing comprehensive and systematic policies with regard to the promotion of science and technology, this law laid out the scope of measures to be taken by the Government. (The outline is shown in Table 4.1.)

It is also noteworthy that this law was initiated by members of Parliament supporting science and technology, and that the law was enacted unanimously, which is rare in the Japanese Diet. The law made the promotion of science and technology one of the nation's principal themes, leading the Government to formulate the Science and Technology Basic Plan and submit an annual "report on measures taken to promote science and technology" to the Diet. It can also be noted that the Basic Law stipulates the responsibilities of local governments in promoting science and technology as well as those of National Government.

In July 1996, the Government decided on the Science and Technology Basic Plan based on Article 9 of the Basic Law, to implement policies for the next decade with special reference to the period covering fiscal years 1996-2000. The plan is composed of two chapters. Chapter 1 describes comprehensive policies to promote R&D. It presents directions for R&D and the improvement of systems and conditions for its promotion, including the expansion of R&D investment by the government. Chapter 2 prescribes concrete measures to be taken for the next five years based on the policies laid down in Chapter 1 (the outline is shown in Table 4.2.)

The most remarkable output of this Basic Plan together with the Basic Law has been the level of government investment in research. The Basic Plan estimates the total science and technology budget for the five years from 1996 to 2000 at ¥17 trillion (Figure 4.1). This was an ambitious plan, particularly since the government's research investments had been ¥12.6 trillion during the five-year period prior to the formulation of the Basic Plan. Despite the

TABLE 4.1 – The outline of the Science and Technology Basic Law

1. Guidelines for S&T promotion:
   - development of researchers' creativity;
   - harmonious development of basic, applied and development research;
   - promotion of S&T in harmony with human life, society and nature.

2. Responsibilities of the National and Regional Governments in promoting S&T.

3. Formulation of S&T Basic Plan by the Government through consultation with the Council for S&T, in order to promote S&T policies comprehensively and systematically.

4. Measures taken by the National Government:
   - balanced promotion of diversified R&D;
   - securing and training researchers and technicians;
   - improvement of research facilities;
   - promotion of information intensive research;
   - promotion of R&D exchanges.

TABLE 4.2 – The outline of the Science and Technology Basic Plan

1. The basic aims of R&D:
   - intensive promotion of R&D applicable to socio-economic needs such as the creation of new industries, solving global problems, the construction of a comfortable community;
   - active promotion of basic research, aiming at the discovery of new laws and principles, the creation of original theories and the prediction and discovery of unknown phenomena.

2. Constructing a new R&D system:
   - constructing a new R&D system to facilitate creative R&D activities;
   - constructing a co-operation and exchange system beyond sectors, regions and nations;
   - impartial evaluations.

3. Expansion of R&D investment by the Government:
   - expected S&T budget total for 5 years from 1995 to 2000 is about ¥17 trillion (about $130 billion).

severity of the fiscal situation, the national budget for science and technology has increased significantly for five consecutive years, notably as a consequence of the Basic Law and the Basic Plan, reaching a cumulative total above the target level of ¥17 trillion. The Plan aims at creating a new R&D system based on the maximum utilisation of the ingenuity of researchers for a new era.

FIGURE 4.1 – Science and technology expenditures during the Science and Technology Basic Plan (5 years from FY1996)

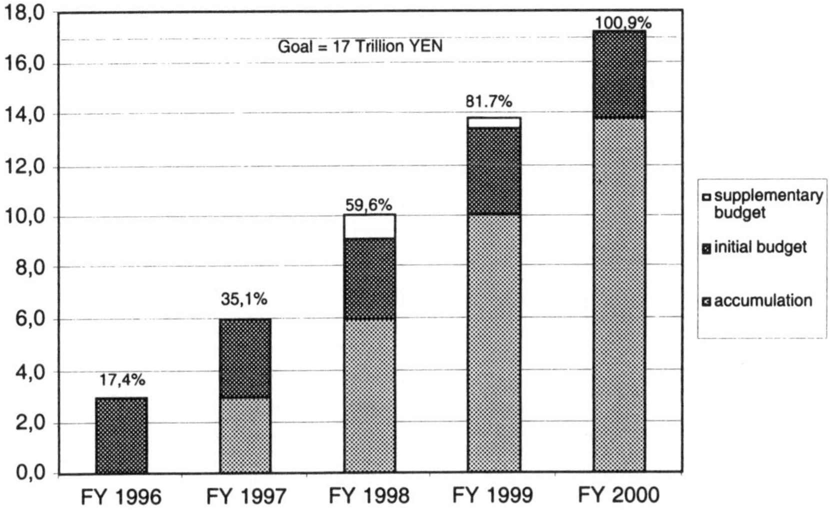

The Administrative Reform Basic Law stipulates, among other things, that the Science and Technology Agency and the Ministry of Education, Science, Sports and Culture (*Monbusho*) are to be merged into one new Ministry of Education, Science and Technology (Figure 4.2). The budget of the new ministry will be about two-thirds of government's total science and technology budget. The law also establishes a new General Council for Science and Technology (GCST) to replace the existing Council for Science and Technology (CST). Both GCST and its secretariat are to be set up in the Cabinet Office.

One of the purposes of the Administrative Reform Basic Law is to increase efficiency in the administration of the central Government by establishing more direct leadership by the Prime Minister, supported by the Cabinet Office with comprehensive co-ordinating functions and directly controlled operations as well as by high-level advisory bodies. Both changes will have the effect of strengthening science and technology policy co-ordination. The GCST will be chaired by the Prime Minister, as the CST is at present. In a new Cabinet, the Minister in charge of GCST can be appointed. Major differences between GCST and CST are as follows:

FIGURE 4.2 – Administrative reform (by the Administrative Reform Basic Law)

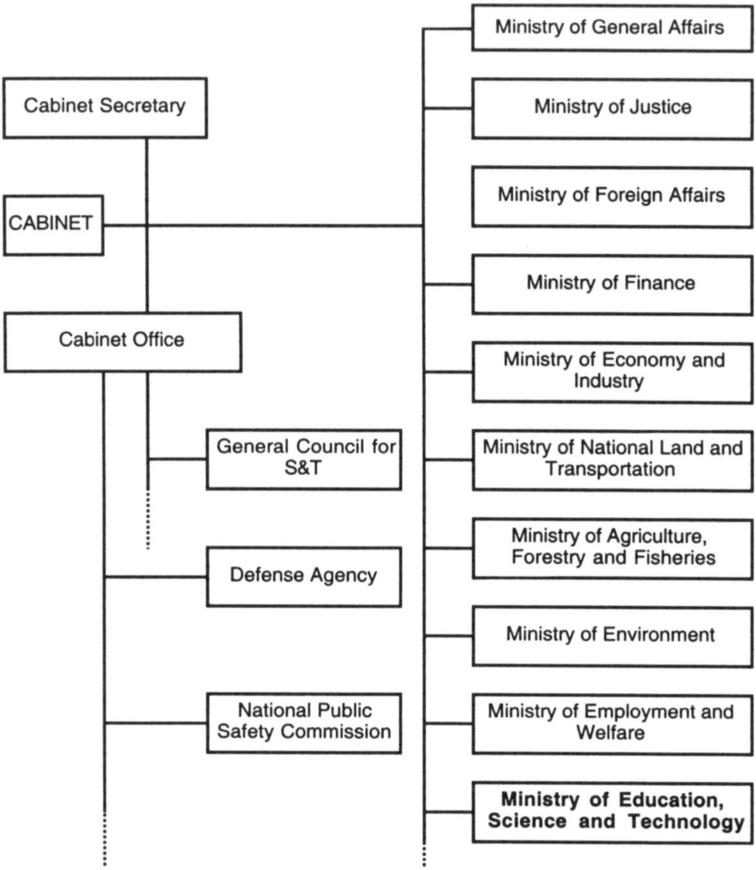

- The GCST will also assume responsibility for the humanities and the social sciences while the existing CST is in charge of policy in science and technology excluding those relevant only to the humanities and social sciences.
- In addition to recommendations in response to requests by the Prime Minister for which the present CST is responsible, the GCST will be able to make autonomous policy recommendations.
- The number of members of GCST will be larger than that of CST covering a wider variety of fields.
- The GCST will have its own secretariat, comprising staff not only from various ministries but also from academia and industry.

- The GCST will not only formulate basic science and technology policy and R&D resources allocation policy but will also conduct evaluation of important projects. The CST is in charge of formulating basic science and technology policy, goal-setting for R&D activity and its promotion.

Moreover, the Administrative Reform Basic Law stipulates the reformation of national research institutes, which are attached to Ministries and agencies. Out of 83 national research institutes, 56 will be transformed into Independent Administrative Corporations in the spring of 2001. Law for each of the Independent Administrative Corporations was enacted by the Diet in December 1999. The transformed research institutes will have more freedom and flexibility in research and organisational management than existing national research institutes.

A brief review of the historical evolution of the structure for science and technology in Japan would show the drastic nature of these recent developments. This is exemplified by the Science and Technology Basic Law, which was initiated by politicians to promote science and technology, and was the first of its kind to be enacted in Japan's history. More than 30 years ago, the Government introduced a draft of the Science and Technology Basic Bill to the Diet for deliberation. However, this bill was not passed by the Diet because of strong opposition by non-government parties, who responded to the negative feelings in academic circles at that time. The change in the Japanese administrative structure would alter two main principles concerning S&T administration, which have existed for more than 40 years.

Firstly, administrative reform will put an end to a unique system in which an agency headed by a Cabinet Minister, the Science and Technology Agency (STA), has been dedicated to formulating comprehensive S&T policies, and to implementing fundamental and infrastructure-type R&D and large-scale projects which are not suitable for implementation in any other Ministry or agency, and to the co-ordination of S&T policies across Ministerial boundaries. Here, it should be noted that S&T policies in Japan are implemented by various Ministries and agencies, according to their respective missions. In the new Government structure, from January 2001, the formulation of comprehensive S&T policies and the co-ordination of overall policies are to be assigned to the Cabinet Office that will be placed above other ministries. From a statutory point of view, more strategic formulation of S&T policies and more top-down type implementation will be realised.

Secondly, inclusion of the humanities and the social sciences in "science and technology" would bring about new perspectives and complexities in the formulation and implementation of S&T policies. The compound word "science and technology" in Japanese, together with the exclusion of those disciplines relevant only to the humanities and the social sciences, have produced a somewhat unique notion which has characterised Japan's remarkable development in science and technology since the end of World War II. The notion is not simply the combination of Science and Technology but an amalgamation of the two notions into an enhanced one. This has unintentionally contributed to the transfer of information and technology from academic circles to industry but has also generated a gap between the natural sciences and the humanities. The need for increased harmony between science and technology and human society requires more involvement of the humanities and social sciences in the formulation of S&T policy. A more harmonised approach to complicated issues, utilising and combining various disciplines in the natural and social sciences including the humanities, can be made as a result of the administrative reform.

In addition to these changes, a new system for S&T policy, in line with the Administrative Reform Basic Law, would induce more explicit distinctions between the different functions involved in the formulation and implementation of S&T policies. This new tendency to separate the two functions would be strengthened by the introduction of nationwide research evaluation into the governmental R&D system that has been made according to the Basic Plan to respond to the increased governmental financial support for R&D despite recent severe financial conditions.

Each ministry or agency responsible for policy formulation has national research institutes whose functions are to implement R&D in the field of its jurisdiction. Although their major role is the implementation of R&D, these research institutes, together with R&D-related public sector corporations, have greatly contributed to the formulation of national S&T policies through the exchange of information and personnel on a regular and/or *ad hoc* basis. Transformation of national research institutes into independent entities, namely Independent Administrative Corporations, would change the basis of the existing relationship between organisations involved in the formulation and implementation of policy into a more contractual one: the policies and missions of ministries would be carried out by the research institutes through R&D contracts. Extensive introduction of a contract-based research system with strict evaluation, which is necessary for the greater accountability and transparency of S&T activities requested by society, would exert a profound influence on the research community and the researchers of the national institutes and universities.

3. PROFILE OF JAPANESE R&D ACTIVITIES

S&T policies that will greatly influence the future level of S&T activities reflect the past R&D activities as well as the socio-economic needs of society. Although it is very hard to grasp precisely the real state of affairs with regard to R&D activities, indicators such as the amount of R&D expenditure and the number of research personnel, including research support staff, together with a description of the R&D system can provide an image of real R&D activities.

Japan's total expenditure on R&D during fiscal year (FY) 1998 stood at ¥16.1 trillion (¥14.8 trillion for the natural sciences), an increase of 2.5% (2.3% for natural science) over the preceding year, reflecting government efforts based on the Basic Plan including supplementary budgets, during a recession with a 2.1% decrease in Gross Domestic Product (GDP) (Table 4.3). This increase occurred over four consecutive years, following a decline in the total R&D expenditures in FY1993 and FY1994 due to lower investment by the private sector caused by an economic slowdown. R&D expenditure for the natural sciences accounts for 91.8% of the total R&D expenditure. The ratio has been around this level (±0.5 points) for the last ten years.

TABLE 4.3 – R&D expenditure

	1965	1970	1975	1980	1985	1990	1995	1997	1998
Industry	252	823	1 685	3 142	5 940	9 267	9 396	10 658	10 800
Research institutes	68	155	421	718	1 101	1 416	1 921	1 927	2 012
Universities	106	217	516	824	1 075	1 406	1 875	1 906	2 012
Total	426	1 195	2 622	4 684	8 116	12 090	13 191	14 492	14 824

R&D expenditure by sector (Natural Science only)
Unit: Billion Yen.

	1965	1970	1975	1980	1985	1990	1995	1997	1998
Government	131	301	721	1 210	1 574	1 990	2 866	2 776	3 031
% of total expenditure	30.8	25.2	27.5	25.8	19.4	16.5	21.7	19.2	20.4
% excluding defence R&D	30.1	24.5	27.0	25.4	18.8	15.7	20.8	18.2	19.7

Government funding (Natural Science only)
Unit: Billion Yen.

TABLE 4.4 – Sectoral share of R&D expenditure by source and performer in Japan

		Industry	Government	Other	Total
1994	Source	68.2%	21.5%	10.3%	100.0%
	Performer	66.1%	9.0%	24.9%	100.0%
1995	Source	67.1%	22.9%	10.0%	100.0%
	Performer	65.2%	9.6%	25.1%	100.0%
1996	Source	69.0%	21.0%	10.0%	100.0%
	Performer	66.7%	8.8%	24.5%	100.0%
1997	Source	69.7%	20.4%	10.0%	100.0%
	Performer	67.7%	8.3%	24.0%	100.0%
1998	Source	68.3%	21.7%	10.0%	100.0%
	Performer	66.9%	8.7%	24.4%	100.0%

Note: "Other" performers include universities and colleges and private non-profit institutes.
Source: Management and Co-ordination Agency, "Report on the Survey on S&T Research".

The ratio of total R&D spending to GDP in FY1998 was 3.26%, up by 0.14 points from FY1997. The ratio fell for four consecutive years in succession, from 2.95 in FY1991 to 2.82 in FY1994, partly due to declining private R&D expenditure, but then jumped up to 2.92 in FY1995 and increased for four consecutive years. Japan continues to maintain the highest level of R&D spending among the major industrialized countries.

R&D expenditure can be characterised by the financing and performance aspects of constituent sectors. Of the total R&D expenditure in FY1998, ¥10 800 billion (66.9%) was spent by companies, an increase of 1.3%; ¥3223 billion was spent by universities and colleges, up by 5.4%; and ¥2117 billion by research institutions, up by 4.6% compared with FY1997. In Japan, about two-thirds of total expenditure has been spent by companies and about 20% by universities for the last several years. The private sector provided ¥12 593 billion in FY1998, accounting for 78.0% of total, while the public sector including local governments provided ¥3499 billion, accounting for 21.7% of the total (Table 4.4).

Shares by type of activity were 14.4% for basic research, 24.6% for applied research, and 61.0% for development research. The structure of R&D expenditure in industry, government, research institutions and universities and colleges is distinctly different. Industry spends more on development due to the nature of corporate activities and the relatively high costs of commercialisation. Universities and colleges give priority to basic and applied research. Research institutions fall between the two categories (Table 4.5). Thus the

TABLE 4.5 – Percentage distribution of R&D expenditure by character of activity in each sector (FY1998)

	Basic research	Applied research	Development	Total
Industry	5.5%	21.9%	72.6%	100%
Private non profit org.	14.4%	21.9%	63.8%	100%
Government	24.5%	29.9%	45.6%	100%
Universities and colleges	54.8%	36.3%	8.9%	100%
Total	14.4%	24.6%	61.0%	100%

Source: Management and Co-ordination Agency, "Report on the Survey on S&T Research".

low proportion of R&D expenditure devoted to basic research is attributed to the low share of government R&D expenditure as well as to the lower share of basic research conducted at universities compared with other selected countries.

As for the number of researchers, countries use different definitions and statistical measures. Therefore, simple comparison may not be adequate, although it is useful for grasping general trends of human resources in research. The number of researchers in the Report on the Survey of Research and Development conducted by the Statistics Bureau of Japan's Management and Co-ordination Agency is counted as follows:

- In companies as well as in private research institutions: the actual number of researchers (head-count) who have two or more years of research experience after graduation from university.
- In universities and colleges: the actual number (head-count) of teachers, students in Ph.D. programmes, medical staff, and so on. Thus, the number of researchers in Japan is inflated because the full-time equivalent (FTE) conversion method is not used.

The number of researchers in Japan as of 1 April 1999 was 639 000 (733 000 including researchers in social science and humanities), showing an increase of 4.0% (including researchers in social science and humanities) from the preceding year.

As for the number of researchers by sector, industry has the most of any sector at 429 000 researchers, 67.2% of the total (58.6% of the total including researchers in social science and humanities), followed by universities and colleges, at 167 000 researchers (256 000 researchers including social sciences and humanities), and research institutions, at 42 620 researchers (47 023 researchers including social sciences and humanities). It is to be noted that the number of researchers at national institutes in Japan as of 1 April 1999 was only 10784 (11471 including researchers in social science and humanities).

In the ten-year period from 1989 to 1999, the number of researchers in companies increased by 46% and the number of researchers at universities and colleges increased by 27%, while the number of researchers at government research institutes showed only a slight increase. As for the number of researchers by industry type, as of 1 April 1999, the electrical machinery sector registered the highest number (155700), followed by the chemicals industry (59600), the transport equipment industry (41300) and the general machinery industry.

4. GOVERNMENT STRUCTURE FOR THE FORMULATION AND IMPLEMENTATION OF SCIENCE AND TECHNOLOGY POLICY

In Japan, science and technology policies are implemented by various ministries and agencies, according to their respective missions. Therefore, almost all Ministries and agencies have their own research institutions and budgets to conduct R&D related to their missions (see Annex). According to policies based on their missions, Ministries and agencies carry out various research and development activities at relevant national research institutes, public research corporations under the jurisdiction of Ministries and agencies as well as at national universities including inter-university research institutes. Sometimes, research and development are carried out through contracts with private companies. These policies have a basis in the Science and Technology Basic Law and the Basic Plan as well as in the recommendations of the CST and other advisory organs, attached to respective ministries and agencies.

The description of the government science and technology budget is presented in Table 4.6. In FY1999, the total science and technology budget of the government is ¥3155 billion. Of that amount, 42.7% is designated for *Monbusho*, most of which finances national universities. It also includes grants to university researchers as well as grants to private universities. Of the total science and technology budget, 24.5% is allocated to the Science and Technology Agency (STA). The STA carries out R&D in nuclear energy and space development and promotes basic research in important areas. 16.2% is allocated to the Ministry of International Trade and Industry (MITI), while the Defence Agency of Japan accounts for 4.6%. It is noteworthy that the ratio

TABLE 4.6 – FY1999 Budget for Science and Technology by Ministries and agencies

	Billion Yen	%
Ministry of Education	1 348.7	42.7%
Science and Technology Agency	773.8	24.5%
Ministry of International Trade and Industry	508.3	16.1%
Defence Agency	146.5	4.6%
Ministry of Agriculture, Forestry and Fisheries	109.2	3.5%
Ministry of Health and Welfare	101.7	3.2%
Ministry of Posts and Telecommunications	74.4	2.4%
Ministry of Transport	23.7	0.7%
Others	94.9	3.0%
Total	3 155.2	100.0%

Unit: Billion Yen.
Note: ¥26 billion appropriated for the Japan Key Technology Center, are included in the budgets of MITI and MPT in duplication (duplications are eliminated in totaling).

of defence-related S&T budget to the total S&T budget in Japan is very low compared with that in other major countries. This ratio increased every year for more than ten years, reaching 6.2% in 1995, and then began to decrease.

5. THE COUNCIL FOR SCIENCE AND TECHNOLOGY AND THE EVOLUTION OF COMPREHENSIVE S&T POLICY

The 1995 Basic Science and Technology Law states that, in formulating the Basic Science and Technology Plan, the Government must first consult the Council for Science and Technology (CST). The main task of the CST is to advise the Prime Minister on the formulation of:

- general science and technology policy;
- comprehensive long-term research goals; and
- basic policy measures for promoting research in important areas.

Under the law establishing the Council for Science and Technology, the Prime Minister must consult the CST in cases where inter-ministerial co-ordination is necessary. If there is a report in response to that consultation, its recommendations must be taken into account.

The CST consists of the Prime Minister, as chairperson, four Cabinet ministers (from the Ministry of Finance, *Monbusho*, the Economic Planning Agency and the Science and Technology Agency), the chairperson of the

Science Council of Japan and five distinguished leaders from the academic and industrial world. These five members are appointed by the Prime Minister with the consent of the Diet. Because it is composed in this way, S&T policies recommended by the CST have been well implemented by various ministries and agencies as if they were Cabinet decisions. In practice, the CST is a quasi-advisory, quasi-executive body.

Plenary meetings of the CST are usually held twice a year with the Prime Minister in the chair. Cabinet ministers other than those listed above may also attend plenary meetings. A panel of the CST, the Committee on Policy Matters deliberates on important policy issues and meets regularly twice a month. The Committee is composed of the members of the CST (except Cabinet Ministers) and several distinguished experts from the academic, industrial and administrative spheres appointed by the Prime Minister. In addition to the Committee on Policy Matters, there are several panels composed of selected non-ministerial members of CST plus additional co-opted experts in various fields to consider important policy issues.

In designing a new framework for the S&T policy, it is necessary and instructive to review the old framework in a positive way. Since a review of the evolution of the whole Japanese government structure in S&T policy would require several volumes, only a brief overview is given here, focusing on the framework for the formulation and co-ordination of comprehensive policies and the role of the CST.

The main organisational framework for the S&T policy-making system in Japan was established during the period from the mid-1950s to the mid-1960s. This reflected a growing recognition of the importance of developing indigenous technological capacity rather than depending on technology transfer from advanced countries.

In 1956, the Science and Technology Agency (STA) was established to promote national science and technology. The responsibilities of the STA were defined as follows:

- to formulate comprehensive S&T policy;
- to co-ordinate the S&T activities of various ministries and agencies;
- to promote the development of comprehensive R&D projects to fulfil national policies on nuclear energy, aerospace, material science and processing, and other areas as required.

However, there remained a perceived need to integrate research at universities with overall national S&T policy, since academic research fell outside the jurisdiction of the STA. In order to address this need, the CST was established in 1959 as the supreme advisory body to the Prime Minister encompassing both academic and government research policy. In the following year, the CST presented a recommendation entitled "Comprehensive and Fundamental Measures for the Development of Science and Technology for the Next Decade". The recommendation proposed a policy to increase human resources available for science and engineering and to intensify R&D efforts in order to catch up with Western advanced countries. This recommendation was the first overall policy framework for science and technology in Japan.

Changes of emphasis in Japanese Science and Technology policy thereafter have been brought about by the recommendations of the CST. In 1971, the CST proposed measures in response to problems encountered in the period of high economic growth such as environmental pollution. The recommendation made in 1977 stressed the need to strengthen responsiveness to sudden changes in the international environment such as the oil crisis. It also stressed the need to develop government policies promoting concern for better quality of life.

In 1986, the Cabinet approved the General Guideline for Science and Technology Policy based on a recommendation made by the CST, recognising that Japan had become one of the technologically advanced countries. This policy was based on three major objectives: the promotion of creative science and technology, the promotion of science and technology in harmony with society, and an emphasis on international aspects of science and technology. In 1992, an updated guideline was formulated by the Cabinet, based on the eighteenth recommendation of the CST. The new guideline sets the following goals: coexistence of humans in harmony with the earth, the expansion of intellectual stock and the construction of a changing society where people can live with peace of mind.

In accordance with this guideline and the Science and Technology Basic Law, based on the recommendation of the CST, the Government established the Science and Technology Basic Plan. In addition to these recommendations on comprehensive policy, the CST has often made more specific policy recommendations in response to and in anticipation of the needs of society as a whole. Specific areas taken up in specific policy recommendations are materials, information/electronics, life science, soft S&T and disaster prevention and safety. Also, the CST has made several recommendations on intersecting matters such as national research institutes, S&T activities in regions, international S&T activities and so forth.

Thus, the CST has played a pivotal role in the evolution of S&T policy since its foundation four decades ago. The major development of Japanese

TABLE 4.7 – Changes in the framework of Japanese Science and Technology policy

1956 – Science and Technology Agency (STA) was established to promote Japan's S&T (except for research in universities).

1959 – Council for Science and Technology (CST) was established to comprehensively formulate and co-ordinate overall science and technology policy for the Prime Minister.

1960 – CST produced the recommendation for Inquiry No. 1: *Comprehensive and Fundamental Measures for the Development of Science and Technology for the next 10 years.*

1971 – CST produced the Recommendation for Inquiry No. 5: *The Fundamentals of Comprehensive S&T Policy for the 1970's.*

1977 – CST produced the Recommendation for Inquiry No. 6: *The Foundation of Japan's Overall Science and Technology Policy Based on Long-Term Prospects.*

1981 – Special Co-ordination Fund for Promoting Science and Technology was established to facilitate the comprehensive promotion of S&T.

1982 – Provisional Research Council for Administration recommended strengthening the co-ordination function in S&T administration.

1983 – Committee for Policy Matters was established within CST.

1984 – CST provided the Recommendation for Inquiry No. 11: *Comprehensive Fundamental Policy for the Promotion of Science and Technology according to the Changing Situations from the Long Term Point of View.*

1985 – Provisional Council for Promotion of Administrative Reform recommended the establishment of a general guideline for Science and Technology Policy and the reform of some bureaux in STA.

1986 – *General Guideline for Science and Technology Policy* based on the recommendation of CST was approved at the Cabinet meeting.

1986 – Three Bureaux in STA were reformed.

1986 – The Law for Facilitating Government Research Exchange was enacted.

1988 – National Institute of S&T Policy (NISTEP) was founded as a core institute to conduct theoretical and survey-based research on S&T Policy.

1992 – CST produced the Recommendation for Inquiry No. 18: *Comprehensive and Basic Science and Technology Policy toward the New Century.*

1992 – *General Guideline for Science and Technology Policy* based on the above recommendation was approved at the Cabinet meeting.

1995 – Science and Technology Basic Law was enacted.

1996 – *Science and Technology Basic Plan* was approved at the Cabinet meeting after consultation with CST.

1997 – *The National Guideline on the Method of Evaluation for Government R&D*, drafted by CST, was approved by the Prime Minister.

1998 – The Administrative Reform Basic Law was enacted.

Note: Events related to the recommendations by CST are in italics.

S&T policy is summarised in Table 4.7. Although the framework for S&T policies has greatly changed, the administrative structure for the formulation and the implementation of S&T policy has remained largely untouched. The legal status and functions of the CST and related ministries and agencies have not greatly changed until today. As stated earlier, however, the CST will be dissolved in order that the new General Council for Science and Technology will play a more consolidated role in promoting S&T in the 21st century. In order to have a smooth transition to the GCST, the CST and its secretariat are working hard with various issues to be solved.

In particular, the CST is preparing a draft of the next Science and Technology Basic Plan, which is expected to be decided by the end of 2000 and will be implemented in April 2001, at the start of the fiscal year. Compared with the present Basic Plan, which has greatly contributed to the expansion of the overall budget for S&T, more attention will be paid to prioritisation of S&T areas in the preparation of the next Basic Plan. In this process, an attempt, which will be described in section 7, has been made to utilise the results of the Technology Foresight Survey conducted by the STA in 1996.

6. OTHER ADVISORY ORGANISATIONS FOR S&T POLICY

The role of advisory councils in the formulation of policies in the field of science and technology is greater than that in any other field, since S&T policy requires an extremely deep and wide range of technical knowledge which is not easily accessible to decisionmakers in the government administration.

In addition to the Council for Science and Technology (CST), several S&T-related advisory bodies are attached to the Prime Minister's Office, the mission of which is to co-ordinate and promote overall policies in those fields where various ministries and agencies are involved. They are: the Atomic Energy Commission, the Nuclear Safety Commission, the Space Activities Commission, the Council for Ocean Development and the Science Council of Japan. Also, four councils are attached to the STA: the Council for Aeronautics, Electronics and other Advanced Technologies, the Resources Council, the Consulting Engineer Council and the Radiation Council (see Table 4.11 in the Annex).

Apart from the councils and commissions in the Prime Minister's Office and STA, there are six S&T related advisory councils attached to respective ministries, namely: the Science Council and the Geodesy Council in *Monbusho*; the Council for Health and Welfare Science in the Ministry of Health & Welfare; the Industrial Technology Council in MITI; the Council for Transport Techniques in the Ministry of Transport; and the Telecommunications Technology Council in the Ministry of Posts and Telecommunications. It should be recalled that S&T policies in Japan are implemented by various ministries and agencies according to their respective missions.

The legal basis for all these councils lies in Article 8 of the National Government Organisation Law. It stipulates that "each administrative organ referred to in Article 3 may establish by law or Cabinet Order collegial bodies to be in charge of deliberation of important matters, review of administrative appeals or other affairs deemed proper to be dealt with by conference of men of learning of knowledge and the like within the scope of its jurisdiction as provided for by law".

The Administrative Reform Basic Law together with other laws related to administrative reform will also change the structure of these advisory organisations. Five councils attached to the Prime Minister's Office, in addition to the CST, will change in the following ways.

The Atomic Energy Commission (AEC), the main function of which is to plan, deliberate and decide important matters in the field of research, development and utilisation of atomic energy, will be attached to a new Cabinet Office. Its secretariat, now in STA, will be attached to the Cabinet Office, consolidated with the secretariat of GCST. The Chair of the AEC is the Minister of State for Science and Technology (*ex officio*), who is joined by four commissioners of the AEC, appointed by the Prime Minister with consent of the Diet. In the new AEC, however, the chair will not be a Cabinet Minister, but will be appointed by the Prime Minister as other commissioners.

The Nuclear Safety Commission (NSC) is also to be attached to the Cabinet Office in January 2001, according to the Administrative Reform Basic Law. A reinforced secretariat for the NSC will be set up in the Cabinet Office independent of the secretariat for CST. Because of the nuclear accident in Japan that occurred at the end of September 1999, the secretariat of the NSC, with an increased number of staff, is to be transferred to the Prime Minister's Office in April 2000 in advance of the Administrative Reform in 2001.

On the other hand, the Space Activity Commission (SAC), the main function of which is to plan, deliberate and decide important matters related to the research and development of space, will be transferred to the new Ministry of Education, Science and Technology in January 2001. The current chair of the SAC is the Minister of State for Science and Technology (*ex officio*) but this will change after the transfer.

The Council for Ocean Development will be abolished in January 2001. Its function will be taken over by a new Council for Science and Technology and Academic Policy, which will be set up in the new Ministry of Education, Science and Technology. As for the Science Council of Japan, it will be attached to the Ministry of General Affairs and its role and mandate will be discussed at GCST.

Three of the four councils currently attached to STA, the exception being the Radiation Council which will be attached to the new Ministry of Education, Science and Technology, will be abolished in January 2001. The functions of these three councils will be taken over by a new Council for Science and Technology and Academic Policy.

Out of six S&T-related advisory councils in other ministries, only the Council for Health and Welfare Science will continue to exist under the new Ministry of Welfare and Labour. The five other councils will be abolished and some of their functions will be adopted in some way or other by new or retained councils.

Given the important roles advisory councils have played in the formulation and implementation of S&T policies during the past decades, the structural change in the advisory council system in January 2001 will exert a profound influence on future S&T policy and the outcome of R&D activities.

7. TECHNOLOGY FORESIGHT ACTIVITIES

In any age of drastic change, technological foresight serves as a basis for the formulation of policy for the future. Since various ministries and agencies are responsible for R&D related to their missions within the government, and since the industrial sector plays a major role in R&D activities in Japan, a variety of technology foresight surveys have been conducted in various organisations. These activities can be categorised into the following four levels. The first is the holistic-level Delphi survey conducted by STA. On the second level are macro-level surveys, carried out by various ministries and agencies in the Government. In this case, survey fields are limited within the mandate of ministries or agencies with a usual time horizon of 10 to 15 years. The third level comprises surveys conducted by groups of private firms or semi-public organisations. On the fourth level are the forecasting activities of private firms to help their business decision.

Among these foresight activities, the STA Delphi survey plays the most important role, in the sense that it serves as a basis for all other foresight activities, covering all science and technology fields with the participation of thousands of experts, and that it is authorised by the Council for Science and Technology (CST), enabling it to directly influence to policy formulation. The budget for the survey was allotted from the Special Co-ordination Fund for the promotion of Science and Technology, over which the Council of Science and Technology has control.

In 1971 the Science and Technology Agency (STA) made public the results of its forecast survey on Japan's scientific and technological developments in the 30 years up to the year 2000. Since then, STA has been conducting a technology foresight survey using the Delphi method almost every five years, in order to obtain a comprehensive overview of future science and technology and the kind of society it will generate, as envisaged by experts engaged in various fields of research and development in Japan. The National Institute of Science and Technology Policy (NISTEP), which is part of the STA, conducted the two most recent STA technology foresight surveys. The results of the last survey, the sixth survey, were released in June 1997.

In conducting this sixth survey, NISTEP established a technology foresight steering committee to examine the overall plan and its implementation guidelines. Also 13 subcommittees headed by members of the steering committee were established in order to set the survey topics in each field, select the survey participants, and analyse the survey results. NISTEP conducted analysis of all fields handled by the survey, and the technology foresight committee prepared a comprehensive report based on the results of this analysis.

The sixth foresight survey covers 1072 technological topics from 14 technology fields, namely: material and processing; electronics; information; life science; space; marine science and earth science; resources and energy; environment; agriculture, forestry, and fisheries; production and machinery; urbanisation and construction; communication; transportation; and health, medical care, and welfare. The numbers of technological fields and topics covered in the survey have been increasing with each survey (Table 4.8).

The forecasted period is 30 years, from 1996 (the year the survey was conducted) to 2025. Like the previous surveys, this survey was conducted using the Delphi method, and responses were consolidated through two rounds of questionnaires. The numbers of responses in the first round was more than 4000 and in the second round more than 3500. The percentage breakdown of respondents differs from field to field, but the overall percentage breakdown of respondents across all fields in the sixth survey is as follows:

TABLE 4.8 – Changes in the coverage of STA's Technology Forecast Surveys

	Survey period	Fields	Topics	Forecasted period	Effective responses
1st survey	1970-1971	15	644	30 years to 2000	2 482
2nd survey	1976	7	656	30 years to 2005	1 316
3rd survey	1981-1982	13	800	30 years to 2010	1 727
4th survey	1986	17	1 071	30 years to 2015	2 007
5th survey	1991	16	1 149	30 years to 2020	2 386
6th survey	1996	14	1 072	30 years to 2025	3 586

36% of respondents were from industry, 37% from university, 15% from public research institutes and 12% from other organisations, including non-profit organisations.

The questionnaires cover seven survey parameters. The first is degree of expertise: respondents were asked to rank their degree of the expertise as high, medium, low or none. The next parameter is degree of importance in Japan. An index of importance was calculated by attaching 100 points to "high" responses, 50 points to "medium", 25 points to "low", and 0 points to the responses of "none". The third parameter is the expected influence of topics upon realisation. Four expected influences are listed: (1) contribution to socio-economic development, such as the development of innovative products, the creation of new industries, expansion of economic frontiers, and development of socio-economic base; (2) the resolution of various problems of a global scale; (3) responses to people's needs, such as the prevention and cure of diseases, improvement of living environment, and support for elderly people; and (4) expansion of human intellectual resources. For this parameter, respondents are allowed to select more than one effect. The fourth parameter is the forecasted realisation time: respondents were asked to select one of a number of listed time horizons, "never", or "don't know". The other three parameters are the countries leading in the topics in question; effective measures the government should adopt in Japan to promote R&D aimed at the realisation of the topic in question, and possible anxieties or potential problems that the realisation of the topics in question could create in Japan.

As for the dissemination of the results, thousands of full copies of the report both in Japanese and in English were provided for interested parties. Various summaries on interesting topics were introduced to the public through a variety of newspapers and magazines, having influence on public perspectives on the future. The Delphi process itself has an important influence on Japanese R&D activities. More than 100 senior researchers are involved in the design of the survey and in the analysis of the results. They have responsible and effective positions in their institutes, universities and enterprises. In addition, more than 3000 researchers are taking part in the survey as respondents. A new aspect for the seventh survey, to be conducted in 2000, is the incorporation of experts in social sciences and society at large in order to capture views from users on the demand side.

8. RESEARCH EVALUATION

Nationwide evaluation of science and technology research in Japanese national research institutes, research programmes and universities is being carried out, and will have a profound influence on these research communities.

The Science and Technology Basic Plan specified the establishment of a national guideline to implement evaluation for governmental R&D. The National Guideline on the Method of Evaluation for Government R&D, drafted by the Council for Science and Technology (CST) and approved by the Prime Minister in August 1997, applies to all government-funded R&D. It therefore applies to R&D implemented by national research institutions, national universities and special public corporations. It also applies to R&D implemented by private sector institutions acting as partners in joint research projects with government funding, to R&D implemented by local government institutions with support from government funding, as well as to R&D implemented overseas funded by the Japanese government (Table 4.9).

The introduction of fair and strict evaluation into the governmental R&D system responds to the two major growing requirements. One is to promote R&D in response to social and economic needs under severe financial conditions, and the other is to gain the public's deep and broad understanding for the promotion of science and technology.

Although several ministries and agencies had been conducting research evaluation before the establishment of the Guideline, more comprehensive and systematic evaluation systems have been established by the Guideline. Each Ministry and agency related to science and technology has set its own guideline for implementing evaluation and/or has requested its R&D institutes and institutions to develop specific methods for implementing evaluation based on the Guideline. An inter-ministerial co-ordination

TABLE 4.9 – Outline of the National Guideline on the method of evaluation for government R&D

Purpose of evaluation:
- effective implementation of R&D evaluation;
- an efficient allocation of limited government R&D funds;
- realisation of open/flexible/competitive R&D environment;
- seeking public support for R&D activities.

Approach to evaluation:
- clarification of evaluation criteria and processes;
- introduction of external evaluation;
- disclosure of relevant information on evaluations;
- appropriate utilisation of evaluation results.

Matters demanding care:
- avoidance of excess burden which accompanies evaluation;
- appropriate consideration of the character of R&D;
- utilisation of numerical indices;
- evaluation of technical examination and R&D which have difficulty producing results in the short term;
- harmonisation of R&D with human lifestyle, society and nature.

Evaluation of R&D institutions:
- national research institutions: evaluation result must be reflected in improvement of management of the institution;
- universities: establish further self-examination/evaluations.

meeting, bringing together 15 Ministries and agencies, concerned with the promotion of research evaluation was organised in January 1998. This helped the Science and Technology Agency (STA) to compile a white paper, entitled "The Status of Research and Development Evaluation", for the first time.

Specific methods for implementing evaluation naturally differ from institute to institute according to their respective missions and the characteristics of R&D performed. The evaluation of some national institutes is implemented by the ministry or agency responsible for them. Most of the other national research institutes, at which direct evaluation by the responsible Ministry or agency is not implemented, have established their evaluation methods respectively. Heads of institutes and researchers have reacted favourably to explicit evaluation, external evaluation by experts and learned people, and so forth, as requested in the Guideline. Two-thirds of national research institutes and almost all programmes have started evaluation activities and almost all of the rest have clear plans to do so. However, intermediate or post-research evaluations are not so popular.

Most institutes and programmes have already introduced external evaluation. The introduction of external evaluation is one of the big changes brought about by the Guideline, and has been adopted favourably by heads of institutes and researchers, as they can obtain suggestions from different points of view. From the evaluators' perspective, the necessary materials and information were available, enabling them to incorporate their own opinions into the evaluation. For the evaluation of some research activities, which was of particular public interest, the public were invited to submit opinions through the Internet. Most evaluation results have been published on the Internet or through press releases. Publication is considered to be useful for promoting R&D.

Selection of research themes and reconsideration of the content of R&D were influenced by the research evaluation. For heads of institutes, the result of evaluation is also useful in resource allocation. Though some researchers share the opinions of the heads of institutes, others entertain some misgivings about being forced to produce immediate results. More than half the heads of institutes and researchers believe that the implementation of evaluation in accordance with the Guideline has induced a change in the behaviour and mentality of researchers.

More than 80% of universities have implemented self-monitoring and self-evaluation, including all the national universities. More than 80% of inter-university research institutes and research institutes attached to individual universities have introduced external evaluation and almost all the results have been made public. Inter-university research institutes had already implemented evaluation of research themes and organisation prior to the establishment of the Guideline, although some of them did not specify evaluation methods. Self-monitoring and self-evaluation at universities have been implemented in accordance with the recommendation by the University Council in 1991. Nearly half of evaluators feel that the utilisation of evaluation results is still inconclusive.

Measures for effective evaluation have not yet been established in Japan. Tools such as a strategic plan for post-research evaluation in the US or foresight for pre-research evaluation in the UK should be developed to enhance the discussion in the evaluation. Ministries and agencies should be heavily involved in the discussion of the strategic plan or foresight, since the evaluation is a management process and routine daily activity is required. Incorporating evaluation into management systems is extremely important and the focus should be placed on how the organisation or theme will be improved rather than focusing on a "go"/"no-go" decision. As science and technology require very high levels of expertise, the participation of experienced experts related to specific R&D areas as well as the participation of experts in evaluation are also needed to promote evaluation. In addition, a good system for fostering such experts is expected.

9. REGIONAL SCIENCE AND TECHNOLOGY POLICY

In Japan, the country is divided into 47 regional units called "prefectures". Each prefecture has its own government with an elected governor and elected assembly. In addition, there are 12 so-called "specially designated cities". These are cities that have been granted special status that allows them to exercise some prefectural prerogatives because of their size. They all have a population greater than one million. So the expression "region" or "regional" here refers to the 47 prefectures and 12 specially designated cities.

The framework of regional S&T policy in Japan has been changing greatly with the merging of two major topical trends: the promotion of "decentralization" and the promotion of science and technology.

In May 1995, the Decentralization Promotion Law was enacted to clarify the respective roles of the central and regional governments, to increase the independence and self-reliance of local public entities, and to cultivate active regional societies.

The Science and Technology Basic Law in 1995 clearly stipulates that the regional governments are responsible for formulating and implementing policies with regard to the promotion of science and technology, corresponding to national policies, and policies for their own initiatives in accordance with the characteristics of their regions.

The idea of promoting S&T at the regional level was further refined in the Prime Minister's resolution of December 1995: "Comprehensive Basic Policy concerning the Activation of Regional Science and Technology Activities". The Basic Policy calls for progress in three broad areas to promote S&T at regional level. Namely:

- promotion of creative knowledge that raises standards of economic, social and cultural activity;
- activation of regional economies through geographical clustering of R&D activities;
- development of human resources and expansion of public awareness to ensure continuous upgrading of local economic and social conditions.

The Science and Technology Basic Plan in 1996 advocates the creation of a new R&D system that stresses the need for greater competition and freedom of movement among researchers, upgrading the R&D infrastructure and technology transfer. In line with these requirements, the Plan calls for regions to "activate themselves as cultivators of economic frontiers" by actively promoting R&D based on local needs and resources.

In this context, the regional governments have already stepped up their promotion of regional science and technology activities, including the adoption of their own regional science and technology basic plans. As of the end of FY1998, 35 of the nation's 47 prefectures had adopted a regional S&T basic plan (Figure 4.3). Most of these plans have been made in the last four or five years.

FIGURE 4.3 – Trend in the number of Science and Technology Basic Plans adopted by regional Governments

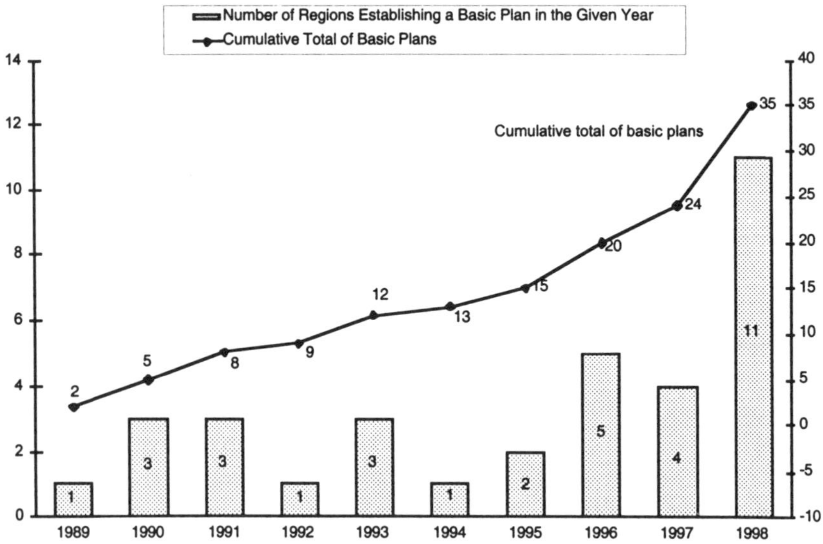

The remarkable feature of local government finance in Japan is the high degree of dependence on the central Government for funds. Because it is local governments that actually deliver the majority of government services to citizens, their share of total public sector expenditures is twice that of the central Government. In other words, on a final expenditure base, the ratio of national to regional is roughly one to two.

However, most local governments supply less than half of their own funds through taxes or other sources of revenue; the rest comes from the central Government in the form of revenue sharing and subsidies. In other words, on a tax revenue basis, the ratio of national to regional is roughly two to one, exactly the reverse of the expenditure ratio.

Regions in Japan differ widely in their collection of tax revenue; the wealthiest prefectures receive from three to four times the tax revenue of the poorest prefectures. To rectify this uneven distribution of revenue, the central government returns a share of revenue back to the regions in the form of "local allocation tax grants".

As for the government expenditure on the promotion of science and technology, regional governments pay out a far smaller portion of the total than the central Government. In 1997, regions spent 862 billion yen on the promotion of science and technology, which was 28.7% of the amount spent by the central government. However, in most other expenditure categories, regional governments pay out considerably more than the central Government. In science and technology, the roles played by central and regional governments, in terms of spending, are the reverse of what they are in most other spending categories. Consequently, there is plenty of room to shift some of the spending authority in S&T from the central to the regional level.

In terms of how funds related to S&T are expended, the largest proportion of regional government expenditures (28.5%) was spent on agriculture in 1997 (Table 4.10), though this represents a significant reduction from the figure of more than 40% in 1990. Commerce and industry, whose proportion in 1997 is 19.2, is also decreasing.

By performer, public testing research centres run by local governments, which provide technical consultation and research support to regional industry, absorb more than half of regional government research expenditures.

TABLE 4.10 – Proportion of R&D expenditure from regional Governments related to science and technology by field

	1990	1992	1995	1997
Commerce and manufacturing	26.3%	26.5%	23.4%	19.2%
Agriculture, forestry and fisheries	40.2%	34.1%	32.1%	28.5%
Environment and health	11.4%	13.3%	15.7%	18.6%
Education	4.1%	8.4%	8.9%	13.0%
Others	18.0%	17.7%	20.0%	20.7%
	100%	100%	100%	100%

A further 25% is expended through institutions of higher learning (i.e. local universities, colleges, special vocational schools, and so on) in 1995. For the central Government, the largest proportion (46.6%) of expenditure is disbursed in the form of subsidies and government investment. This is followed by national universities, which receive nearly 36% of total science and technology expenditure.

In line with the Science and Technology Basic Plan and the Basic Policy for Promoting Regional Science and Technology, ministries and agencies of the central Government have stepped up their support of regional research and technology development. Among the major agencies promoting regional science and technology activities, STA is by far the most significant source of funds, followed by the Ministry of International Trade and Industry (MITI).

Although policies administered by the central Government reflect the needs and resources of the respective ministries promoting them, they embrace the common objective of raising the standards of R&D and boosting innovative potential based on distinctive qualities of each region.

Among the more innovative programmes currently being implemented are those involving regional collaborative research and technology transfer. A new aspect of these programmes is the recognition that geography is an important factor in the process of technological innovation. Consequently, the ministries and agencies sponsoring these programmes generally require interested parties to organise a research consortium or other collaborative body whose members come from the same region. The thinking behind this approach is that any ensuing technological innovation will have important spillover effects in the regional economy.

The passage of the University Technology Transfer Promotion Law in spring, 1998 has further favoured regional co-operative R&D project. The purpose of this legislation is to encourage the technology transfer from national universities to industry by providing assistance to universities establishing "technology licensing offices" (TLOs). TLOs have only recently gained attention in Japan. As a result of this law, interest in technology transfer has surged.

These new policy initiatives aimed at promoting regional science and technology mean that in Japan we are facing a paradigm shift in our regional science and technology policy. This shift moves away from a promotional system directed from the centre, and toward a system in which centre and regions collaborate to promote S&T. This is still very much work in progress, but a primary goal is to strengthen the ability of regions to use science and technology to differentiate themselves from other regions and develop their distinctive characteristics.

10. INTERNATIONAL ACTIVITIES IN SCIENCE AND TECHNOLOGY

Science and technology contribute to the creation of intellectual property which is a common property of all mankind, as well as to the solution of global-scale issues, such as problems of the global environment, energy and resources, not to mention the development of industries and economies. As described in the Science and Technology Basic Plan, international deployment of these science and technology activities is important, for Japan to play an active role in global society and to contribute to further national development of science and technology. Japan has been making efforts to address these global-scale issues utilising multinational frameworks, such the summit meetings of major nations. Co-operative activities with nations of Europe and the US are being promoted effectively by sharing and mutually complementary activities, which take into consideration the fact that the potential of each of the co-operating partners is almost equal.

Japan has made science and technology co-operation agreements with 24 countries, including the US and European countries. Japan has also been promoting science and technology co-operation in the arena of the United Nations, the Organisation for Economic Co-operation and Development (OECD), the Asia-Pacific Economic Co-operation (APEC), the Association for Science Co-operation in Asia (ASCA) and the International Science and Technology Centre (ISTC).

For the purpose of further expansion of international research exchanges, the Science and Technology Agency (STA) has established a new international research co-operation system, that includes both the early stage of formulating ideas for research co-operation and the later stages of implementing them. In order to develop new concepts and ideas, which will be the buds of future science and technology in basic research areas, International Co-operative Research Projects have been carried out in close co-operation with foreign research institutes in developed countries.

To promote international research exchanges, three programmes have been established. The support programme provides foreign researchers and their family with accommodation in Tsukuba Science City, and support such as Japanese language training, daily living counselling in English and daily living information brochures in English. The research information programme provides foreign researchers and research institutes with information on Japan's research activities. The co-operative research fellowship programme began in fiscal year 1993 to send researchers of national research institutes to research institutes of Asia-Pacific countries for the purpose of research co-operation. In fiscal year 1988, STA established the STA Fellowship Programme, whereby overseas researchers are invited to Japan's national institutes.

The Human Frontier Science Program is an international co-operative programme, initiated by Japan, which promotes collaborative basic research on elucidating the sophisticated and complex mechanisms of living organisms.

Japan also participates in large international co-operative programmes such as the International Space Station Programme with the Japanese Experiment Module (JEM) or the International Thermonuclear Experimental Reactor Project (ITER).

11. CONCLUSION

The change in the framework for S&T policy in Japan is inexorable and intrinsically intertwined with the change in the entire Japanese socio-economic system within a dynamic global environment. The direction of change is one in which various functions involved in the formulation and the execution of S&T policy are becoming more distinctive and transparent requiring more output-oriented and contract-based practices. This change requires prioritisation of policies, clarification of the roles of the main actors, including researchers, and strict evaluation of outcomes. These practices have been uncommon in Japan, but are now elaborated as challenges for administrative reform in the field of science and technology as well as in other fields of governmental services.

Another challenge facing the Japanese S&T community is the enhancement of the mobility of researchers. Some efforts have been made to allow researchers in national research institutes, as well as professors in national universities, to undertake work for private companies under certain conditions. The transformation of many national research institutes into "independent administrative corporations" which is expected to take place in April 2001 will also contribute to the enhancement of mobility. High mobility is needed not only in terms of organisational affiliation but also in terms of academic discipline. A key concept here is that of "trans-border" activity, which includes crossing the "borders of disciplines", the "borders of organisations" and the "borders of nations", without which real mobility could not be attained. The international mobility of researchers in Japan will be one of the most important challenges for the future of Japanese S&T systems.

The mentality of researchers is also undergoing profound change, not only due to the disintegration of the lifetime employment and seniority systems, but also because of the introduction of strict research evaluation. The more direct linkage between the results of research evaluation and the treatment of researchers will make researchers more conscious of research output. Furthermore, the Administrative Reform Basic Law stipulates the necessity of policy evaluation in each ministry. R&D policy is no exception to this rule. In a society where evaluation of output is prevailing, whether a person's mentality is "risk-taking" or "risk averse" is a very important factor to determine the characteristics of society. In order for Japan to revive and maintain its vitality in the future, a large portion of researchers who have become evaluation-conscious should have a "risk-taking" mentality. Also there should be chances for former losers to make new attempts. These changes in mentality and practice that have not been common in Japan will be further important challenges.

All these challenges can be summarised as "drifting away from homogeneity". Introduction of prioritisation at the policy level, and the process of differentiation at the R&D level in a highly competitive world have been transforming a traditional homogeneous society based on rather stable organisations into non-homogeneous (but not necessarily heterogeneous) one, which Japan has never experienced before. Faced with these challenges, great efforts are needed to change the Japanese S&T system, to enable it to become effective and attractive in a highly competitive world. The profound change Japan is facing could be turned into an opportunity to embrace a radically new system for research, development and innovation. The new system will be characterised by flexibility, competition and an open research environment. The emphasis will be put on creativity and responsiveness to social and economic needs, as well as responsiveness to opportunities arising from scientific discovery itself. The emergence of a knowledge-based society in the 21st century will surely offer Japan the challenge of basing its vitality on science and technology with the broadest possible perspective.

ANNEX: THE ADMINISTRATIVE ORGANISATION OF JAPANESE SCIENCE AND TECHNOLOGY

The administrative organisation of Science and Technology in Japan is based on a departmental approach. Typically, a ministry or agency: (1) is advised by one or more councils, (2) has national research institutes, and (3) has authority (sometimes with other ministries) over a set of institutions, centres or societies. Table 4.11 provides a list of the advisory bodies related to each ministry and agency, while major national research institutes are presented in Table 4.13. Table 4.12 lists the various other structures, over which the different ministries have authority.

The Prime Minister's Office plays a specific role, having under its authority three councils and three commissions, which advise him. The three councils are: the Council for Science and Technology, the Council for Ocean Development and the Science Council of Japan. The three commissions are: the Atomic Energy Commission, the Nuclear Safety Commission and the Space Activities Commission.

Two ministries and one agency play a major role in S&T. These are STA, MITI and *Monbusho*.

- The Science and Technology Agency, STA has four advisory councils and six national research institutes attached to it. It has authority over other six major institutions, which are in charge of managing sets of programmes for the agency, namely: JAERI, RIKEN, NASDA (authority shared with the Ministry of Transport and the Ministry of Posts and Telecommunications), JNC, JAMSTEC and JST.
- The Ministry of Education, Science, Sports and Culture, *Monbusho*, is primarily responsible for 99 national universities, which include 53 research institutes and nine national inter-university research institutes. It has authority over the Japan Society for the Promotion of Science (whose main role is to contribute to the advancement of science). It is advised by the Science Council and the Geodesy Council.
- The Ministry of Trade and Industry, MITI, embraces the Patent Office, and the Agency of Industrial Science and Technology, AIST, which in turn has 15 national research institutes (eight national, as listed in Table 4.12, and seven regional ones). It has authority over NEDO and (with the Ministry of Posts and Telecommunications) over JKTC. It is advised by the Industrial Technology Council.

The departmentalisation of S&T goes far beyond these two ministries and one agency. More than ten other ministries and agencies include

50 national research institutes (excluding regional stations), while four of them have established an advisory body (those marked with an*).

- The Ministry of Health and Welfare*, MHW, has 11 national research institutes.
- The Ministry of Agriculture, Forestry and Fisheries*, MAFF, has 16 national research institutes, plus 13 regional stations. It shares, with the Ministry of Finance, control over the Bio-oriented Technology Research Advancement Institution.
- Attached to the Ministry of Posts and Telecommunications*, MPT, is the National Research Institute on Communication. MPT has control over TAO and shared authority over JKTC and NASDA.
- The Ministry of Transport*, MOT, has shared authority over NASDA and JAERI while five research institutes on shipping, electronic navigation, port and harbour, traffic safety and nuisance, and meteorological research are attached to it.
- The Ministry of Finance has the Brewing Research Institute.
- The Defence Agency has the Technical Research and Development Institute.
- The Environmental Agency includes two NRIs on environmental studies and Minamata disease.
- The Ministry of Labour has research institutes concerned with industrial safety and industrial health.
- The Ministry of Construction, MOC, has research institutes for building research and public works.
- The Ministry of Home Affairs has the Fire Resistance Research Institute.

TABLE 4.11 – List of Japanese advisory bodies, by Ministry or agency

Ministry	Councils
Prime Minister	Council for Science and Technology
	Atomic Energy Commission
	Nuclear Safety Commission
	Space Activities Commission
	Council for Ocean Development
	Science Council of Japan
STA	Council for Aeronautics, Electronics and other Advanced Technologies
	Resources Council
	Consulting Engineering Council
	Radiation Council
Monbusho	Science Council
	Geodesy Council
MITI	Industrial Technology Council
MAFF	Agriculture, Forestry and Fisheries Research Council
MHW	Council for Health and Welfare Science
MOT	Council for Transport Techniques
MPT	Telecommunications Technology Council

TABLE 4.12 – Other research institutes (per Ministry affiliation)

Ministry in charge	Acronym	Title	Employment 1999
STA	JAERI	Japan Atomic Energy Research Institute	2360
STA	RIKEN	Institute of Physical and Chemical Research	631
STA	JNC	Japan Nuclear Cycle Development Institute	2687
STA	NASDA	National Space Development Agency of Japan	1087
STA	JST	Japan Science and Technology Corporation	454
STA	JAMSTEC	Japan Marine Science and Technology Center	223
Monbusho	JSPS	Japan Society for the Promotion of Science	72
MHW	OPSR	Organisation for Pharmaceutical Safety and Research	51
MAFF	BRAIN	Bio-oriented Technology Research Advancement Institute	104
MITI	NEDO	New Energy and Industrial Technology Development Organisation	1155
MITI**	JKTC	Japan Key Technology Center	66
MPT**	JKTC	Japan Key Technology Center	66
MPT	TAO	Telecommunications Advancement Organisation of Japan	101

** joint authority

TABLE 4.13 – Main National Research Institutes

Ministry in charge	Acronym	Area of activity	Employment 1999
STA	NAL	Aerospace	425
STA	NRIM	Metals	413
STA	NIRIM	Inorganic Materials	163
STA	NIRIM	Radiological Science	388
STA	NIED	Disaster Prevention	117
STA	NISTEP	S&T Policy	46
Monbusho	KEK	High Energy Physics	754
(only national	NIPR	Polar	135
inter-university	ISAS	Space and Astronomical	154
research	NIG	Genetics	120
institutes)	ISM	Statistical Mathematics	85
–	NAO	Astronomy	285
–	NIFS	Nuclear Fusion	252
–	ONRI	Molecular Science, Biology	125
–	NACSIS	Science Information System	407
AIST, MITI	NRLM	Metrology	195
AIST, MITI	MEL	Mechanical Engineering	240
AIST, MITI	NIMC	Materials and Chemical	407
AIST, MITI	NIBH	Bio-Science	229
AIST, MITI	GSJ	Geological Survey	324
AIST, MITI	ETL	Electro-technical	623
AIST, MITI	NIRE	Pollution and Resources	284
AIST, MITI	NAIR	Advanced Interdisciplinary	56
Defence Agency	TRDI	R&D, Test and Evaluation	1178
Environmental Agency	NIES	Environmental Studies	270
MHW	NIPH	Public Health	144
MHW	NIID	Infection Disease	414
MAFF	ABR	Agro-biological Resources	226
MAFF	NIAES	Agro-environmental Science	212
MOT	SRI	Ship	249
MOT	PHRI	Port and Harbor	199
MOT	MRI	Meteorological	179
MPT	CRL	Communications	426
MOC	PWRL	Public Works	468
MOC	BRI	Building	170

REFERENCES

Fujigaki, Y. and A. Nagata, 1998, Concept evaluation, *Science and Public Policy*.

Government of Japan, 1996, *Science and Technology Basic Plan*, Tokyo: Science and Technology Agency.

Kawasaki, M., 1994, Historical Development of Japanese Science and Technology Policy in Conjunction with socio-economic Policy in G. Aichholzer and G. Schieustock (eds), *Technology Policy: Towards or Integration of Social and Ecological Concerns*.

Kuwahara, T., 1999, Technology Forecasting Activities in Japan, *Technological Forecasting and Social Change*, Vol. 60, No. 1.

Law No.130 of 1995: The Science and Technology Basic Law.

Mori, W. and T. Ochiai, 1997, Science and Technology in Japan in W.T. Golden and J.T. Ratchford (eds), *Technology in Society*, Vol. 19, No. 3/4.

National Institute of Science and Technology Policy, 1997, *Science and Technology Indicators: 1997*, Tokyo: NISTEP Report No. 50.

National Institute of Science and Technology Policy, 1997, *Study of Regional Science and Technology Promotion (Third Survey) – Analysis of science and technology policies by local governments*, Tokyo: NISTEP Report No. 51.

National Institute of Science and Technology Policy, 1997, *The Sixth Technology Survey – Future Technology in Japan*, Tokyo: NISTEP Report No. 52.

Sato, Y., 1998, The Changing Framework for the S&T Policy in Japan, Presentation to NISTEP 10th Anniversary Conference, Tokyo, 8-9 October.

Sato, Y., 1998, The Structure and Perspective of Regional Science and Technology Policy in Japan, Presentation to The Fourth Regional Science and Technology Policy Research Symposium, Chapel Hill, 21-24 November.

Science and Technology Agency, annual editions, *White Paper on Science and Technology*, Tokyo.

ACKNOWLEDGEMENT

The author would like to express his sincere thanks to the staff of the National Institute of Science and Technology Policy (NISTEP) for their assistance in collecting data and preparing tables and figures for this chapter. Also, the author appreciates the studies conducted by them and discussions with them while he was Director-General of NISTEP. In addition special thanks are given to Professor Larédo, whose kind advice and patience led the author to complete this work.

CHAPTER 5

THE RESEARCH, DEVELOPMENT AND INNOVATION SYSTEM IN KOREA

SUNG CHUL CHUNG

1. INTRODUCTION

Over the past three decades, Korea has made phenomenal economic achievements.

Since 1962, when the First Five-Year Economic Development Plan was launched, Korea has maintained an average economic growth of over 8% per annum, transforming itself from a poor agrarian society into one of the fastest-growing industrial economies. By 1995, Korea had grown to be the eleventh largest economy and twelfth largest trading country in the world, with a per capita GNP of over $10 000 dollars.[1]

Korea's economic achievement has received great attention from many development economists as well as policy practitioners because Korea has accomplished its development under most unfavourable socio-economic circumstances. Korea inherited an economy distorted by 36 years of colonial rule (1909-1945) and, what is worse, the industrial base was totally flattened during the Korean War (1950-1953). So, when it started its drive towards industrialisation in the early 1960s, Korea faced all the problems that poor, developing economies have to suffer.

1. Due to the depreciation of the Korean currency and the negative economic growth caused by the financial crisis in Asia, GNP per capita has gone down to the level of $6000 in 1998.

What, then, are the factors that have made such a tremendous economic transformation possible? Korea's economic success can be attributed to an assortment of diverse factors, but the most pronounced of these appear to be the outward-looking development strategy pursued by the Korean government, high-quality human resources, and bold adoption of foreign technologies for industrial development.[1] Indeed, Korea's economic development has been based on bold introduction of advanced technologies from foreign sources for assimilation and improvement, while promoting the development of domestic capacity for technological development (Amsden, 1989 and Kim and Dahlman, 1992). So, we can say that Korean industrialisation has been a process of learning or industrialisation based on learning rather than on invention or innovation.

This suggests that the research, development and innovation system in Korea has evolved largely in response to industrial development. In other words, the current research and innovation system in Korea is the result of the planned and unplanned efforts towards industrialisation over the past four decades. In the early stage of its development, the system worked as a facilitator for learning about foreign technology (Amsden, 1989). It was only in the early 1980s that Korea embarked on serious efforts for indigenous research and development to support the development of high-tech industries. Amazingly, however, it did not take long for Korea to attain technological competitiveness in such technology areas as memory chips, LCD and telecommunications that have led economic growth over the recent years. Some say that Korea achieved within 30 years what it took advanced industrial countries more than a century to attain.

This chapter provides an overview of the research, development and innovation system of Korea, identifies the strengths and weaknesses of the system, and puts forward policy prescriptions for future developments. To this end, it begins with an analytical review of the Korean R&D and innovation system, focusing on the major features of the system. Section 3 presents the national priorities in research and development, including target technologies for development, and the process through which such priorities are set. Section 4 deals with public sector research and development, focusing on the roles of the government research institutes, universities and other pertinent agencies. Section 5 discusses industrial research and development and

1. K. Kim (1987), D.M. Leipziger and V. Thomas (1993), L. Kim (1993), A. Amsden (1989) and many others provide similar explanations with varying emphases.

government promotion policies. In the concluding section, current policy challenges are discussed and policy prescriptions are offered.

2. RESEARCH AND DEVELOPMENT ACTIVITIES: AN OVERVIEW

This section provides an overview of the research and development (R&D) system in Korea. Here, R&D system is defined as the mechanism through which R&D is financed and performed. In order to put the discussions in a context, this chapter begins with a review of past and present policies for industrialisation and science and technology development. Then, R&D activities on the aggregate national level are analysed and evaluated.

2.1 – Industrialisation and technology development

Considering the relationship between industrialisation and science and technology policy, it is quite natural that Korea's scientific and technological developments directly mirror its stages of industrialisation. The 1960s were a period of technological learning for the development of such light industries as shoes, clothing, textiles and some intermediate goods for import substitution as well as export. During this period, the major sources of technological learning were OEM (original equipment manufacturing) production arrangements, technical training as part of turnkey base plant importation and DFI (direct foreign investment) firms. Korean firms benefited most from the OEM production arrangements, because they offered opportunities to work with foreign buyers who provided everything from product designs and materials to quality control at the final stage of production. This was especially so in the case of garment and electronic industries.[1]

During the 1970s, Korea made massive investments in machinery and chemical industries. For the development of its chemical industries, Korea relied largely upon turnkey base importation of plants, which included technical training programmes as part of the packages. In the case of heavy machinery, international licensing was the major channel for technology acquisition (Chung and Branscomb, 1996). To compensate domestic industries for their weaknesses in technology, the Government created Government Research Institutes (GRI) in the fields of heavy machinery and chemicals.

---

1. For discussions on this, see Hobday (1995).

As a whole, Korea relied on foreign sources rather than domestic R&D for the technologies required for industrialisation in the early stages of development. Korean firms brought in mature technologies that were readily available from foreign sources and adapted them for domestic use without investing much in R&D. Until the early 1980s, R&D efforts, especially in the private sector, were negligible. The ratio of R&D investments to sales remained far below 0.5% in the early 1970s. And it was only in the early 1980s that major efforts for indigenous R&D were launched in Korea.

The changes in the economic environment in the early 1980s stimulated Korea's investment in domestic R&D. On the one hand, industrial development reached such a stage that Korean industries could no longer rely on imported technology and cheap domestic labour for competition in international markets. On the other hand, since Korea was seen as a potential competitor in international markets, advanced countries became increasingly reluctant to transfer technologies to Korea. Therefore, it was inevitable that Korea would develop an indigenous base for R&D.

To meet the challenge, it required highly trained scientists and engineers as well as the financial resources to support R&D activities that are by nature uncertain and risky. Korea was fortunate in this respect, because it already had a fairly large pool of well-trained scientists and engineers and also because the large conglomerates were financially able to venture for new technologies.

Scientific and technological human resources

As a matter of fact, Korea's modernisation started with education. Due to the Confucian tradition that places a very high value on education and scholarship, coupled with government investments in education that rose from a mere 2.5% of the total budget in 1951 to over 25% in the 1990s, Korea has been able to make remarkable achievements in human resource

TABLE 5.1 – School enrolment rate by age group

	1953	1960	1970	1980	1990	1996
Elementary (age 6-11)	59.6	86.2	102.8	101.1	100.7	100.5
Secondary (age 12-14)	21.1	33.3	53.3	94.6	98.7	99.0
Secondary (age 15-17)	12.4	19.9	29.3	68.5	86.9	88.7
Tertiary	3.1	5.0	8.7	16.0	37.7	48.8

Unit: %.
Source: MOST (various years).

TABLE 5.2 – Number of researchers

	Research inst.		University		Industry		Total	
1975	5 300	(na)	2 300	(na)	2 700	(na)	10 300	(na)
1980	4 600	(484)	8 700	(2 877)	5 100	(56)	18 400	(3 417)
1985	7 200	(1 105)	14 900	(6 640)	19 000	(251)	41 100	(8 005)
1990	10 434	(2 933)	21 332	(21 332)	38 737	(1 139)	70 503	(17 662)
1995	15 007	(4 659)	44 683	(27 073)	68 625	(3 373)	128 315	(35 105)
1996	15 503	(4 938)	45 327	(27 434)	71 193	(3 734)	132 023	(36 106)

Unit: Persons (scientists and engineers only).
Notes: In measuring manpower, Korea follows the OECD method. In the brackets are the numbers of researchers with Ph.D. degrees.
Source: MOST (1997a), KITA (1997).

development. Enrolment at various levels of formal education has increased rapidly. Table 5.1 shows that in 1970 the school enrolment as a percentage of the population of corresponding age group reached 100% for elementary education. The enrolment rate for middle and high schools rose to 99% and 86% respectively in 1994. What is more surprising is that the enrolment rate for tertiary education jumped to 49% in 1994 from a mere 3% in 1953.

This has served as an excellent base for the development of scientific and technological human resources. The number of research scientists increased more than twelve times from 10 300 to 132 000 during the period 1975-1996. The number of researchers with Ph.D. degrees grew even faster: over the period 1980-1996, the number of Ph.D.s increased more than tenfold from 3417 to 36 106. Most pronounced is the increase of Ph.D.-level researchers in the private sector (see Table 5.2). The number of research scientists per 10 000 of the population increased from 4.8 to 29 during the same period. The latter figure is the highest among developing countries and almost the same level as that of advanced industrial countries, such as the UK and France.

R&D investments

The government initiated domestic R&D efforts by launching a national R&D programme in 1982 and by various policy measures to promote and facilitate private R&D activities. Private industries also responded to policy by investing very heavily in R&D. R&D investment has since undergone

FIGURE 5.1 – R&D investment trend (1982-1998)

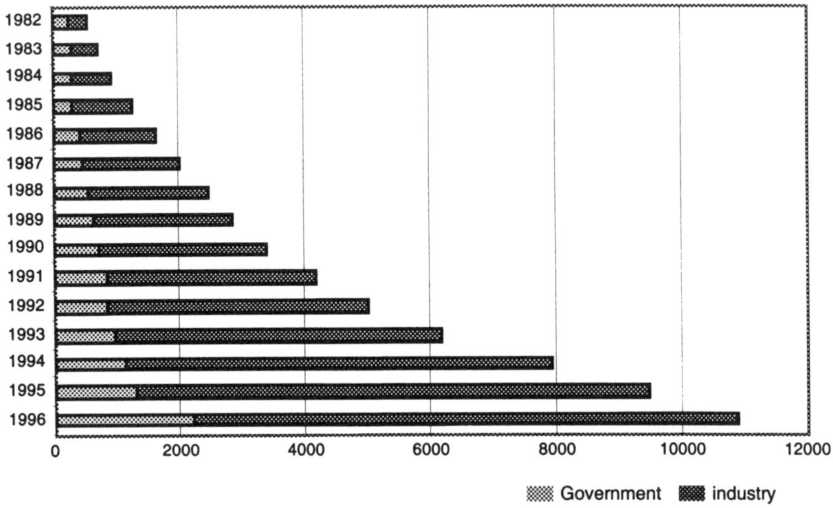

Unit: Billion Won.
Note: The upper portion of the bar refers to industry contribution, and the lower part to that of the government.

a quantum jump. Korea invested only $526 million (368.8 billion Won, 0.81% GDP) in R&D in 1981, but the figure rose to $13.5 billion (10 878 billion Won, 2.8% of GNP) in 1996. Over a period of about 15 years, investment in R&D increased almost 26 times, with an average annual growth rate of about 21.6% (see Figure 5.1). Such a phenomenal increase is largely attributed to private industries. In 1981, the Government accounted for 53.5% of the nation's total R&D investment, but the Government's share declined to 19.4% in 1990. The Government's share further decreased to 16% in 1994, but the tendency was reversed to some extent subsequently and the figure rose to 22% in 1996.

There may be many factors that have contributed to the rapid increase in private sector R&D investments, but basically such an increase has been possible because Korean firms have been put under market pressure for technology development, to which they have been able to respond financially. The Government contributed to this development in two indirect ways. First, the outward-looking development strategy (export drive) of the Government drove domestic industries towards international markets.

Second, the Government industrial policy that favoured large firms gave birth to a unique form of business organisation in Korea: the "Chaebol", which is similar to the "Zaibatsu" of Japan before WW II. Chaebols enjoy greater financial affluence owing to the economies of both scale and scope of their business operation. Chaebol companies, which are usually large international operators, have deeper pockets and are able to engage in risky and expensive R&D projects that would be unthinkable for small and medium-sized firms. This is explained by the fact that the top 20 firms account for about 57% of the total industrial R&D investments in Korea.[1]

R&D actors and targets

In the earlier stages of development, public research institutions,[2] including the government Research Institutes (GRIs), played the major role in R&D, simply because universities and industries lacked R&D capability. Up until 1980, more than 50% of R&D investment had been allocated to public research institutions. In 1980, 62% of R&D investment went to public institutes, 28.8% to private industrial laboratories, and the remaining 9.2% to universities. But since then, the composition has changed rapidly. It took less than ten years for private industries to take the lead in R&D activities. In 1990, the share of private industries in gross national R&D expenditures reached 74%, with public R&D institutes and universities accounting for only 18.5% and 7.6% of the total R&D performed. In 1996, private industrial laboratories accounted for 73.3% of R&D performed, while the share of GRIs and universities remained at 17.4% and 9.3%, respectively.

Collaboration between the major R&D actors has not been so active in Korea. As a matter of fact, the lack of partnership relations within the R&D community has been pointed out as one of the serious weaknesses of the Korean R&D system. A Ministry of Science and Technology (MOST) analysis shows that about 35% of national R&D was conducted by research consortia comprising industries, universities and research institutes. This suggests that inter-sector research co-operation is quite active in Korea. However, these statistics are very misleading, because the members of the consortia work separately for different sub-projects under the same umbrella project.[3] Korea has not been so successful in developing well-functioning inter-sector research networks.

---

1. Section 4 discusses this issue in more detail.
2. This includes national laboratories and GRIs. GRIs are funded by the Government, but are not government organizations. See section 4 for more details.
3. MOST (1998a) pp. 70-71.

From the apparent prominence of private industries in R&D, it can be inferred that the lion's share of R&D investments in Korea has been devoted to applied research and technology development. As shown in Table 5.3, the share of basic research had declined over time until 1993, when this trend weakened.

Since 1984, basic research has never received more than 18% of the national R&D investment and this share decreased to 13% in 1996. In 1996, almost 60% of total R&D investment was spent on technology development and 27% on applied research.

TABLE 5.3 – R&D investment structure

1965	1965	1970	1975	1980	1985	1990	1996
R&D Expenditure (billion Won)	2.1	10.5	42.7	282	1 237	3 350	10 878
Government	1.9	9.2	30.3	180	307	651	2 398
Private	0.2	1.3	12.4	103	930	2 699	8 467
Gov: Private	93:3	90:10	71:29	64:36	25:75	19:81	22:78
R&D/GDP (%)	0.26	0.38	0.42	0.77	1.58	1.95	2.79
R&D performers							
Industry	-	-	28.9	28.8	65.4	74.9	73.2
University	-	-	5.1	9.2	10.3	7.6	9.4
GRI	-	-	66.0	62.0	24.3	18.5	17.4
R&D by nature (%)							
Basic research	-	22.9	-	-	16.8	16.1	13.2
Applied research	-	31.3	-	-	29.2	24.4	26.9
Development	-	45.8	-	-	54.0	59.5	59.9
Industrial R&D intensity (%)	-	-	-	0.47	1.23	1.72	2.40

Source: MOST (1997a, 1998b), KITA (1998).

R&D outputs

Evaluations of the results of R&D efforts in Korea are mixed. There is criticism that Korea leads other countries in terms of R&D inputs, such as human and financial inputs, but lags way behind in outputs. Other critics maintain that R&D results have not been effectively linked to industrial uses. All in all, the major criticism is that Korea's R&D investments have not been efficient enough to be economically justified. Most of the criticisms are targeted at public research. However, these criticisms are based on anecdotal evidence rather than on formal analyses.[1]

Despite such criticisms, one cannot deny the positive contributions that R&D activities have made. Rapid growth in R&D investment led to a remarkable increase in patent registration. The number of patents granted by the Korea Industrial Property Office (KIPO) increased from 1808 in 1981 to 16 516 in 1996, with an average annual growth rate of 14.7%. What is more encouraging is the growth of patents granted to Koreans. The number of patents granted to Koreans was only 232 (12.8% of the total patents registered) in 1981, but this figure rose to 14 497 (58.9% of the total) in 1997, recording an average annual growth rate of 24% (Table 5.4).

Furthermore, Korea ranks sixth in the world in terms of the number of intellectual property applications (including patents and utility models), behind only Japan, the US, Germany, the UK and France (KITA, 1997). US patents are sometimes used as an indicator of the international technological competitiveness of a nation. The number of US patents granted to Koreans was only five in 1969, but grew to 413 in 1991, and to 1175 in 1995, placing Korea ninth (Table 5.5) in terms of the number of patents. According to a patent analysis by the US Department of Commerce, Korea has established world prominence in such technology areas as information and telecommunications, pharmaceuticals, advanced materials and automotive technology (Albert, 1998). These indicate that Korea has been rapidly gaining in technological competitiveness.

Another important development is the remarkable increase in the number of scientific publications in internationally recognised academic journals. The number of scientific publications by Koreans reported by the Science

TABLE 5.4 – Number of patents granted

	1981	1985	1990	1995	1997
Total number of patents	1 808	2 687	7 620	12 512	24 579
Granted to Koreans (%)	12.8%	13.0%	33.5%		58.9%

Source: MOST (1998b).

1. There are a couple of studies on the economic effects of R&D investments in Korea, such as S.C. Chung and J. Jang (1993) and M.I. Nadiri (1993). The results in general are in support of R&D investments, but the studies do not provide any direct justification of R&D investments, such as cost and benefits of R&D.

TABLE 5.5 – Number of US patents granted to Koreans

	1969	1992	1993	1994	1995
Number	5	543	789	941	1 175
Rank in the world	-	11	-	-	9

Source: USPTO.

TABLE 5.6 – Number of SCI publications

	1973	1980	1985	1993	1997
Number	27	171	1 227	2 997	9 124
Rank in the world	-	-	37	27	17

Source: MOST (1998b).

Citation Index (SCI) increased from a mere 27 in 1973 to 171 in 1980, 1227 in 1988 and 9124 in 1997, rising from thirty-seventh in the world in 1988 to seventeenth (Table 5.6). Although Korea still lags far behind advanced countries in scientific publications, it recorded the highest growth (24.2% per annum during the period 1973-1997).

R&D efforts have also contributed to the development of high-tech industries in Korea. Based on in-house R&D, Korean industries have recently emerged as world leaders in semiconductor memory chips, cellular phones and LCDs, and have also established themselves in world markets in areas such as shipbuilding, home appliances, automobiles and telecommunications.

2.2 – Strength and weakness of the R&D system

Korea has made enormous developments in science and technology over the past three decades. By making large-scale, continuous investments in human resource development and R&D, Korea succeeded in building up an R&D system on barren land. The very strength of the Korean system is its dynamism, fuelled by the strong commitment of the Government to "technology-based national development" and private industries' efforts for competitiveness. Despite its short R&D history, Korea has already harvested rich crops from its endeavours in the forms of patents, scientific papers and exports of technology-intensive products such as semiconductors, cellular phones, LCD and automobiles.

Yet there are problems, too. R&D activities in Korea have grown very rapidly, led by private industries under the active promotion policy of the Government. Despite this growth, however, the size of R&D investment in Korea is only comparable to that of a single major business corporation in advanced countries. General Motors and Siemens, for example, spend as much as Korea does on R&D. In other words, even though Korea spends a larger share of its GDP on R&D than other countries, it still lags far behind advanced industrial countries in terms of its absolute R&D expenditure. The challenge is how to overcome the disadvantage of being a small economy.

Second, discussions so far show that Korea has approached the level of an advanced country in terms of scientific and technological inputs, but they also suggest that Korea still has a long way to go to reach advanced countries in terms of R&D productivity. The most important source of inefficiency is the lack of sufficient interactions and exchanges between the major actors in innovation, such as universities, research institutes and industries. Inter-sector mobility of scientists and engineers is extremely low in Korea.[1]

Third, a fundamental problem of the Korean system is its weakness in basic sciences, because scientific capability determines the technological potential of a nation. Since the Korean system has been overly devoted to industrial development, it has not been successful in nurturing a base for scientific research. The lack of a strong scientific base already functions as a limit to technological progress in Korea. Weakness in science results not just from a funding policy that favours technology development but also from a weak university research capability. Therefore, strengthening the university research base poses one of the major policy challenges.

Fourth, excessive reliance on private industries for R&D investments and activities has made the R&D system very vulnerable in two ways. The system places too much emphasis on applied research and development. This has led to a failure to build up the strong foundation required for the long-term development of science and technology. In addition, the R&D system responds too sensitively to changes in the economy and business environments. For instance, large Korean enterprises responded to the financial crisis of 1997 by cutting their R&D spending by about 14%, destabilising the whole system. If the crisis had continued for several more years, the R&D system might have collapsed.

---

1. Traditionally Koreans tend to regard it a virtue to serve at one job as long as possible. "Life-time employment" is also part of the culture.

3. NATIONAL PRIORITIES IN SCIENCE AND TECHNOLOGY

Since the early 1960s, the Government has played a key role in Korea's development. The Government first initiated science and technology development as part of the national economic development plan, and has led development, not just as rule-setter but also as target-setter and financier. As discussed in the preceding section, science and technology policy in Korea has been closely linked to industrial development, and thus policy priorities have been adjusted in response to the changes in industrial development targets. In the 1960s and 1970s, the government set specific policy goals and led private industries in pursuing the goals.

As industrial development has proceeded, it has become increasingly difficult for the Government to intervene in economic as well as R&D activities because of the increased scale and complexity of industrial activities. Therefore, the pattern of government intervention in science and technology has been changing from direct involvement as a target-setter and "commander-in-chief"-style leader to indirect involvement as a facilitator and promoter.

This is a natural path of change in view of the growth of private industries in R&D as well as management and information capabilities. Korean industries are now more capable than the Government at understanding and responding to market signals.

The problem under these circumstances is the possibility that the business interests of private sectors may not coincide with national interests that are represented by the Government. This section discusses the mechanism through which policy priorities are set, and outlines current policy priorities in research and technology.

3.1 – Mechanisms for policy and budget co-ordination: How are policy priorities set?

Various ministries and agencies conduct science and technology activities, including R&D, independently of each other, which very often leads to duplication and/or inconsistency among government policies and R&D programmes. Without question, duplication and inconsistency result in inefficiency and wasted resources. To reduce such inefficiency, national science and technology policies are reviewed for harmonisation through two major processes: one is a policy co-ordination process and the other a budgetary co-ordination process.

Science and technology policy co-ordination has been an issue of hot debate in Korea over recent years. Even though MOST is legally the central agency for science and technology policy co-ordination, it has not been able to function as an effective co-ordinating agency, basically because it has neither financial resources nor political power. In the past, MOST conducted its role as a policy co-ordinator through the National Science and Technology Committee, which was chaired by the Prime Minister. However, this committee failed to function as a system of co-ordination, because of weak commitments by the Prime Minister's Office as well as by the member ministers. Consequently, the committee was temporarily replaced by the Science and Technology Ministers' Committee, chaired by the Deputy Prime Minister (the Minister of Finance and Economy) with MOST as its secretariat (1996-1998).

After these experiments, it was concluded that co-ordination can be achieved only through effective priority-setting and resource allocation. Therefore, the Special Law for Scientific and Technological Innovation (1997, amended 1999) stipulated that the Government set up a National Science and Technology Council (NSTC) as a body responsible for policy co-ordination. The President chairs NSTC, and its members include ministers engaged in science, technology or research policies, the Chairman of the Presidential Council on Science and Technology (PCST) and three non-government experts from industry and the science and technology community. NSTC deliberates the major policy agenda on science and technology and sets policy directions, including priorities for national R&D resource allocation. For this purpose, NSTC evaluates the performances of national R&D programmes. MOST functions as the secretariat for the NSTC (Figure 5.2).

To help set R&D policy and priority, the Technology Development Promotion Act (1967, amended in 1992) and the Special Law for Scientific and Technological Innovation (1997) require MOST to conduct long-term technology forecasting every five years. The first round of technology forecasting using the Delphi method was undertaken in 1994 (Shin and Kim, 1994). The second-round survey has recently been launched. Taking into consideration the basic directions of national development, social and economic environments, domestic technological capability and resources, they set science and technology policy directions and technological priority areas for R&D.

Sectoral R&D policies are basically the responsibility of individual ministries. In formulating sectoral policies, individual ministries usually rely on *ad hoc* committees or advisory groups comprising scientists, engineers

FIGURE 5.2 – The Korean policy process

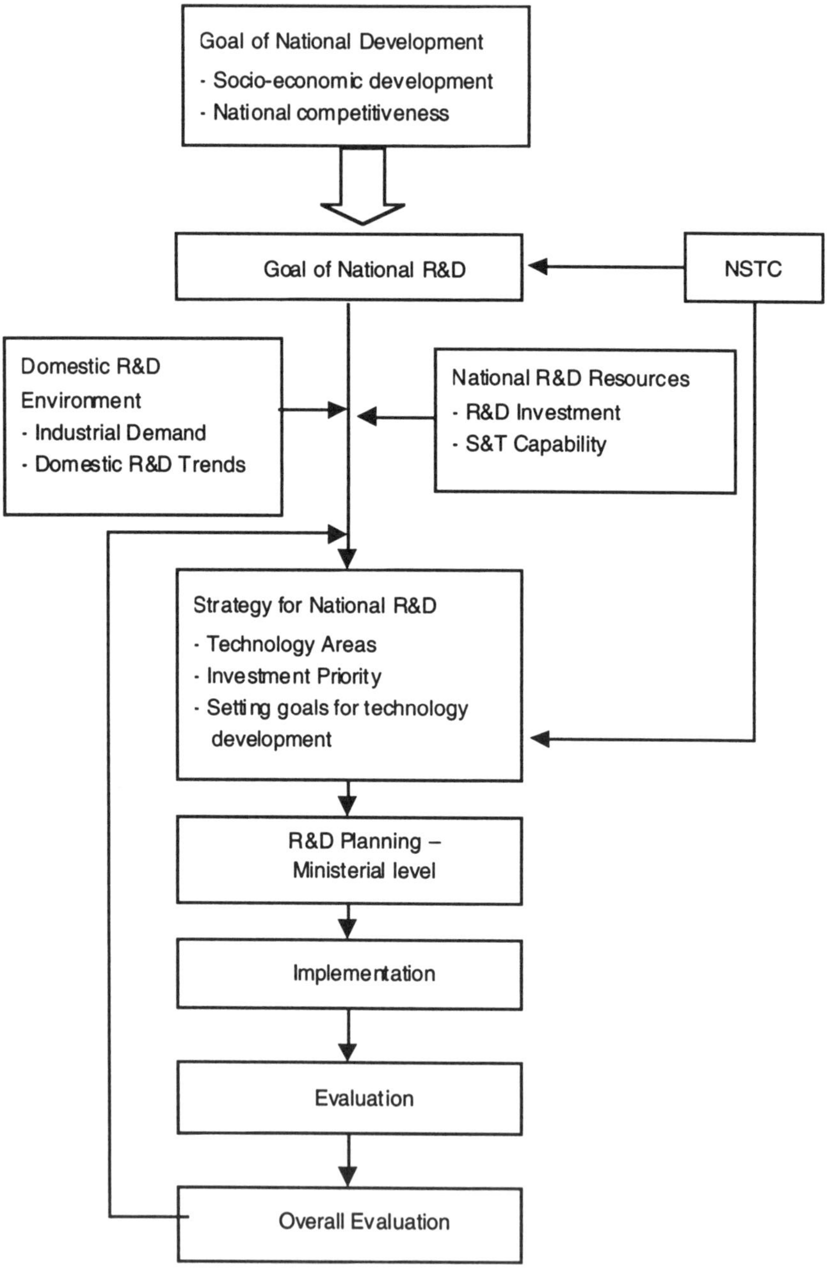

and business people.[1] NSTC reviews the sectoral policies of individual ministries to ensure that they are consistent with other national policies, say, education, industrial, social and economic policies, and do not duplicate other policies or programmes. Through such a process, NSTC tries to minimise possible duplications among ministerial programmes, and ensuing waste of resources. NSTC also evaluates the performances of R&D programmes, the results of which are reflected in budget allocation processes.

Indirect policy co-ordination is conducted by PCST, which advises the President on the issues of science and technology. The members of PCST represent the science and technology and industrial communities.

With respect to budgetary co-ordination, the Ministry of Planning and Budget (MPB) exercises substantial influence. MPB not only draws up annual budget guidelines but also allocates budgets. In the past, the budget authority played the single most important role in science and technology budget allocation, but it is likely that NSTC will play the central role in science and technology system.

3.2 – Major programmes and target technologies[2]

With the emergence of a knowledge-based economy, the capability to generate, diffuse and utilise new scientific knowledge and information has become the key to national welfare and security. Reflecting such changes, the Korean Government has been reformulating its basic policy directions based on the following premises:

- Knowledge will be the dominant factor in national development in the new century.
- Science and technology are the major instruments for sustainable growth.
- A strong and efficient national system of innovation is required to realise the potential of the emerging knowledge-based economy.
- Science and technology have to contribute to not just economic development but also the enhancement of quality of life as well as national prestige.

1. A kind of technology foresight exercise was conducted in the early 1990s for the formulation of the HAN project, but there has not been any such major exercise since then. Individual ministries conduct technology foresight activities at sector levels.
2. This chapter draws largely upon MOST (1999).

Under these premisses, the Korean Government pursues nine major policy goals. They are (1) early development of the core strategic technologies,[1] (2) increasing government R&D investment to 5% of the total budget and enhancing the efficiency of the investment, (3) the development of high-quality scientists and engineers, (4) strengthening basic scientific research at universities, (5) establishment of a scientific and technological information system, (6) strengthening policy support for the development of knowledge industries, (7) promotion of international science and technology co-operation, (8) promotion of regional scientific and technological activities, and (9) nurturing a science-oriented culture in Korean society.

To attain each of these policy goals, the Government implements specific programmes or projects at ministerial or inter-ministerial levels. The policy goal of foremost importance to the Korean government is the development of those technologies that are critical for future industrial development.

"The Review of the FY 2000 National R&D Budget"[2] of NSTC (March 1999) states that the national R&D policy goal is "to develop world-level capability in strategic areas of technology in the early 21st Century". It further says that the Government should pursue the development of key technologies for future socio-economic development, along with an effort to solve the short-term technological problems of industry.

In accordance with this, the Government pursues the following as major programmes for technology development:

- *Industrial Technology Development*: this aims at developing technology to enhance short-term industrial competitiveness. Target technologies include machinery and parts, semiconductor, precision equipment, non-memory chips, ASIC, LCD, and so on.
- *Highly Advanced National Project (HAN Project)*: this is a ten-year inter-ministerial programme designed to develop core technologies for the future development of Korean industries. The HAN Project aims to develop future-oriented technologies, such as broad band ISDN, next-generation automobile, TFT-LCD, advanced materials for electronics, advanced manufacturing system, environmental technology, fuel battery, next-generation nuclear reactor, nuclear fusion, etc. (cf. later section)

---

1. Core strategic technologies refer to the target technologies of the major programs.
2. This is a document reviewed and approved by NSTC in March 1999.

– *Development of Strategic Technologies*, which include technologies such as semiconductors, information and telecommunications, biotechnology, new materials, space technology.

3.3 – Problems

The major problem here is related to the process through which priorities are set. As described earlier, NSTC plays the central role in setting national science and technology policy directions and priorities. But NSTC may not function effectively and fulfil the mandates, unless measures are taken to get individual ministries actively involved in the activities of NSTC. Even though MOST functions as the secretariat of NSTC, there are doubts about the qualification of MOST as an objective and fair navigator of NSTC, on the ground that MOST has conflicts of interest as one of the government agencies engaged in science and technology policy and R&D activities. Other ministries and agencies are not so co-operative and even reluctant to release some of the R&D data required by MOST for the evaluation of government R&D programmes. As a matter of fact, MOST had to rely on the data covering only 55% of total government R&D expenditure in assessing the performances of the government R&D programmes in 1997 (MOST, 1998a). This shows that the system is not working properly.

4. PUBLIC SECTOR RESEARCH

In the early phase of industrial development, science and technology policy in Korea was geared to helping industries absorb and implement foreign technologies. However, entering into the 1980s, the Government recognised the need for government R&D effort as a means to build up indigenous technological capability required for Korea's transition to an industrial economy. MOST launched the National R&D Programme in 1982 on the basis of the Technology Development Promotion Act of 1962.

The National R&D Programme was the first public sector R&D programme in Korea, and accordingly, it placed emphasis on developing an indigenous R&D capability as well as encouraging and stimulating private sector R&D activities. This section describes how public sector R&D activities have evolved in Korea, and analyses the current structure of public sector R&D programmes. As the Government accounts for almost 100% of public sector R&D, discussions are limited to the government R&D programmes.

4.1 – Evolution of public sector R&D

As mentioned, the first public sector R&D programme in Korea, in a formal sense, is the National R&D Programme initiated by MOST in 1982. In the initial year of the programme, 18.7 billion Won was invested and since then, investment has increased at an annual rate of about 25%, reaching 398.1 billion Won in 1995. Noteworthy is the rapid growth of private participation in the National R&D Programme. In 1982, private industries accounted for 28.9% of the total investment in the National R&D Programme, but their share increased to 52.2% in 1995. The number of companies participating in the National R&D Programmes also increased from 86 to 980 during the same period. Over the 13-year period (1982-1995), a total of 2 trillion Won was invested in the National R&D Programme, of which 1.2 trillion Won (60%) came from the Government and the remaining 0.8 trillion Won (40%) from private industries. So, it can be said that the remarkable expansion of the National R&D Programme was made possible not only by the firm commitment of the Government but also by the active participation of private industries.

Together with the increases in investment, the nature of the National R&D Programme has also changed over time. In its formation stage (roughly during the first four years, 1982-1985), the main objective of the programme was to absorb and improve foreign technologies. In other words, the focus of the programme was placed on the development of technologies required for the local production of major products, parts, components and materials considered essential to industrial development at that time. During this period, the National R&D Programme relied totally on a bottom-up approach for project selection, and priority was given to those proposals involving private enterprises.

The subsequent five years, 1986-1990, can be dubbed the "take-off" stage, because the objective of the programme was gradually switched from simple internalisation of foreign industrial technologies to development of core technologies that private industries were not able to tackle due to technical and financial risks. The programme also aimed at building up a technological base for the development of high-technology industries. Accordingly, project selection was partly linked to long-term technology development plans.

At about the end of the take-off stage (around the late 1980s and early 1990s), other ministries began to establish their own R&D programmes in order to solve the problems in the areas of their purview. The Ministry of

TABLE 5.7 – National R&D Programme: inputs, 1982-1995

	1982	1985	1990	1995
Total Investment (billion Won)	18.7	43.1	200.1	398.1
Government (%)	71.1	69.6	60.5	47.8
Private industry (%)	28.9	30.4	39.5	52.2
Number of researchers (persons) (1)	2 700	4 500	5 000	6 500
Number of enterprises (units) (2)	86	240	263	930
Average project size (million Won) (3)	150	180	200	280

Notes: (1) Researchers involved in the national R&D Programme; (2) Private enterprises that participated in the national R&D Programme; (3) Average budget per project.
Source: MOST (1997b).

TABLE 5.8 – Changes in the national R&D Programmes by stage

	Formation stage (1982-1984)	Take-off stage (1985-1990)	Maturing stage (since 1991)
NRDP objective	Internalisation of foreign technologies	Development of core technology	Creative research and future-oriented research
Planning	No planning: bottom-up	Based on loose long-term plan	R&D planning and technology foresight
Main actors	Government R&D institutes (GRI)	Main: GRI Minor: Universities and industries	Main: GRI, with increased role of universities and industries

Source: MOST (1997b).

Commerce, Industry and Energy launched the Industrial Base Technology Development Programme in 1987 and the Alternative Energy Development Programme in 1988. The Ministry of Information and Communication created the Information and Communications Technology Development Programme in 1989, and several other ministries followed these in subsequent years. Up until 1987, the National R&D Programme of MOST was the only government-sponsored R&D programme, and actually, MOST was the sole player in public sector research. But with the emergence of other ministries on the scene, the role of MOST has been gradually reduced. Now, the share of MOST in government R&D expenditures stands at only 33% (1996).

As government R&D investment grew, questions arose as to the immediate and long-term values of the R&D results: "what good do the government R&D programmes do for the future of the nation?" At the same time,

TABLE 5.9 – Government R&D Programmes by Ministry

Ministry	Year initiated	Programme	Budget (1996)
MOST	1982	National R&D Programme	354.3
MOCIE	1987 1988	Industrial Base Technology Development and Alternative Energy Development Programmes	293.4
MIC	1989	IC Technology Development Programme	207.3
MOE	1992 1996	Environmental Engineering Technology and Environmental Basic Technology Development Programmes	19.5
MOCT	1994	Construction Technology Programme	13.3
MOAF	1994	Agricultural Technology Programme	55.0
MHW	1995	Health and Medical Technology Development Programme	7.2
MoEd	1983	Basic Scientific Research Support Programme	90.0

Unit: Billion Won.
Source: MOST (1997b).
Note: MOST: Ministry of S&T; MOCIE: Ministry of Commerce, Industry and Energy; MIC: Ministry of Information and Communications; MOE: Ministry of Environment; MOCT: Ministry of Construction and Transportation; MOAF: Ministry of Agriculture and Forestry; MHW: Ministry of Health and Welfare; MoEd: Ministry of Education.
Defence R&D (W271.2 billion, 1996) is excluded.

the private industrial sector, as a major financial contributor, became increasingly sceptical about the economic values of R&D results. On the other hand, as individual ministries created their own R&D programmes, the problem of inter-ministerial resource allocation emerged as an important policy issue. In other words, the diversification of government R&D programmes brought up a new set of problems, such as duplication of research efforts, delineation of R&D areas between different ministries, inter-ministerial R&D priority-setting and efficient allocation of the R&D budget.

All these issues boil down to the question of how to allocate limited resources to a selection of areas. The question is not only technological but also economic and political in that the government R&D programme is justified only by the taxpayers' consent to the investment.

To deal with the issues, in the mid-1980s the Government adopted the concept of technology planning and evaluation in implementing the government R&D programmes. In other words, in setting priorities for technology development, the Government employed a strategic approach based on long-term planning. Industries and academia participated in the process so that the interests of private industries and academia could be reflected in the planning process of government R&D programmes. It was during this period that industry–academia–GRI collaborative research was first undertaken as part of government R&D programmes. The International Co-operative Research Programme was also launched during this period.

Yet, it was not until 1992 that a Korean system of public sector R&D management took form. In that year, the Government launched the Highly Advanced National (HAN) Project, which is a ten-year inter-ministerial R&D programme to develop core technologies for the industrial development toward the 21st century.[1] The HAN Project is the first government R&D programme that was developed through a full cycle of the planning process including technology foresight activities and inter-ministerial consultation.

Through these stages, the government R&D expenditure grew very rapidly from 264 billion Won in 1982 to 2411 billion Won in 1996. As a result of such growth, the role of MOST has been reduced from the central and sole player in the public sector R&D to just one of the major players.

4.2 – Current structure of government R&D programmes[2]

In 1997, the government budget for R&D was 2994 billion Won or about 2.8% of the total government budget. This is approximately 4% of the corresponding US budget (1996), 11% of that of Japan (1996), and 25% of that of Germany (1996). Of the total budget, MOST's share was 33%, MOD

1. The HAN Project is broadly categorised into two groups. One is the product technology group, which focuses on the development of specific products, particularly high-technology products, that will lead Korea's industrial development toward the new century, such as new agro-chemicals, ISDN, HDTV, ASIC, flat panel displays, biomedical, micro-machine, next-generation vehicles and express rail systems. The other category emphasizes the development of core generic technologies that are indispensable for the long-term development of the Korean economy, namely: next-generation semiconductors, advanced materials, advanced manufacturing systems, new energy, next-generation nuclear reactors, advanced super-conducting tokamak, ergonomics and dual-use technologies.
2. This part draws largely on MOST (1998). Due to the lack of co-operation from some of the ministries, only 55% of the total expenditure or 1644 billion Won were covered in this analysis.

TABLE 5.10 – S&T budgets by Ministry, 1997

Ministry	% of the total budget
MOST	33.1%
MOCIE	18.1%
MOD	17.6%
MoEd	9.2%
MOAF	10.8%
MOMF	3.5%
MHW	3.4%
MOE	1.4%
Others	2.9%
Total	100.0%

Source: MOST (1998a).

18%, MOCIE 16%, MOEd 9%, MOAF 7% and Ministry of Maritime and Fisheries Affairs 3.5%. Consequently, a large share of the science and technology budget is under the command of a few Ministries, MOST, MOD and MOCIE.

R&D expenditure by technology area

Of the total R&D expenditure analysed (997 billion Won) covering 710 projects, 19% was for bio-science, 15% for transportation, 14% for information and communications, 10% for nuclear energy, 9% for machinery, 8% for chemicals, 7% for energy resource and 6% for environment. Given this structure of R&D investment, it is not so clear if the government R&D programmes are focused on certain strategic areas of technology, although it seems that the 1997 government R&D programmes placed priority on such areas as bio-science, information and communication, and public welfare (transportation).

R&D expenditures by industry area

Large shares of national investment went to information and electronics (17.6%), machinery and metals (16.6%), automobiles and shipbuilding (14.6%), and medicine and agro-chemicals (10.5%), but the R&D investment in information and electronics still understates the reality because the R&D activities financed by other public funds (approximately 300 billion Won) were excluded from the analysis.

TABLE 5.11 – R&D investment by area

Technology area	% of the total investment (997.5 billion Won)
Bio-science	18.9%
Transportation	15.2%
Information and communication	13.7%
Nuclear energy	9.7%
Machinery	9.3%
Chemicals	8.1%
Energy and resources	7.2%
Environment	6.4%
Materials	5.1%
Others	6.4%
Total	100.0%

Source: MOST (1998a).

Budget allocation to R&D performers

In Korea, GRIs conduct the lion's share of government R&D programmes. According to the MOST analysis of the 1997 government R&D expenditure, (1644 billion Won), 48% of the total government R&D expenditures went to GRIs, 35% to research consortia,[1] 7% to national laboratories, 6% to universities and only 2% to private industry. This seems to support the allegation that government R&D programmes are just for GRIs. However, these statistics should be read very carefully, since they understate the share of universities, and do not capture the R&D funds that entered universities via GRIs. Another source of this understatement is the fact that the salaries of professors engaged in R&D come from other sources and are not counted as university R&D expenditures.[2] This implies that the statistics understates the share of universities, while overstating that of GRIs. Readers may also be misled by the large share (35%) of research funds allocated to research consortia to think that industry–university–research institute (usually, GRIs) research co-operation is quite active in Korea.[3]

1. Research consortia refer to collaborative research group comprising industries, universities and research institutes.
2. MOST solved this problem by adopting the OECD method in measuring R&D expenditure. The salaries of GRI researchers and current expenditures of GRIs have been included in the R&D expenditure.
3. This has been discussed in the previous section.

TABLE 5.12 – Major national R&D performers

Performer	GRIs	R&D consortia	National laboratories	Universities	Others
% of the total budget allocated	48.5%	35.3%	7.5%	6.4%	2.3%

Source: MOST (1998a).

Other statistics published by MOST provide a different picture: in 1997, GRIs received 48% of the government R&D layouts, national laboratories 15%, universities 20% and private industries 14% (MOST, 1999). The difference between the two sets of data appears to explain the degree to which the share of universities was understated by the statistics. Still it cannot be denied that Korean universities are not so actively involved in R&D as their foreign counterparts are.

Socio-economic objectives of the national R&D programmes

Finally, an international comparison has been made as to which social aspects the national R&D programmes address. It has been found that the Korean national R&D programmes place more focus on defence (18.1%),[1] industrial development (17.5%), energy (11.4%), agriculture, fisheries, etc. (9.4%), and health and social welfare (4.6%). This is in sharp contrast to what is observed in some of the advanced countries. In the US, for example, 66% of national R&D programmes are devoted to defence technology, while only 0.6% of national R&D programmes are for industrial development (1994). In the case of Germany, about 53% of national R&D efforts are directed to knowledge generation, and about 13% for industrial development. This clearly shows that the Korean national R&D programmes are geared to industrial development.

4.3 – Public research organisations

The primary funder of public R&D activities is the Government. Individual ministries fund R&D activities in the areas of their purview through their funding agencies. Consequently, public R&D funds flow

1. Defence research has not been included in the analysis because of the lack of data.

from the Government to R&D performers – industries, universities and GRIs – via intermediary funding agencies that actually plan, manage and evaluate R&D projects. These constitute the major public research organisations in Korea together with national laboratories, GRIs and universities that carry out about 84% of the government R&D activities in terms of expenditure.

Funding agencies

Funding agencies conduct the planning, management and evaluation of public R&D programmes. They interact directly with R&D performers on behalf of the ministries they represent. There are two types of funding agencies: one mandated to support university research, and the other managing government R&D programmes.

There are two agencies that belong to the first category. The Korea Science and Engineering Foundation (KOSEF), a funding agency under MOST, supports basic scientific and engineering research at universities. The Korea Research Promotion Foundation is also a major source of funds for university research. Unlike KOSEF, it serves broader constituents, including not only science and engineering but also humanities and social sciences. It is under the Ministry of Education.

Individual ministries also operate funding agencies to manage their R&D programmes. For example, the Korea Institute of Science and Technology Evaluation and Planning (KISTEP) manages the R&D programmes sponsored by MOST, and the Institute for Industrial Technology Evaluation and Planning (ITEP) plays the same role for the Ministry of Commerce, Industry and Energy. Several other ministries operate similar organisations for the same purposes.

National laboratories

National laboratories are government organisations whose functions include research and testing. Researchers at the national laboratories are civil servants. As of 1997, there were 77 national laboratories, 43 of which are under the central government and the remaining 34 under local governments. They are mostly for research, testing and technical extension services in the areas of agriculture (46), manufacturing (15), and medicine (8). The national laboratories employ 4604 researchers, of whom 1178 are of doctoral level. They account for about 15% of the government R&D expenditure. In 1997, they spent 309 billion Won or 81% in agriculture of the total R&D expenditure of 381 billion Won.

Government Research Institutes (GRI)

GRIs are the major players in public R&D. They operate with the financial assistance of the Government, but are independent non-government organisations operating under the provisions of the civil laws and the Law for the Creation and Promotion of the Government Research Institutes (1999). Therefore, GRI researchers are not government employees.

There are 32 GRIs and they conduct 48% of the government R&D activities in terms of expenditure and 12.4% of the gross national R&D. In 1997, GRIs' R&D expenditures amounted to 1511 billion Won, of which 1252 billion Won (83%) came from the government. GRIs employ 8360 scientists and engineers, of whom 3316 have Ph.D.s, 4011 masters, and 984 bachelors. Of the total R&D expenditures by GRIs, 60% was devoted to technology development, 25% to applied research and the remaining 15% to basic research. This shows that the major role of GRIs still has not changed much despite the changes in their operating environments. Over 80% of the research funds came from the Government, while the inflow of funds from industry was less than 20%, signifying the fact that despite the overwhelming industrial research orientation of GRIs, industry–GRI research interactions are not that pronounced.

The Korea Institute of Science and Technology (KIST), the first GRI in Korea, was established in 1966 as an integrated technical centre to assist industrialisation by finding solutions for simple technical and practical problems and helping internalise imported foreign technologies. With the development of heavy machinery and chemical industries in the 1970s, the demand for technical support, such as those provided by KIST in the 1960s, increased in various industries. To meet the demand, the Government spun off from KIST a number of specialised institutes in the priority industrial areas, namely, electronics, telecommunications, energy, machinery, chemicals, shipbuilding and marine resources. These institutes operated under the patronage of the ministries that were responsible for the development of the respective industries. To accommodate these institutes as well as private R&D laboratories, the Government started the development of the Daeduk Science Town in 1970s.

Entering into the 1980s, the operating environments of GRIs changed very rapidly. Private industries began to establish in-house R&D systems in order to build up technological capability to cope with increasing market competition. Universities, which had been heavily teaching-oriented, also launched various efforts to develop research capability as a means to upgrade the quality of education. These two developments squeezed the position of

GRIs in the Korean R&D system. GRIs, that had been virtually the sole player in R&D in the 1970s, have been rapidly losing ground.

The major criticisms have been that they have not contributed as much as they have spent, and that GRIs have overly expanded their range of activities as a means to claim more resources, causing duplications among themselves and leading to severe wastes of resources. Another strand of criticism is that the demand for the services of GRIs has gone and so government support for them is no longer justifiable. In response, in the early 1980s, the government downsized GRIs and merged some of them to reduce duplications, and then put them under the single jurisdiction of MOST, in the hope of promoting inter-institutional management of research personnel and resources. Along with this, they launched the National R&D Programme, and enacted the Law for the Promotion of National R&D Organisations (1982), which provided the legal base for GRIs. GRIs were transformed from industrial technical research centres into institutions for government R&D.

After almost twenty years of operation since then, the criticisms against GRIs have not been toned down and they have had to undergo another round of major reform and reorganisation. Two reasons for the reorganisation are that resource wastes resulted from the inter-institutional barriers to mobility existing between and among GRIs, and that GRIs tend to work for the interests of their patrons (ministries) rather than those of the nation. So, they downsized GRIs in terms of both budget and personnel and put them under three newly-created research councils (see Table 5.13). The research council reports directly to the Prime Minister's Office. Some GRIs were allowed to stay under MOST or other ministries if that was deemed necessary for efficient operation. For example, the Korea Atomic Energy Research Institute is under MOST even after the reorganisation.

It is too early to evaluate the reorganisation, but, despite its significance, the fundamental problem of GRIs still remains unsolved: GRIs do not have clearly defined missions by the government. Even though the charter of each GRI contains statements on the missions of the institute, the statements are too broad or too ambiguous. The broadness and ambiguity of the missions lead to duplications and overlaps in R&D among GRIs and make it practically impossible to evaluate the performances of GRIs on the basis of their missions.

Another possible source of problems is their relationships with the research council and the Government. Research councils manage GRIs, appoint the directors of GRIs, and evaluate the performances of GRIs at the end of each fiscal year. On the other hand GRIs receive funds directly from

TABLE 5.13 – Research councils and government research institutes

	Personnel	Budget
Korea Research Council of Fundamental Science and Technology		
Korea Institute of Science and Technology	781	100 549
Korea Research Institute of Bio-science and Bio-technology	267	35 215
Korea Basic Science Institute	174	39 263
Korea Astronomy Observatory	95	11 003
Korea Research Council of Public Science and Technology		
Korea Research Institute of Standard and Science	339	50 499
Korea Ocean Research and Development Institute	227	43 583
Korea Institute of Geology, Mining and Minerals	340	41 511
Korea Institute of Energy Research	250	40 732
Korea Institute of Construction Technology	286	11 858
Korea Railroad Research Institute	180	15 088
Korea Institute of Industry and Technology Information	207	12 009
Korea Aerospace Research Institute	249	72 562
Korea R&D Information Center	40	9 107
Korea Research Council of Industrial Science and Technology		
Korea Institute of Oriental Medicine	40	4 756
Korea Institute of Industrial Technology	225	21 650
Electronic and Telecommunications Research Institute	1 561	250 812
Korea Food Research Institute	199	13 567
Korea Institute of Machinery and Metals	520	65 734
Korea Electro-technology Institute	191	51 183
Korea Research Institute of Chemical Technology	366	47 771

Note: 1999 budget in million Won.

individual ministries on an individual project basis and the budgets for current expenditure from the budget authority through the Prime Minister's Office. The research councils have neither research funds to support GRIs nor any voice in the government funding process. The only role the research councils play in the budgeting process is to collect budget requests from GRIs and review them for consistency before submitting them to the Prime Minister's Office. This means that the research councils have to manage GRIs with only sticks because they do not have any carrots in their hands. The GRIs' relationship with the Government has also become more complex, because the Prime Minister's Office is responsible for their operation, while their research activities are funded and assessed by other ministries and funding agencies.

Universities

There are 156 four-year universities[1] in Korea, of which 24 are national universities, two public, and 130 private. There are also 669 graduate schools, almost all of which are part of the universities, except for a few independent institutions. Private universities are basically operated on endowments, fees and donations, but the government complements private university foundations by providing various forms of financial support. Private universities account for over 76% of the total university enrolments.

Universities are a rich pool of high-quality scientists and engineers. They have highly-qualified research manpower of over 45 000, of whom 27 000 hold Ph.D. degrees, 17 000 master's degrees. This constitutes approximately 34% of the total research manpower of Korea, but Ph.D. level research scientists and engineers are extremely concentrated in universities (76%) (see Table 5.2).

Even though universities command the largest pool of qualified scientists and engineers, they account for only 9.4% of gross national expenditure on R&D (1019 billion Won, 1996), which is smaller than the share of GRIs (12.4%). This reflects the situation of Korean universities, especially their research environments. First, professors cannot afford to engage in serious research because of excessive teaching obligations. Amazingly, the student–professor ratio (number of students per professor) at Korean universities exceeds 34:1. Even at national universities, which are in much better situations, the ratio is 29:1 (Ministry of Education, 1998). Second, as Korean universities place more emphasis on teaching at undergraduate levels, graduate programmes have not been well developed. And therefore, university education is not linked to research. Naturally, universities are very poor in research facilities. Kim, an engineering professor at one of the top engineering schools in Korea, describes the situation as: "only a few academic institutions in Korea provide anywhere near sufficient research support. In most universities, a newly recruited professor must start with an empty laboratory space and very meagre university support for expendable supplies and equipment" (Kim, 1996). Third, university professors have never

1. Education system in Korea is very similar to that of the US with 12 years of Elementary and High schools and four years of university education.

TABLE 5.14 – Basic statistics on universities in Korea (1997)

		Total	National	Public	Private
Number	Universities Graduate schools	156 669	24 123	2 10	130 536
Enrolment ('000)	Universities Graduate schools	1 478 77	330 20	18 2	1 130 55
Professors	Number ('000)	45	11.9	0.44	33
Student / professor ratio	Percentage	34.5%	29.4%	43.2%	35.9%

Source: Ministry of Education (1998).

TABLE 5.15 – R&D structure at universities

Research manpower	Persons	45 327	Ph.D.: 27 434
R&D expenditure	Billion Won	1 019	9.4% of GERD
Source of fund	Government Industry Foreign	505 510 3	49.6% 50.0% 0.3%
Type of University	Public Private	390 629	38.2% 62.8%
Nature	Basic research Applied research Development	452 271 296	44.4% 26.6% 29.0%
Areas	Exact sciences Engineering Agriculture Medicine Others	238 470 151 105 54	23.3% 46.1% 14.8% 10.3% 5.2%

Source: KITA (1998).

been put under pressure for research. Once employed, they are automatically tenured. Even though there are some requirements they have to satisfy in order to stay at the teaching job, these could be met without serious research efforts.

University R&D activities are directed more toward basic research than other sectors. Of the total R&D expenditure, 45% is devoted to basic research, 27% to applied research, and 29% to development. Interestingly, universities rely more on private industries (50%, 510.4 billion Won) than the Government (49.5%, 505.4 billion Won) for research funds. This is a good contrast to the case of GRIs, which focus on applied and development research (85%) but depend heavily upon government for research funds (83%).

The composition of university R&D expenditure by area shows that over 46% of the funds were spent on engineering research, while scientific research received only 23%. Other major areas are agriculture (14%) and medical science (10%).

Even though private universities exceed national, public universities in the size of student enrolments, their R&D activities lag way behind. The share of private universities in university R&D expenditures was 62% in 1996, which is much lower than their share in student enrolments – 76%. This suggests that research environments at private universities are much worse than they are at national, public universities.

Overall, universities in Korea have not been so important an actor in R&D as their foreign counterparts are. Various factors may be behind this, but the fundamental reason is the extreme teaching-orientation of Korean universities. To reorient Korean universities toward more research-oriented institutions, the Government has taken various measures, including the "Brain Korea 21" programme, which supports selected universities in their transformation into research- and graduate education-oriented institutions. For this programme, the Ministry of Education has planned to pour in 90 billion Won every year for nine years from 1999.

5. INDUSTRIAL RESEARCH AND DEVELOPMENT

In a market economy, technological innovation in private industry is the key element for sustaining economic growth. Since the early 1980s, private industries have been leading technological innovation in Korea: they have been the prime sources of R&D funds as well as the key performers of R&D. The Korean government has also been promoting industrial R&D through various policy measures. Under the provisions of the Industrial Technology Development Promotion Law and other measures, private firms are given tax incentives and financial supports for their R&D activities, along with tariff deduction on imported R&D equipment and R&D subsidies. This section describes industrial R&D activities in Korea, and the government policy measures to promote industrial innovation.

5.1 – Industrial R&D

Up until the early 1980s, private industrial R&D activities were negligible. Private industries depended heavily on foreign technologies, and their major technological efforts were directed toward adapting and assimilating imported technologies. After 1982, private investment in R&D began to increase sharply. Before 1982, the industrial R&D intensity (the ratio of R&D expenditures to sales) remained below 0.5. In industrial R&D intensity first exceeded 0.5%, in 1982 and it has been increasing steadily since, although the ratio of technology imports to total R&D expenditures has stayed stable. After 1992, the ratio of technology imports to R&D expenditures fell to 20%, and R&D intensity exceeded 2%, reaching 2.4% in 1996. These changes show that Korean industries rely increasingly on in-house R&D for technology development. This turnaround has of course been made possible by the massive R&D efforts that started in the early 1980s.

As discussed earlier, as the industrial structure became more technology-intensive, Korean firms had to compete with advanced countries in high-tech product markets and needed to build up technological capabilities to survive international competition. In 1982, private industries invested 268.7 billion Won in R&D that accounted for 50.4% of the gross national investments in R&D. Since then, industrial R&D investment has undergone a quantum jump, recording an average annual growth rate of over 24% over the period of 1982-1996. In 1996, private industries invested 8466.7 billion Won or 77.8% of gross national investments in R&D (see Figure 5.1 and Table 5.3).

Structure of R&D activities

In terms of R&D performance, private industry accounted for 73.2% of the nation's gross R&D expenditures in 1996. This means that private industries are the dominant players in R&D in Korea. In 1980, there were only 321 industrial R&D laboratories with 5100 researchers, of whom only 56 were Ph.D. degree holders. After 16 years, the number of industrial R&D laboratories grew to 2435, employing 71 193 researchers, including 3374 Ph.D. degree holders (1996).

Of the total industrial R&D expenditure in 1996, manufacturing industries accounted for 84.5%, construction 5.5% and services 3%. Manufacturing R&D was led by communication equipment (38.8%), followed by transportation equipment (30.4%), chemicals (9%), machinery (5%) and basic metals (3%). With communication equipment, transportation

TABLE 5.16 – Industrial R&D, 1996

Research manpower	Persons	71 193	54% of the total R&D manpower
Out of which	Ph.D. diploma Master degree	3 374 19 702	4,7% 14,9%
R&D investment R&D expenditure	Billion Won Billion Won	8 467 7 964	77.8% of the total 73.2% of the total
Size of firms	Large firms SMEs	7 256 707	91.1% 8.9%
Purpose	Product development Process development	6 251 1 713	78.5% 21.5%
Nature of R&D	Basic research Applied research Development		8.1% 26.3% 56.5%
Industry	Manufacturing Construction Services Other	6 734 438 245 69	84.6% 5.5% 3.1% 0.8%

Source: KITA (1997).

equipment, and chemical industries explaining almost 70% of manufacturing R&D, there is a very high industrial concentration of manufacturing R&D activities.

The average R&D intensity of Korean industry was 2.1% in 1996, and that of manufacturing industry was 2.6%. Those industries that invested the largest shares of their sales in R&D include: communication equipment (5.5%), medical and precision equipment (4.6%), transportation equipment (4.2%) and office equipment (4%). In contrast, it turned out that agriculture, construction, pulp and paper industries, food and beverages, textile industries invested less than 1% of their sales in R&D, while agricultural industries invested less than 0.5% of their sales in R&D (see Table 5.16).

Quite naturally, industrial research focused on development (65.6%) and applied research (26.3%) with the remaining 8.1% devoted to basic research. In the case of office equipment, medical and precision equipment, and miscellaneous industries, more than 80% of the R&D expenditures were used for development. In contrast, the food and beverage, pulp and paper, and

TABLE 5.17 – Manufacturing R&D, 1996

Industry	R&D expenditure (billion Won)	R&D intensity (%)
Food	118	0.6
Textile	51	0.9
Pulp and paper	26	0.8
Chemicals	601	1.8
Non-metal minerals	99	1.2
Basic metals	205	0.8
Fabricated metals	46	1.1
Machinery	338	1.7
Office equipment	165	4.0
Electrical machinery	144	1.6
Communication equipment	2 614	5.5
Medical and precision equipment	55	4.6
Transportation equipment	2 057	4.2
Others	35	1.7
Total	6 734	2.6

Source: KITA (1998).

TABLE 5.18 – R&D concentration, 1996

	Top 5	Top 10	Top 20
Agriculture	98.9	100.0	100.0
Mining	100.0	100.0	100.0
Manufacturing	38.2	51.5	63.9
Communication equipment	79.1	89.8	93.5
Transportation equipment	75.2	87.9	92.2
Construction	47.5	70.2	87.6
Services	28.7	42.6	56.9
Others	78.3	89.9	96.9
All industries	32.3	44.5	56.5

Source: KITA (1998).

transportation equipment industries invested relatively more in basic research. And most of the industrial R&D activities were geared to developing new products rather than new production processes. Of the total industrial R&D,

78.5% was for new product development, while only 21.5% was devoted to the improvement or development of production processes.

Industrial R&D activities in Korea are highly concentrated in large enterprises, which account for 91% of R&D expenditure and 83% of research manpower. Therefore, large firms spend more than double on R&D per researcher than small and medium-sized firms (SMEs) (123 million Won for large firms and 58 million Won for SMEs).[1] More notable is the fact that the top five firms command more than 32% of the total industrial R&D expenditure, the top ten firms 44%, and the top 20 firms 56%. In agriculture and mining industries the top ten firms explain almost 100% of the total R&D expenditure, and among manufacturing industries, the top ten firms account for more than 80% of the total R&D in such industries as basic metals, communication equipment, transportation equipment and miscellaneous industries. Such an extremely high concentration of industrial R&D expenditure is a direct reflection of the large-firm oriented industrial structure of Korea. The top five firms in terms of R&D investment in 1996 are: Samsung Electronics (1276 billion Won), Hyundai Motors (620 billion Won), LG Electronics (446 billion Won), Hyundai Electronics (380 billion Won) and LG Semiconductors (370 billion Won). Industrial R&D concentration has been increasing over time. For example, the share of the top 20 firms increased from 49% in 1992 to 56% in 1996. The exceptionally high concentration of R&D activities is also a consequence of the Chaebol system.

Technological competence of Korean industries

As a result of massive R&D activities, Korean industries have made remarkable achievements in technological development. It is not easy to assess the technological competence of industries, because of data requirements. Here, we rely on patent data from the Korea Industrial Property Administration and the US Patent and Trademark Office to obtain a rough understanding of the technological competence of Korean industries.

The R&D efforts of Korean industries produced enormous amounts of industrial property. The number of Korean patents granted to Koreans increased more than 25 times from a mere 300 in 1982 to 8300 in 1996. In 1982, Koreans' share in the total number of patents registered was only 11.5%, but the share increased to 50% in 1996. In 1994, Korea ranked fifth

1. In Korea, firms employing less than 300 employees are considered as SMEs.

in the world in terms of the number of world patents and utility models applied. More remarkable, Korea ranked seventh in terms of the accumulated number of US patents registered over the period 1982-1996. The US Department of Commerce notes: "what is most noticeable here is the strong growth in US patents for Taiwan and Korea. At current growth rate, both will overtake the UK in a relatively short time" (Albert, 1998, p. 20).

Of the Korean patents granted in 1996, electronics and communications took up 44%, chemical and textiles 22% and machinery 16%. This result is consistent with the data on Korean industrial R&D expenditure, which are concentrated in a few industries such as communications and transportation. This is also well reflected in Korea's US patents. In US patent counts, over the period 1992-1996, Korea ranks fourth in information technology, fifth in advanced materials, sixth in transportation and logistics, and seventh in automotive, pharmaceuticals and biotechnology. In the US patent analyses for the period 1992-1996, it has been assessed that in overall technological strength, Korea ranks third in information technology, fifth in advanced materials, sixth in automotive and seventh in pharmaceuticals and biotechnology (Albert, 1998).

5.2 – Policy incentives for industrial R&D

Incentive measures for industrial technology development have their legal bases in the Science and Technology Promotion Law and the Industrial Development Promotion Law. Overall, Korea employs grants and tax credits as the major instruments to promote industrial R&D. How effective those measures have been in promoting industrial R&D is an issue of great concern in Korea. There have been a few attempts to evaluate the effectiveness of policy programmes (Lee and Jang, 1998). While, in order to evaluate policy programmes, quantitative assessments on each of the measures are the most desirable, full-scale quantitative analysis has not been practically possible due to the lack of data.

Past studies on this issue have produced some useful qualitative assessments of policy measures. The studies seem to concur that tax incentives are the most effective tools for the promotion of industrial R&D, and that the Korean government offers so diverse and complex measures that it is not easy for industries to fully understand the support programmes. On the other hand, some industries take advantage of the complexity of the programmes in order to get multiple benefits. The incentive programmes offered by the Korean government are shown in Table 5.19.

TABLE 5.19 – Incentive programmes for firms in Korea

Tax incentives

- Reserve Funds for Technology Development: private enterprises that reserve funds for technology development, technical information, R&D manpower and facilities, etc. for three years are given tax deductions. The maximum tax-deductible reserve is 5% of the total sales.
- HRD Expenditure: private enterprises may take advantage of a tax deduction of up to 15% of their total expenditures on human resource development, including technical training and in-house technical college.
- Research Facilities: a tax deduction is allowed for up to 10% of the investments in research facilities by private enterprises. A depreciation of research and test facilities at the rate of 90% a year is also allowed.

Subsidies

- R&D Subsidy by Government: private enterprises are eligible for a maximum subsidy of 50% of the total R&D expenditure, if they are involved in national R&D programmes. The government also provides financial support of up to 80% to large enterprises and up to 90% of the total cost to individuals or small enterprises to help commercialise new technologies.
- Support by Public Corporations: government-owned corporations such as the Korea Electric Power Corporation (KEPCO) and the Korea Telecommunication Corporation (KT), extend to private industrial R&D centres or research co-operatives financial support of up to 80% of the total R&D investment for the development of new products.

Financial support

- Financial Institutions for R&D: financial institutions, such as the Korea Development Bank, the Citizens National Bank and the Industrial Bank of Korea, provide private industries with long-term, low-interest loans for the development of new products, process technologies and the commercialisation of new technologies.
- Venture Capital: venture capital, such as that available through the Korea Technology Banking Corporation (KTB), offers various forms of financial support for technology development activity by the private sector. Financial support takes the form of equity investments, debenture purchases, conditional loans, technology development loans, leasing and factoring services.

5.3 – Structural weakness of industrial R&D system

Despite the remarkable performance of Korean industries in technological development, Korean industries harbour fundamental structural problems that have to be corrected to sustain the technological dynamism. The first structural problem is the extremely high concentration of R&D activities. High concentration means that only a few large firms are actively involved in R&D, while others are not. If this persists for long, this will dichotomise Korean industries into technologically advanced and retarded firms and sectors. This will reduce inter-firm and inter-industry interactions that are the key elements of innovation. The second problem is the weakness of SMEs in R&D, because SMEs are the origin of the technological vitality of an economy. Even Chaebol companies would not be able to sustain competitiveness without technologically strong domestic SMEs. The third problem is the insufficient interaction between industries, universities and GRIs. This lack of interaction between R&D performers extends the distance between laboratories and the market.

6. CURRENT ISSUES: IMPACTS OF THE ECONOMIC CRISIS

As the consequence of the financial crisis of 1997, the Korean economy recorded a negative growth, of –6.5%, in 1998. The impacts of the economic setback have been wide and deep in many sectors of the economy. Most affected were the future-oriented investment activities, especially R&D.

According to the survey by STEPI (Oct. 1998), large enterprises reduced R&D investment by 13.6% and R&D personnel by 6.0%, while SMEs downsized their R&D activities by 5.6% (investment) and 2.8% (R&D staff).

Most of the reductions were for mid- and long-term R&D projects rather than short-term projects. Therefore, industry drastically reduced its funding for university as well as public R&D: particularly notable is that large enterprises reduced their funding for university research by 22.7%. Universities responded to the survey in a very similar way: they answered that research fund inflow from industry decreased by 25.6%. Considering that universities rely on industries for more than 50% of their R&D funds, this was a very serious blow to university research in Korea that is still at an infant stage. According to the survey of universities, total R&D fund inflow into universities decreased by 12.4%.

The magnitude of the impacts varied widely across universities. Some universities were hurt more severely than others by the reduction. For instance, mid-level universities had to suffer almost 20-30% reductions in R&D funding from industry, while the impacts on top universities were rather minimal (–3.6%). As the survey shows, the financial crisis resulted in a very severe shrinkage in R&D activities, especially in private industry and

TABLE 5.20 – Private industries' R&D responses to the economic crisis

	Large enterprises	SMEs
R&D personnel	-6.0%	-5.6%
R&D investment	-13.6%	-2.8%
R&D fund outflow towards		
- Universities	-22.7%	-5.4%
- Public R&D institutes	-9.2%	-41.0%
- Others	-11.1%	-14.6%
- Overseas	-18.7%	-31.7%

TABLE 5.21 – Impacts on university research

Source of fund	University rank	Changes (97-98)
Public sector	Top 10	-1.5%
	11th to 30th	-11.3%
	31st to 80th	-21.8%
	81st and over	17.0%
	Average	-7.2%
Private sector	Top 10	-11.1%
	11th to 30th	-37.7%
	31st to 80th	-38.1%
	81st and over	-23.1%
	Average	-25.6%
Total	Top 10	-3.6%
	11th to 30th	-17.6%
	31st to 80th	-27.3%
	81st and over	0.1%
	Average	-12.4%

universities. The impacts were so deep and wide that many worried that if this situation continued for several more years, Korea's R&D system might collapse.

In contrast, public sector R&D was not very much affected, as the Government maintained its commitment to the development of science and technology. Such a commitment is based on the recognition that science and technology constitute the core of the nation's capability to grow. Reflecting this, R&D investment suffered relatively minor cuts in the course of the 1998 budget revision, made by the new Government in response to the crisis.

Almost all of the budget items suffered 10% or greater reduction from the original budget approved by the National Assembly.

As a result of the budget revision, government R&D investment was cut by 7.4%. It is, however, encouraging that according to the 1999 budget, the Government plans to increase its investment in science, technology and information by more than 4%. This suggests that public sector R&D efforts will remain strong despite the economic crisis.

Two sets of policy issues emerged in the course of the crisis. One is the set of short-term issues, such as reduced jobs for researchers and under-utilisation of research facilities that resulted from the crisis. The other is related to the structural weakness of the R&D system, such as the instability of the R&D system and the inefficiency of public sector R&D that were revealed in the course of the crisis.

To deal with the short-term issues, the Government took various measures to create temporary jobs to help retain scientists and engineers in laboratories or similar environments until the economy recovers. This included such measures as the S&T D/B Development Programme and the Research Intern Programme. In addition, in order to stimulate industrial R&D, the Government launched new programmes, including direct financial assistance for industrial research and support for new start-ups. Anyhow, these problems will be cured as the economy revives.

More serious is the structural weakness of the R&D system, because it cannot be cured automatically, even when the economy recovers its dynamism. The government undertook a massive reorganisation of GRIs in order to enhance the productivity of public sector R&D (see section 3), but despite the reorganisation, the fundamental problem has not been solved because the missions of GRIs have not been specified and clarified. The broadness and ambiguity of the missions have always been the source of problems: duplications, overlapping and so on. Consequently, providing well-defined missions, which are consistent with the direction of national development, to GRIs is the first action the Government should take to make GRIs more productive.

What was revealed during the economic crisis was the extreme instability of the Korean R&D system. The instability stems basically from its excessive reliance on industries for R&D investment. As profit-seekers, industries are bound to respond much more sensitively than the Government to changes in the market. As they did in 1998 in Korea, industries tend to respond to major cutbacks in the market by cutting future-oriented investments such as R&D in an abrupt and substantial way. Such cuts in industrial R&D

spending may disrupt the whole R&D system, when the system depends too much on industrial R&D. Indeed, if the situation in Korea had continued for several more years, the nation's R&D system might have collapsed. Excessive reliance on industries also makes the R&D system unstable, because under such a system, they stress applied research and development too much and tend to neglect basic research, failing to develop a strong foundation for the future development of science and technology. The policy prescription is simple and clear. The Korean government has to increase its share of investment in R&D in order to make the R&D system more balanced and stable.

REFERENCES

Albert, M.B., 1998, *Global Patenting Trends in Five Sectors*, Washington, DC: US Department of Commerce.

Amsden, A., 1989, *Asia's Next Giant, South Korea and Late Industrialisation*, Oxford: Oxford University Press.

Chung, S.C. and L.M. Branscomb, 1996, Technology Transfer and International Cooperation, in Branscomb and Choi (eds), *Korea at the Turning Point*, Westport: Praeger.

Chung, S.C. and J. Jang, 1993, *Gisul Gaebal Tujaeu Kyungjejok Hyogwa Bunsuk* (The Economic Effects of R&D) (Korean), Policy research 93-04, Seoul: STEPI.

Hobday, M., 1995, *Innovation in East Asia: The Challenge to Japan*, Aldershot: Edward Elgar.

Kim, K., 1987, *Korea in the 1990s: Making the Transition to a Developed Economy*, Mimeo, Seoul: The Ilhae Foundation.

Kim, L. and C. Dahlman, 1992, Technology Policy for Industrialisation: An Integrative Framework and Korea's Experience, *Research Policy* 21, 437-452.

KITA, 1997, 1998, *Major Indicators of Industrial Technology*, Seoul.

Lee, W.Y. and J. Jang, 1998, *Sanup Gisul Gaebal Jiwon Jedoeu Jonghap Pyonggawa Gaesun Bangan* (Evaluation of the Government Supports for Industrial Innovation and Policy Implication) (Korean), Policy research 98-10, Seoul: STEPI.

Leipziger, D.M. and V. Thomas, 1993, *The Lessons of East Asia, An Overview of Country Experiences*, Washington, DC: World Bank.

Ministry of Education, 1998, *Statistical Yearbook of Education*, Seoul.

MOST, 1996, 1997a, *Report on the Survey of R&D in Science and Technology*, Seoul: MOST.

MOST, 1997b, *Tukjung Yongu Gaebal Saub Shihaeng 15 Nyun* (Fifteen Years of the National R&D Programme) (Korean), Seoul: MOST.

MOST, 1998a, *97 Nyun Gukka Yungu Gaebal Saub Chosa, Bunsuk, Pyungga Gyulgwa* (Analysis of the 1997 National R&D Programmes) (Korean), Seoul: MOST.

MOST, 1998b, *1997 Report on the Survey of R&D in Science and Technology*, Seoul: MOST.

MOST, 1999, *Jungchaekeu Gibon Banghyang-gwa Chujin Jollyak* (The Basic Direction of Policy and Strategies) (Korean), A Document submitted to the National Science and Technology Council, March 1999.

NSTC, 1999, *2000 Nyundo Gukka Yungu Gaebal Yesan Sajun Jojung Gyulgwa* (Adjustment of the FY 2000 R&D Budget) (Korean), an agenda document for the meeting of NSTC of March.

Nadiri, I.M., 1993, *Output and Labor Productivity, R&D Expenditure and Catch-up Scenarios: A Comparison of the US, Japanese, and Korean Manufacturing Sectors*, Mimeo, New York University: Department of Economics.

OECD, 1996, *Review of National Science and Technology Policy, Republic of Korea*, Paris: OECD.

OECD, 1998, *Basic Science and Technology Statistics*, Paris: OECD.

Shin, T. and H. Kim, 1994, Research Foresight Activities in Korea and Technological Development: S&T Policies in National R&D Programme, *Technology Forecasting and Social Changes*, 45, 31-45.

CHAPTER 6

THE EMERGENCE OF A NEW EUROPEAN UNION RESEARCH AND INNOVATION POLICY[1]

PARASKEVAS CARACOSTAS AND UGUR MULDUR

INTRODUCTION

Europe is on the move. The advent of a common currency, the Euro, has given a new impetus to European economic and political integration. Growth and job creation are at the centre of the debate between Heads of state and governments. There is strong political support in Europe for research, innovation and learning as key dimensions of a strategy for economic growth and social cohesion.

But what does a European Union mean? What are the foundations, the achievements, the perspectives of research and innovation policy institutions, which translate the political projects of 15 European countries to share their resources and efforts for the common good? What is specific in the objectives, decision-making procedures and ways of operating government research activities at a transnational, European level of governance?

The purpose of this chapter is to provide an overview on the recent evolution of European Union (EU) research and innovation policy in its political and economic context. It is organised in five parts. The first recalls briefly the main elements of the co-evolution between EU research and innovation policy and the European integration process. The second draws a synthetic picture of the EU as a global actor in research and innovation. Thirdly, the

1. The views expressed here are those of the authors and may not in any circumstances be regarded as stating an official position of the European Commission.

evolution of research and innovation policy institutions, resources and achievements are presented. The fourth part attempts to review the rationale and shaping of the current phase of this EU policy characterised by a problem-solving and systemic approach, and to assess how EU research and innovation policy has coped with the double trend of globalisation and subnational regionalisation. The fifth part explains how the policy process works at EU level. Finally the conclusions sum up the achievements of the EU in the 1990s and mention some of the challenges ahead.

1. RESEARCH AND INNOVATION POLICY AS A KEY DIMENSION OF EUROPEAN INTEGRATION

A research and innovation policy carried out at European level necessarily implies that it is part of the broader European integration process. The analysis should therefore start by describing the main phases of this process (Section 1.1), analyse the co-evolution involved (Section 1.2) and explain the "subsidiarity" principle which helps to define what needs to be done in common by the 15 member states (Section 1.3).

1.1 – From Rome to Amsterdam: "an ever closer union among the peoples of Europe"

In the aftermath of World War II it was crucial to have peace in Europe. At that time, Jean Monnet had the idea of pooling European countries' economic interests. His aim was political, rather than economic. In his memoirs he points out that "the political unity of tomorrow will depend on making the economic union effective in everyday activities".

A lot has been done since then. In 1957, the Treaty of Rome set up the European Economic Community. In 1987, the Single Act gave fresh impetus to the internal market, after which the Maastricht Treaty of 1993 paved the way for Economic and Monetary Union. It also introduced the pillar of European foreign policy and the desire to create an area of free movement in safety. On 17 June 1997 the leaders of every member state of the European Union agreed on a new Treaty for Europe, the Treaty of Amsterdam. After the official signing of the texts, the new provisions were publicly debated in each country and submitted to the electorate for approval, either by referendum or by parliamentary decision. The new Treaty entered into force on 1 May 1999.

What is European integration? What is the European Union? The European Union is the product of a political choice: to stop unending wars between neighbours with a different culture, wars that led Europe to lose its world leadership. This political project, unique in the world, of institutionalising co-operation between 6, 9, 10, 12, today 15 and tomorrow more than 20 states has been realised by the creation of multiple economic interdependencies. Its motors now are the creation of an Economic and Monetary Union, increasing co-operation in the fields of internal affairs and justice as well as of foreign and security policies. The direction is clear, and is specified in the first articles of the new Amsterdam Treaty:

"This Treaty marks a new stage in the process of creating an ever closer union among the peoples of Europe, in which decisions are taken as openly as possible and as closely as possible to the citizen. The Union shall be founded on the European Communities, supplemented by the policies and forms of co-operation established by this Treaty. Its tasks shall be to organise, in a manner demonstrating consistency and solidarity, relations between the member states and between their peoples". (Article 1).

"The Union shall set itself the following objectives: to promote economic and social progress and a high level of employment and to achieve balanced and sustainable development, in particular through the creation of an area without internal frontiers, through the strengthening of economic and social cohesion and through the establishment of economic and monetary union, ultimately including a single currency in accordance with the provisions of this Treaty; (...)" (Article 2)

The EU is therefore based on Treaties and secondary legislation derived from the Treaties (for example, Directives which are similar to framework laws at national level, and action programmes decided for funding specific co-operation activities). The most recent Treaty, the Treaty of Amsterdam, streamlined the previous legal bases of the EU in two Treaties:

- the Treaty on the European Union (which covers matters such as common foreign and security policy or common internal affairs and justice policy);
- the Treaty instituting the European Community (which covers the rest, i.e. the Community policies and institutional framework).[1]

1. We will use loosely throughout this chapter, the expressions "European Union", "Union", "European Community", "Community" to designate activities carried out by the European Union.

1.2 – The co-evolution of European integration and research policy

Research and innovation policy started fully in the mid-1980s, although the European Community had been active in this domain since its creation. Without specific legal basis – apart from the measures foreseen by the EURATOM (European Atomic Energy Community) Treaty, still valid today, from one article in the EEC (European Economic Community) Treaty concerning agricultural research and one article in the ECSC (European Coal and Steel Community) Treaty concerning research in the field of coal and steel – these activities were developed on the basis of Article 235 of the EEC Treaty, which allows actions to be taken when it is necessary to realise one Community objective.[1]

In 1987, the Single European Act introduced into the EEC Treaty a new Title relative to research and technological development (RTD). Its articles define the objective – "to strengthen the scientific and technological bases of European industry and to favour the development of its international competitiveness" – as well as the shape of Community action in this domain. The Maastricht Treaty, in 1993, by extending the objective of Community RTD to the promotion of activities "deemed necessary in nature of other Chapters of the present Treaty", has provided an explicit basis for the research activities in all fields where the Union is competent. It also strengthened the democratic legitimacy of EU research activities by stipulating that multi-annual RTD Framework Programmes would be adopted by a co-decision process, involving both the Council of Ministers and the European Parliament.

Finally the Amsterdam Treaty, signed in 1997, brought a significant improvement in the decision-making process for the adoption of Framework Programmes, by substituting the rule of unanimous voting with qualified majority voting.

Table 6.1 (in Annex) summarises the main elements of the co-evolution of EU research and innovation policy and the general European integration process. It is clear that the focus of the Second and Third Framework Programmes on industrial competitiveness, as an objective and an emphasis on generic (key) technologies, as a privileged means of achieving the objective, is related to the preoccupations of the 1980s concerning European integration through the achievement of a Single European Market. Similarly,

---

1. An history of European research institutions can be found in Guzzetti's work (1995), analysed by Caracostas and Soete (1997) in a "systems of innovation" approach.

the current emphasis in FP5 on societal objectives, matching the needs of society with the potential of research and industry, cannot be understood without taking into account the current stage of European integration: a stage where political union is on the agenda through EMU, enlargement, a pressure for a stronger common foreign and security policy, and an increasing focus on a citizen's Europe.[1]

1.3 – The European Union is not a state: the principle of "subsidiarity" determines the activities common to the 15 member states

How to define what the 15 member states need to do together? How to avoid too much European legislative and other activity when common objectives could be achieved by doing less at EU level? The subsidiarity principle is intended to ensure that decisions are taken as closely as possible to the citizen and that constant checks are made as to whether action at Community level is justified in the light of the possibilities available at national, regional or local level. Specifically, it is the principle whereby the Union does not take action (except in the areas which fall within its exclusive competence) unless it is more effective than action taken at national, regional or local level. It is closely bound up with the principles of proportionality and necessity, which require that any action by the Union should not go beyond what is necessary to achieve the objectives of the Treaty.

The main grounds for opting for Community action are described in the second paragraph of Article 3b of the EC Treaty, which states that the Community shall take action "only if and in so far as the objectives of the proposed action cannot be sufficiently achieved by the member states and can therefore, by reason of the scale or effects of the proposed action, be better achieved by the Community". Three criteria can be used for judging whether this condition has been fulfilled:

---

1. Caracostas and Muldur (1997) distinguish three phases of research and innovation policies in industrial countries after World War II. Phase 1 (1950-1975) is characterised by the pairing of defence objectives and basic science as a means of achieving them. During phase 2 (1975-1995) the emphasis is put on the objective of reinforcing industrial competitiveness where key technologies seem to be the central focus. Phase 3, starting in the mid-1990s, is characterised by a priority on societal objectives and an emphasis on innovation in order to meet these objectives. Phases 2 and 3 seem to correspond also quite well with the evolution of the EU research and innovation policies from the mid-1980s to the current trends.

- the issue under consideration has transnational aspects which cannot be satisfactorily regulated by the member states;
- action by member states or lack of action would conflict with the requirements of the Treaty;
- action at Community level would produce clear benefits.

Article 163 of the EC Treaty sets out some general objectives for research policy at European level:

- to strengthen the scientific and technological bases of Community industry;
- to encourage it to become more internationally competitive;
- to support other Community policies.

Four types of activity are envisaged (Article 164):

- research, technological development and demonstration programmes;
- international scientific co-operation;
- dissemination and optimisation of results;
- training and mobility of researchers.

The Treaty foresees (Article 166) that all EU activities in the field of research are included in a multi-annual (five-year) Framework Programme, a strategic decision defining the science and technology priorities, the main activities to be carried out, the means and modalities of EU funding of these activities. The Framework Programme is implemented by specific programmes specifying the broad strategic lines it defines for a five-year period.

The specific nature of research activities carried out at EU level can finally be illustrated by the type of projects that are supported. EU research policy:

- is not a policy of subsidies; it co-funds projects mobilising national public and private funds; the focus is on multinational and multi-actor co-operations;
- is not a redistributive policy; the projects are selected on the basis of specific scientific excellence and socio-economic relevance criteria;
- is not a policy of permanent funding of national research organisations; only co-operative projects of a limited duration are concerned.

Two main types of activities are funded by the EU:

- indirect activities (such as shared cost research and demonstration projects, access to scientific infrastructures; SME-related measures; training grants; network funding, studies and conferences);
- direct activities carried out by the EU Joint Research Centre (research and scientific and technological support to EU policies).

2. THE EU AS A GLOBAL AND DIVERSE ACTOR IN RESEARCH AND INNOVATION

In order to understand the changes in EU research and innovation policy, it is necessary to get a clear picture of the European macroeconomic context (Section 2.1) and of the situation of Europe as a global research player (Section 2.2). This global comparison can then be supplemented by an analysis of European diversity in research and innovation (Section 2.3).

2.1 – The macroeconomic context: the challenges of unemployment and ageing

Between 1986 and 1990, prior to the Maastricht Treaty, the average growth rate of the European economy was 3.3% a year, real interest rates were falling, the average rate of gross capital formation was close to 6%, consumption by private households rose by 3.7% a year and unemployment was either stable or declining in all the member states. During the crisis of 1992-1994, in the period immediately following the Gulf War and German reunification and marked by monetary upheavals, the European economic machine shuddered to a halt and the Union lost some 5.5 million jobs, i.e. more than half the number it had created during the previous eight years.

Thanks to member states' efforts to respect the convergence criteria of the Treaty, new albeit tentative signs are emerging of a return to growth and the creation of jobs in the near future. The world economic climate is favourable to the development of European investment and exports; the basic parameters of the European economy are good: low inflation, moderate wage demands, good investment returns. Monetary policies have become more flexible, long-term interest rates are down, confidence in exchange rates has returned, budget deficits have been cut. These recent trends in the European economy explain why a majority of the member states were able, from 1997, to meet the five convergence criteria and were ready to be part of Economic and Monetary Union (EMU) on the basis of the original timetable, i.e. starting on 1 January 1999.

However, even the most optimistic forecasts for growth and employment hold out no hope of a rapid reduction in unemployment levels (see Figure 6.1). It thus behoves us to ensure that the routes to long-term development and growth, which the European economies follow, are secure, which means that short-term political and economic priorities must not take precedence over long-term imperatives.

FIGURE 6.1 – Pattern of unemployment 1975-2000

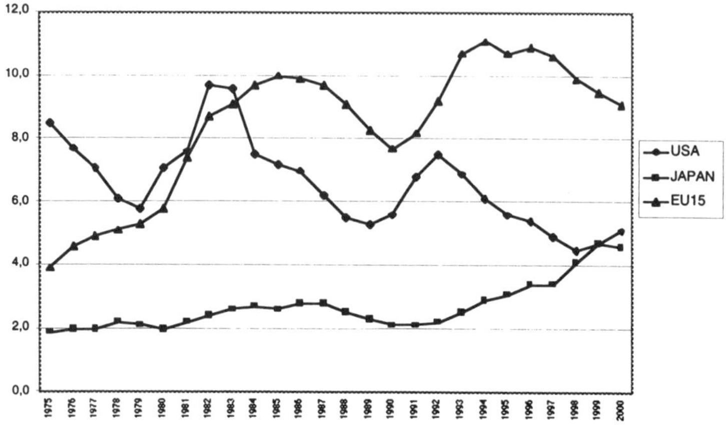

Source: EUROSTAT (NEWCRONOS database).

There is another challenge which many experts see as irreversible at the human level and which shows even more that Europe must innovate: the gradual ageing of Western societies. Europeans, who have colonised other continents throughout the ages, will increasingly be a very small minority, less than 5% of the world's population, which is expected to stabilise at around 8 billion around the third decade of the 21st century. Many recent studies by international organisations also suggest that Europe's population will also be one of the oldest in the world (World Bank, 1994; Eurostat, 1994; and United Nations, 1994).

Studies by the Commission clearly show that it would doubtless be unrealistic to hope that unemployment rates will fall significantly as a result solely of changes in the population structure.

On the other hand, if proper account is not taken of it straight away, this ageing of the population could lead to major dysfunction in the economies and societies of Europe. For example, the ability to sustain the European social model (which includes institutionalised solidarity between generations) might also be jeopardised by the gradual ageing of European societies.

In order to maintain its current levels of industrial production and average *per capita* incomes, Europe would need to double, even treble, the growth of its productivity in the EU over the next few decades. But a productivity increase of this kind will only be possible if Europeans invest hugely, and now, in research and new technologies, at the same time improving the training and skills of their workforce and organising work in their companies more effectively.

2.2 – Europe, research and innovation: insufficient investments, inefficient allocation

The stagnation in the total R&D expenditure observed in most industrialised countries since the mid-1980s, and indeed the relative decline in the case of some countries, together with the marked rise in research and innovation costs, might lead in the next few decades to a slowing of technical progress which could damage the growth of productivity and jobs in the European Union.

Europe invests relatively less than the US and Japan in RTD and human resources

Measuring total R&D expenditure against the total population and the GDP of each country, we find that the EU lags behind its main competitors. Compared with the total population, it is even more obviously behind: the European Union invests 346 in purchasing power standard ECU (PPS) *per capita* in R&D, that is to say almost half of what the US and Japan spend, (752 and 648 PPS *per capita* respectively). Of the member states, only Sweden has *per capita* R&D expenditure close to that of Japan and the US. Using this indicator, there would appear to be wide differences between the member states. Analysed over a 15-year period, they show that the gaps in R&D investment *per capita* between the three world economic powers widened in favour of Japan and the US.

Comparative analysis of R&D expenditure compared to GDP in the three economic powers also confirms that the European Union lags behind. In 1997 total R&D expenditure (in ECU PPS) was 1.82% of EU GDP compared with 2.73% for the US and 2.89% for Japan. If total defence-related R&D expenditure is excluded, the gap clearly widens in favour of Japan: between 1993 and 1997 civilian R&D expenditure calculated as a percentage of GDP averaged 1.7% in the EU, 2.1% in the US and 2.7% in Japan.

The wealth of human resources devoted to R&D and, particularly, the number of researchers and engineers available, is a key element in predicting how well each national system of innovation will be able to respond to the challenges of globalisation and the fledgling information society. The only way to ascertain whether the EU has enough scientific and technological human resources is also to compare it against its main competitors. It is precisely here that one of the EU's major weaknesses becomes apparent: its proportion of researchers (total number of R&D researchers and engineers per thousand employees) was 5.0 in the EU in 1996, as against 7.4 in the US in 1993 (the latest year available) and 8.5 in Japan.

The Union's disadvantaged position is even more striking when one compares the number of researchers working for European companies with that of their US and Japanese competitors: companies in the 15 member states employ a total of just 407 000 R&D researchers and engineers compared with 919 000 and 404 000 in US and Japanese firms respectively. These figures, as a proportion of each country's working population, highlight the scale of this competitive disadvantage for the EU, which has 2.4 in-house researchers per thousand employees compared with 6.7 in the US and 6.0 in Japan.

The EU thus needs to create more research jobs in R&D and particularly in the business enterprise sector. To make up the ground currently lost to their competitors, European companies would need to create 600 000 jobs for R&D researchers and engineers straight away. Bearing in mind associated jobs (such as technicians and administrators), the shortfall in European R&D jobs could easily be put at close to one million.

In Europe, the trend in both state and private enterprise is towards less rather than more research effort. The EU's investment in research and innovation does not match either that of its main international competitors or its own economic, social and political aspirations. Throughout the post-war period the proportion of total R&D funded by the US Federal Government was greater than that funded by US companies and the proportion of public funding in most of the EU member states. In the 1980s, the percentage of R&D funded by the Federal Government was always higher than the average for the 15 member states of the European Union. Towards the end of Ronald Reagan's second Presidential term, the gap between the scale of public funding in the US and the European Union had actually widened: in 1987, 49.1% of total R&D expenditure was funded by the US Federal Government as against just 42% in the EU and 19.6% in Japan. Not until the beginning of the present decade did the total proportion of public funding in the US drop to a level close to that generally seen in Europe.

This indicator of public assistance for R&D is slightly misleading, however, given the importance of the public R&D sector in Europe, which is of course government-funded. From the point of view of competition it makes better sense to compare the levels of public funding of R&D carried out by industry. According to OECD data, in 1989 the percentage of expenditure on R&D carried out by companies and funded by the state was 28% in the US compared with 14% in the Europe of 15. Admittedly, the level of public funding of R&D carried out by companies fell more in the US than in Europe.

But it is still much higher than that in the Union: in 1997 this percentage was 15.2% in the US and 9.5% on average in the EU (13.1% in France, 9.7% in the United Kingdom, 11.9% in Italy and 8.9% in Germany).

This discrepancy should not be underestimated: because R&D expenditure by US companies is higher on average by 41% than that of European companies (139 billion ECU in the US against 85 billion ECU in the Union in 1997), this represents a colossal infusion of US federal funds into industry every year.

Between 1990 and 1996, the US Federal Government spent 137 billion ECU on supporting R&D by US companies (not counting non-federal public aid). In comparison, 64 billion ECU was spent over the same period by the 15 member states and the European Union Programmes (Third and Fourth Framework Programmes plus the proportion of the Structural Funds spent on financial research infrastructures) to fund R&D activities by European companies. This difference of 73 billion ECU represents a substantial comparative advantage for US firms in world technological competition.

Europe is less effective than its competitors in exploiting its excellent scientific potential to best economic and social advantage

The "European paradox"

As we have seen, every year the European Union invests a smaller percentage of its GDP in R&D than its main competitors, and its human resources as a proportion of its total population are far smaller. Some will say it is hardly surprising that it should underperform the US and Japan in science, technology and commerce. However, when we compare the EU's scientific and technological results against its own investment in this area, we can also identify worrying trends, which need to be corrected quickly.

For example, Europe's scientific performance in relation to its investment in this field is excellent and on a par with, if not better than, that of the US and Japan. However, its technological and commercial performance in the leading-edge sectors, in relation to its investment in industrial R&D, has unfailingly declined since the mid-1980s. These facts have been encapsulated in the notion of a "European paradox" analysed in the first European Report on Science and Technology Indicators (1994) and mentioned in the European Commission's Green Paper on innovation (1995) (see Figure 6.2 below).

This paradox reflects, first of all, one of the striking weaknesses of the European innovation system. Although the scientific base is excellent, the technological and economic results are not only poorer than those of Europe's main competitors, they are also disproportionately small compared with the scale of its own investment in this area.

Scientific performance. Studies of scientific writings show that the scientific performance of the EU is by and large excellent, on a par with that of the US but broadly better than that of Japan and the other industrialised countries. The total EU share of world scientific publications has been relatively stable since 1980. Scientific publications by European researchers account for more than 30% of world publications. Over the same period, the US share has remained at around 35%. That of Japan is rising slightly but it is well below that of European and US researchers (9% in 1995).

Technological performance. Specialists regard patents as one of the best indicators of technological performance in the industrialised countries. The following indicators are calculated from data supplied by the US Patent Office (USPO) and the European Patent Office (EPO), which are the two most extensive patenting systems in the world. The EU's share of patents granted in the US has fallen back markedly in recent years: from being more than 20% in the early 1980s, it fell to less than 16% in 1995. After experiencing a decline for most of the early 1990s, the EU shares of patents in the US for almost all the technology sectors showed a modest increase in 1995. That of the US, however, which had fallen from 60% in 1980 to 52% in 1990, grew during the first half of this decade and by 1995 was 55% of the total. The shares of almost all the Asian countries also rose in the same period. Japanese companies, which in 1980 accounted for only 12% of patents granted in the US, accounted in 1995 for 21% of all US patents. Over the last 15 years the Europeans have fared little better at home, since the share of the EU in patents filed in Europe decreased overall from about 57% at the beginning of the 1980s to 44% in 1995, whilst the US improved its share, which rose over the same period from 26% to 34% of all patents filed in Europe. The Japanese for their part have increased their share from 9% of all patents filed in Europe in 1981 to 14% in 1995.

Trade performance in high-tech fields

The EU's commercial performance improved considerably in recent years: its market share of exports of goods and services was one of the highest in the world and its export–import balance has been in surplus for a number of years. Unquestionably, the high-tech products and services, which play a part in intermediate consumption in virtually all sectors of the economy, help indirectly to boost the good results recorded by these sectors in foreign trade.

But the direct contribution of European high-tech industry to improvement of the Union's commercial performance, measured in terms of the proportion of total EU manufacturing exports provided by high-tech products, was relatively small compared with those in the US and Japan. It is a well-known fact, however, that international trade in high-tech products is the most dynamic component of world foreign trade. Between 1985 and 1995, world exports of high-tech products grew, in constant ECU terms, by 50%. The highest rates of growth were recorded by Asian countries like China, Thailand, Indonesia, Malaysia and Singapore.

The EU is the only economic superpower to show a permanent deficit in its world trade in high-tech products, of more than 14 billion ECU year on year. The European Union's trade deficit with Japan and the US, its two main trading partners, is very much greater in this area.

2.3 – The internal diversity of the EU research landscape: weakness or strength?

It will be noted, however, that the EU is treated above as a single unit. The contrasts are far greater when one analyses the situation in terms of member states individually. We see then that this phenomenon, the "European paradox", is an aggregation of widely differing national situations: scientific and technological performance contrasts very starkly in countries like the UK, Denmark, Ireland, Greece, Sweden, Spain, Belgium and Portugal. France and Italy each have a scientific and technological performance proportionate to that of the other but lower than the European average.

In the four other member states, Germany, the Netherlands, Finland and Austria, the existence of this paradox between scientific and technological performance cannot really be proven since these four member states achieve very good results, in terms of patents, compared with their companies' R&D investment and their scientific performance (Figure 6.2). Japan stands out from the other countries in combining a mediocre scientific performance with a technological performance which is broadly better than theirs.

The great diversity of the Union countries in terms of the resources spent on research, the structure of these resources, the scale of relations between universities and industry or of the institutional and organisational architecture of the public system of support for research and innovation (centralisation, federalisation, regionalisation) may be explained by a long history of national institutions and traditions. Each national system of innovation in Europe has different sectoral specialisations, a variety of corporate

FIGURE 6.2 – The European paradox, manifest at member state level: a potential strength

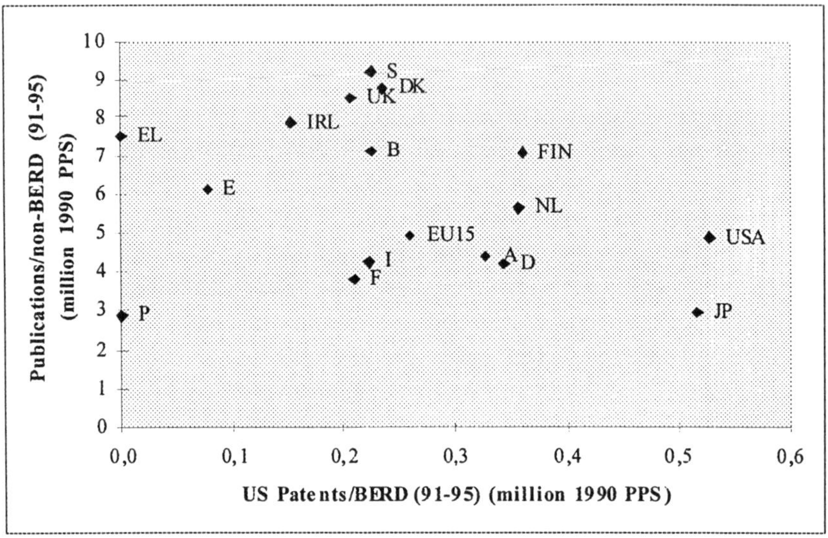

Source: European Commission, Second Science and Technology Indicators Report, 1997.

governance rules and routines, varying institutional mechanisms and highly contrasting performances. The Second European Report on Science and Technology Indicators (European Commission, 1997) provided a good summary of the most recent and comparable data available in the various member states.

The analyses of Amable, Barré and Boyer (1997) have provided for a typology of "social systems of innovation". We shall quote just a few elements of it. On the basis of a sample of eight European countries (France, Italy, Germany, United Kingdom, Netherlands, Sweden, Finland and Norway) and four non-European countries (US, Canada, Australia and Japan), the authors painted a vast comparative canvas of the industrialised economies using 179 parameters. The results are as follows:

- The European countries studied belong to three of the fundamental types of social systems of innovation: the "market" system in the case of the United Kingdom, the "government-regulated" system in the case of France, Italy, Germany and the Netherlands, the "social democrat" system in the case of Sweden, Finland and Norway (which is a member not of the European Union but only of the European Economic Area);

– They also fall into different technological and industrial specialisation groups: France and the United Kingdom into the group which is drawn towards high-tech and the state-controlled sectors (aerospace, pharmaceuticals, defence), the Scandinavian countries into the group based on resource-intensive industries; Germany is rather more like Japan in having a specialisation in medium technology sectors; plus a specialisation in automobiles and machine tools;
– Their performance in terms of innovation or macroeconomic growth varies greatly even though, since the early 1990s, most of them have experienced slow growth and increased unemployment.

According to the authors: "[this diversity] is a strength in terms of innovation and production since in response to international or internal changes Europe can find from within itself the impetuses for fresh competitiveness simply by developing those aspects of its systems which are best adapted to the economic conditions and the imperatives of the time ...".

3. THE EMERGENCE AND IMPACT OF A EUROPEAN SYSTEM OF RESEARCH AND INNOVATION

How has the European integration process contributed to the emergence of a "European system of innovation"? What is the relative financial importance of EU efforts devoted to R&D and innovation? What is their impact on member states' RTD fabric? This section attempts to provide some answers to these questions (respectively Sections 3.1, 3.2 and 3.3).

3.1 – An embryonic European system of innovation: gradually taking shape and accelerating over the last 15 years[1]

Simply to note this fact of European diversity would be to ignore more than 40 years of European integration as regards research and innovation policies. It is broadly agreed (see Caracostas and Soete, 1997) that this process of Europeanising national systems of innovation is largely initiated by the development of a number of simultaneous or successive trends towards

1. In this section the term "institution" is defined as "the humanly devised constraints that shape human interaction" (North, 1990) or, as Edquist and Johnson put it (1997), "a set of common habits, routines, established practices, rules, or laws that regulate the relations and interactions between individuals and groups".

convergence and thus of a stratum of activity which adds to the national measures taken by governments or the private sector. Seven areas can be cited in which public action now takes place at a European level, thus also influencing the capacity for learning and innovation of social and economic players and institutions operating at national level:

- The European Single Market project, launched by the European Economic Community in the late 1950s, became a reality in the early 1990s as the machinery of national markets was deregulated through mutual recognition and to a lesser extent harmonisation of the legislative and regulatory framework of the member states (European Commission, 1997b).
- Globalisation and the development of world trade policy institutions (World Trade Organisation and its agreements) also make it necessary to strengthen the policy instruments shared by the European nations so that they can all have equal weight in negotiations with third countries.
- The Single Market entails a strengthening of European competition policy (scrutiny of state aids, regulation of mergers and takeovers, scrutiny of cartels and inter-company agreements) now extending into negotiations at world level (within the WTO).
- Enlargement of the European Community in the 1980s to take in the less industrialised countries of Southern Europe led to a swift increase from 1988 onwards in instruments of redistribution (of the federal type) between the continent's wealthy and poorer regions (cf. the major objective of "reinforcing economic and social cohesion" and the Structural Funds which co-finance national measures to help countries catch up and convert sectors of their economy). This new redistributive function gave considerable impetus to institutional and organisational learning in the so-called "cohesion countries" (Ireland, Greece, Spain and Portugal) which had to adapt to new rules of partnership with the Community in order to benefit from the Structural Funds.
- The emergence of a European policy of spatial planning is reflected in public–private joint financing of some of the large trans-European networks in transport, telecommunications and energy, which are intended to speed up the access of economic players and citizens in member states to the potential of the Single Market.
- The rapid development, quoted above, in the 1980s and 1990s of European measures and institutions for research and innovation

policy, the Framework Programmes of research and the programmes carried out under them undeniably helped, together with intergovernmental initiatives (EUREKA, Airbus, Arianespace, and so on) to create the nucleus of an embryonic European system of research and innovation.

– In the run-up to the 21st century, Economic and Monetary Union is the most powerful force for "Europeanising" the national policies governing member states' systems of innovation since the Single Market came into being in the early 1990s, as it provides new opportunities for financing research and innovation activities and calls for an accompanying "common European research space".

On the other hand, three components, of critical importance to the working of a national system of innovation, were less structured at European level until the mid-1990s:

– Financial institutions specialising in funding innovation: such institutions are now developing (for example EASDAQ).
– Intense social dialogue and regular discussions between social players: the focus on employment brought about by the Amsterdam Treaty, the recent initiation of the "Luxembourg process" and the decision on the "employment pact" in June 1999 demonstrate that a real European employment strategy is emerging.
– European legislation and policy on education (still a matter of national sovereignty): the inclusion in the Union Treaty of a chapter on training and the fact that numerous initiatives of European co-operation in these areas have been launched justify a measure of optimism.

Thus one may conclude that we have an increasingly unified European market (with an Economic and Monetary Union covering 11 states), which is governed by common rules of competition and trade policy and accompanied by fledgling structural policy institutions (research, innovation, social and regional development). This fledgling European system of innovation frames (and is framed by) specific rules at a supranational level, rules which are both formal (Treaties, European Programmes, Directives, and so on) and informal (networks of players, shared habits, *de facto* standards).

One can hypothesise that the heart of this system is made up of the organisations and institutions which govern the creation, dissemination and use of knowledge and know-how co-operatively across borders, in other words the international rules for research, innovation and the development of capabilities and skills.

Until the mid-1990s the institutional rules and initiatives concerned with the heart of a fledgling European system of innovation were not sufficiently integrated as part of an overall vision. The growing influence of new approaches to innovation and the revival of political debate on sustainable growth and employment in Europe, in the wake of the European Commission's 1993 White Paper on growth, competitiveness and employment made it possible, for the problem to be readdressed from a different angle from 1995 onwards (see Section 4.1. hereafter).

In parallel to the development of EU institutions in the field of RTD, there has been, since the 1950s, a multiplication of European research organisations working on the basis of "intergovernmental" rules and associating, for each of them, a different set of European countries.[1] Amongst them, one needs to quote: the EUREKA initiative launched in 1985 and financing pre-competitive RTD projects according to a bottom-up industrial co-operation approach; COST (Co-operation in Science and Technology) co-ordinating national projects in a "variable geometry" fashion; ESA (the European Space Agency) which funds both compulsory and *à la carte* co-operation projects in the field of space research and technology; CERN (the European research centre for particle physics) and the European Science Foundation which groups 65 organisations of 22 countries and manages programmes and networks.

The European Union is involved in many of these ventures: for example, it is participating in nearly 40 EUREKA projects; the Commission hold the technical secretariat of COST and it co-operates with ESA in the Galileo project aiming at developing a European satellite navigation system. Other European (intergovernmental) organisations such as the European Molecular Biology Laboratory (EMBL) participate in EU research programmes. A future challenge for EU research policy will be to strengthen the links between EU activities and these diverse organisations in order to consolidate further and "rationalise" the emerging European system of research and innovation[2] (see conclusions below).

---

1. For a complete review of these schemes, see Chapter 10, Second European Report on Science and Technology Indicators.
2. For a discussion of this topic, see Caracostas and Muldur (1997), Chapter 6, section 6.3.

3.2 – Resources: important and increasing financial efforts

Community research and technological development activities (RTD) developed rapidly from the mid-1980s onwards, especially after the introduction of the first Framework Programmes. Annual RTD expenditure (according to the Frascati manual definition and not including the Structural Funds) rose from 573 million ECU in 1985 to 2664 million ECU in 1997. This was equivalent to 4.9% (1997) as opposed to 1.6% (1985) of the total budgetary appropriations for R&D by the 15 member states. The budget for the Fourth Framework Programme on RTD was increased subsequent to enlargement of the EU from 12 to 15 member states. This percentage might be estimated between 5.5% and 6.5% by the end of the Fourth Framework Programme, i.e. in 1998, based on the predictions for public R&D appropriations in member states, which have entered a period of restrictions. The Fifth Framework Programme, with a budget of nearly 15 billion Euro (an increase of 4.5% over the Fourth Framework Programme) should normally make it possible to continue this positive evolution from the point of view of European integration.

If compared to the national research budgets, which are not allocated to the permanent funding of research organisations, i.e. to "mobilising" or "incentive" (programme or project) funding, the percentage of EU financing might be much higher. Thus one can say that, even in large member states, the importance of European research programmes is non-negligible, especially in relation to emerging or new fields of research.

Two other indicators allow us to trace the pattern of Community expenditure on RTD in recent years. The first relates this spending to the total GDP of the 15 member states. The level of Community expenditure on RTD (not counting the Structural Funds) remains low, despite a rapid rise between 1985 and 1997: Community expenditure on R&D, which was 0.016% of European GDP in 1985, had increased to 0.037% of European GDP by 1997.

The second indicator is arrived at by comparing Community expenditure on RTD (with or without the proportion of the Structural Funds allocated to R&D) with the total Community budget. This indicator shows us that the funds for the RTD Framework Programmes (Frascati definition) accounted for 3.0% of the total Community budget in 1998 as against 2.0% in 1985.

The average percentage covered by the Fourth RTD Framework Programme for the period 1995-1998 is estimated at about 3.1% of the total

EU budget. If the proportion of the Structural Funds earmarked for R&D is taken into account, this percentage rises to 4.9%.

3.3 – Impacts (1984-1994): the creation and consolidation of transnational research networks

The advantages of European research co-operation (1984-1994)

The Framework Programme is an effective instrument for encouraging companies and public and private research centres to invest, co-operate and create more jobs in the field of R&D. There are many advantages (economic, as well as scientific and technological) in developing European co-operation amongst companies, universities and public and private research centres. The great majority of these advantages have been put under the microscope and analysed in depth in numerous official and academic reports (see, for an extensive review, the Second European Report on Science and Technology Indicators, 1997). We shall confine ourselves here to a brief presentation of the various advantages which often follow from the original nature of the Framework Programme's features.

A large majority of R&D laboratories, in both industry and universities, state that their staffing levels have risen as a result of projects funded by the European Union: 79% of participants in Ireland, 74% in Belgium, 71% in Greece, 67% in Denmark, 54% in Portugal, 49% in Spain, 36% in France, and so on. It is also clear that additional personnel taken on for the project were kept on for the longer term even after the Community project had been completed.

About 8% of total personnel engaged in R&D in the 15 member states, i.e. 125 000 people, are involved in one way or another in research projects funded by the Framework Programme. If account is also taken of other Community activities, notably the Structural Funds, we easily arrive at a figure of 150 000 people.

Of all the benefits derived from participation in Community projects, European researchers most frequently mention the acquisition of new knowledge and capabilities as a result of co-operation. It is, however, extremely difficult to set a value on this contribution which the Framework Programme makes to European R&D. Together with specific measures under the Human Capital and Mobility programme, all projects under the Framework Programme are thus contributing to the development of human resources in science and technology both quantitatively and qualitatively.

Between 1987 and 1995, Community R&D initiatives led to the forging of over 150 000 co-operative links between large companies, SMEs, universities and public and private research centres in the European Economic Area. This dynamic of scientific and technological co-operation, which the Framework Programmes created amongst European players in R&D, strengthened the fabric of the European research and innovation area and improved its scientific and technological performance. In surveys conducted after the programmes were completed, the majority of participants stated that their co-operation would continue.

Studies of the Second Framework Programme's impact on the scientific and technological fabric of member states also point to the important effect of Community funding on national financial incentives (as opposed to "institutional" funding which secures the existence of large public bodies and universities). An analysis of the impact of the Framework Programme on the French research system (Larédo, 1995) demonstrated how European programmes affect national research. The author showed that they helped to create new types of transnational networks between industrial firms and universities in an area which he termed "basic technological research". This type of research lies somewhere between basic research and industrial research.

A more recent study of the strategies of the 500 top European industrial companies and their participation in the Framework Programme and EUREKA (Bossard Consultants, SOGES, Management Partner, 1996) showed that Community and EUREKA funding is only a relatively small part of these companies' research budget. But its impact at micro-level on certain units or divisions of these groups is considerable and may influence the way in which their strategy operates (these groups being, naturally, broadly free to design their overall strategy as they please). The principal achievement of these European co-operative ventures seems to have been that they strengthened research partnerships and networks in Europe.

The Davignon Panel: unanimity explains a lack of strategy in previous Framework Programmes

The only overall assessment of the Framework Programme completed recently is the one finished in February 1997 by the panel of independent experts chaired by Viscount E. Davignon (Davignon, 1997). It covered Community research activities over the period 1991-1995 and involved nearly 170 European experts. This panel based its work, among others, on the

BOX 6.1 – The conclusions of the Davignon Panel

The analysis of the Davignon Panel highlighted both the strengths of Community partnerships and their weaknesses due to the institutional and decision-making framework in which they are devised and operated. Some of the main conclusions in this assessment were as follows:

- "no areas of major concern were noted regarding the quality of the research being undertaken in the programmes …
- (the Framework Programme) is not fulfilling its promise. It lacks focus and is underachieving. This is not the fault of individuals but of a structure which inhibits the formulation of real strategy and makes effective implementation difficult.
- … the programme is not flexible enough to respond to new challenges and opportunities.
- … Nor is it clearly related to the goals and objectives of the European Union.

We agree with the Commission that it is time for a major change, for a leap forward as qualitative and fundamental as the creation of the Framework Programme itself".

assessment reports for the 18 specific programmes implemented under the Fourth Framework Programme and for the Joint Research Centre.

Thus the Davignon report, whilst praising the quality of the research undertaken under the Framework Programmes and the network of capabilities set up as a result of them, puts its finger squarely on the major weaknesses of the institutional machinery of Community research: the detailed substance of Community measures is the subject of an unwieldy procedure of co-decision by the European Parliament and the Council of Ministers, the latter giving its final assent only after the member states have voted unanimously to approve them (see Section 5.2 below).

This method of decision-making slows down the implementation of a common, truly European research strategy, often resulting in unsatisfactory compromises on objectives and priorities. It makes it very difficult to adapt European initiatives regularly to changes in the scientific, technological and socio-economic scene and prevents the European executive from modernising and streamlining the operational management of these initiatives.

In publishing its proposals for the Fifth Framework Programme of research on 9 April 1997, the European Commission broadly took account of the conclusions and recommendations of the assessment reports five years on from the Fourth Framework Programme and more especially those of the Davignon Group's report.

4. TOWARDS A SYSTEMIC AND PROBLEM-SOLVING APPROACH

Against the background of the evolutions in the years 1984-1994, it is possible to recall the initiatives which, during the mid-1990s, aimed explicitly at speeding up the consolidation of a European research and innovation space at macro-level (Section 4.1) and to assess the specific measures taken at meso- (FP5, see Section 4.2) and micro- (key actions, see Section 4.3) levels.

4.1 – Consolidating the European research and innovation area at macro-level

In the last three years the Commission has taken a number of initiatives, which, together with the proposals for the Fifth Framework Programme, reflect an overall view of the challenges which Europe will face in these areas in future. Three of these initiatives can usefully be analysed in terms of what they contribute to the redrawing of European policy on research, innovation and training.

Speeding up the shift towards a learning society

The White Paper on education and training ("Teaching and learning. Towards the learning society", 1995) made a number of proposals to back up national education and training policies.

The idea that it is necessary to "revalue basic culture", "*i.e.* the ability to grasp the significance of things, to understand and create", derives from both a political approach (in a society and an economy based on information and knowledge one must not create a divide between those who know and those who do not know) and a socio-economic approach (basic culture is the foundation of a lifelong learning process). The White Paper perceived a systemic weakness in the economic valuation of practical and occupational skills other than those sanctioned by a diploma, and proposed to improve workers' geographical and occupational mobility by removing administrative, legal and fiscal obstacles and trying out new methods of validating acquired, especially informal, skills. The suggested approach to improving this validation process is an original one: firstly to set up European networks of research centres and vocational training centres, enterprises and professions which will identify the areas of knowledge which are most in demand, the key skills which are essential. This might ultimately lead to a "personal skills record card" which each individual could show to provide proof of his knowledge and know-how anywhere in the European Union.

The method of European networks, tested in the fields of education and training under the COMETT and ERASMUS programmes, and after them the SOCRATES and LEONARDO programmes, could also be applied to centres providing training and learning apprenticeships.

The White Paper contained other ambitious proposals for improving the economic evaluation of "intangible" investment in training, notably the suggestion that companies should list sums earmarked for training as intangible assets on their balance sheets. On the subject of employee demand for training, it suggested developing "training savings plans" for persons who wish to update their skills or return to training after a break in their studies.

Finally, one of the many avenues suggested is the idea, bearing in mind the need to redirect the European system of innovation towards sustainable growth based on services, that encouragement should be given to training in new service industry jobs and the training of engineers and technicians for the tertiary sector.

Reducing obstacles to innovation in Europe and fostering mobility

The Green Paper on innovation, published by the Commission in December 1995, prompted a vast debate on innovation issues in all the countries of Europe.[1] On the basis of this the European Commission drew up an Action Plan for Innovation, at the end of 1996, which sets out the measures to be taken at national and Community levels. Three objectives were singled out:

- to develop a true culture of innovation;
- to adapt the administrative, legal, financial and fiscal environment;
- to strengthen the links between research and innovation.

Society's valuation of the need for "intangible" investment is reflected in public attitudes to technological change and innovation generally. The level of technical competence exhibited by the population influences its perception of technological change in the company context. If young people are well trained in the implications of science and the new technologies this may help to attract young people into careers in science.

---

1. More than 40 000 copies of the Green Paper were distributed. Conferences and debates, involving 5000 people, were held in all the member states plus Norway and Iceland. The Commission received several hundred contributions, primarily from the business world.

The first strategic axis of the Action Plan (to promote a culture of innovation) set out the new measures to improve society's perception of innovation, the mobility of researchers, students and teachers and comparisons of situations and performances.

The second axis concerned the implementation of a legal, administrative, financial and fiscal framework favouring innovation. The European Commission has since been holding targeted debates on the questions of the Community patent, the impact of legislation on innovation and the financial and fiscal instruments needed to consolidate the European system of research and innovation. Services were set up to assist enterprises and universities in matters of intellectual property, access to venture capital and information on aids to innovation. The third axis of the Action Plan, i.e. better structural links between research and innovation, was broadly reflected in the Commission's proposals for the Fifth Framework Programme. But it is clear that the other Community instruments (the Structural Funds in particular) should give greater encouragement to innovation.

Lastly, the Green Paper on "Education, training, research: obstacles to transnational mobility" (1996) analysed the potential advantages and obstacles to transnational mobility on the part of pupils, students, young people, assistants, researchers, teachers and trainers or voluntary workers and suggested concrete measures to address these issues.

4.2 – The Fifth Framework Programme: a twofold systemic approach at meso-level

Public support for research often takes the form of research programmes. In the case of the EU policy, there is firstly a Framework Programme, i.e. a strategic document (and decision) describing the broad strategic priorities (each of them is implemented through a specific programme), the overall EU budget to be spent on research for a five-year period, the breakdown of this amount between the above-mentioned priorities and the ways and modalities according to which funding is made available to projects.

The number of specific programmes, their articulation into an overall strategy, funding for each of them, reflect the policy debate between the decision-makers (the Council of Member States Research Ministers and the European Parliament) (see Section 5 below). Since 1984, the EU has decided upon five Framework Programmes:

- the first (1984-1987) with a budget of 3.7 billion ECU;
- the second (1987-1990): 5.4 billion ECU;

– the third (1990-1994): 6.6 billion ECU;
– the fourth (1994-1998): 13.2 billion ECU;
– the fifth (1998-2002): 15 billion Euros.

On the basis of the Commission's own assessment and of the conclusions of the Davignon Panel, it has been decided that the EU research policy will concentrate its efforts on a more limited number of programmes. The Fifth Framework Programme's architecture is therefore simplified and its priorities reflect a shift towards tackling major societal problems.

A simplified architecture

The architecture of the Fifth Framework Programme crystallises as well the "twofold systemic policy" of research and innovation needed if the emerging European system is to be both redirected and consolidated.

It was decided jointly by the Council and the European Parliament in December 1998 and reflects a desire to focus European public activity on a limited number of objectives: strategic priorities provide for seven major programmes. Taking a twofold "systemic" view, the first four of these (covering the first type of activities mentioned in Article 164 of the EU Treaty) are designed to provide primarily a catalytic and mobilising force, helping to redirect the European system of research and innovation towards addressing problems the solution of which would support socially and ecologically sustainable growth underpinned by an economy and a society which are information-based. The titles of these four thematic programmes are:

– Quality of life and management of living resources;
– User-friendly information society;
– Competitive and sustainable growth;
– Energy, environment and sustainable development.

In addition to these there are three horizontal programmes, designed to strengthen the structure-building function of European activity (i.e. the consolidation of a truly European system of research and innovation). These three programmes correspond to the second, third and fourth activities described in the Treaty and are aimed at:

– Confirming the international role of Community research;
– Promotion of innovation and encouragement of participation of SMEs;
– Improving human potential and socio-economic knowledge-base.

Each of these horizontal programmes, whilst pursuing its own specific objectives, will ensure that all the activities pursing those objectives in thematic programmes are carried out in a manner which is co-ordinated across the whole of the Framework Programme. The functions of a catalyst, a mobilising and structuring force, are thus organically linked and integrated within the Fifth Framework Programme, allowing structured pursuit of the twofold objective of helping to both redirect and consolidate the European system of research and innovation.

The evolution of priorities (1982-2002): towards tackling societal problems

The Framework Programmes are the medium-term planning instrument for Community RTD policy. They determine the scientific and technological priorities as well as the financial scope of Community activity in the field of RTD. Changes in the structure for allocating budget appropriations also reflect the structure of priorities given to the various research fields. Large consultation exercises underpin the changes. The decision-making procedure for the Framework Programmes involves the Commission, the Council of Ministers and the European Parliament and the Economic and Social Committee. Throughout the process there is widespread consultation with the scientific and industrial community (see section 5 below for an analysis of the policy process).

The differences in grouping and content of research fields and specific programmes covered by the five Framework Programmes make it difficult to discern changes in priorities. A diachronic analysis of the priorities put forward in them has nevertheless been made possible thanks to harmonisation work carried out by the Commission.

An analysis, by harmonised groups of activities and their relative funding levels, shows seven major trends. This analysis was undertaken for each type of activity as a function of its relative importance in terms of total Community funding. In absolute terms all research domains have received increased funding, because Community funding has increased since the beginning of the 1980s, when it was at a level about 600 MECU per year, to around 3 BECU per year in the 1990s.

The analysis of how priorities have changed from one Framework Programme to another shows that there has been substantial change: a cutback, initially, in the relative importance of energy and an intensification of efforts devoted to information technologies, followed, secondly, by greater diversification in an attempt to take account of industrial technology, quality of life (biotechnology, health) and society's needs. At a more detailed level, the evolution can be described as follows (on the basis of the structure of the Fifth Framework Programme):

- The share of quality of life research and life sciences is growing fast. Of little importance in 1982 (3% of the research budget), this sector has gained in importance and now accounts for 13% of the Fourth Framework Programme. The allocation of funds to this priority will increase by 32% in real terms in the Fifth Framework Programme. The boosting of efforts in this field reflects the emphasis placed on the contribution of research to the improvement of the quality of life and to implementing other common policies (particularly reconversion of agriculture and fisheries, health, and safe and healthy nutrition).
- Rapid growth and decline of information and communication technologies: after showing rapid growth during the 1980s (42% of the total budget of the Second Framework Programme), the relative share of the activities related to the information society is showing a significant downward trend during the 1990s. In the Fifth Framework Programme, the priority allocated to this line is declining slightly (less 10% in real terms and a share of 27% vs. 28% previously). The emphasis will be on system and services for the citizen, new methods of work and multimedia content while essential technologies and infrastructure still represent an important component.
- Competitive and sustainable growth: There has been a steady increase in Community funding in both absolute and relative terms for this priority. Its relative share increased from 9% to 16% between 1982 and the Fourth Framework Programme. This theme (which covers also transport-related research) will represent a stronger priority in the Fifth Framework Programme, both in relative (19%) and absolute (an increase of 12% in real terms) terms. The key actions "innovative products, processes and organisation" and "new perspectives for aeronautics" represent more than 50% of the resources allocated to this theme.
- Less emphasis on energy research, but relatively more on non-nuclear sources: in the phase preceding the First Framework Programme, energy played a leading part in research activities. This can be explained in particular by the important nuclear research based on the EURATOM Treaty. The share of "energy" dropped from 66% in 1982 to 16% of the total budget for the Third Framework Programme. The Fourth Framework Programme provided for a slight increase in its relative share (to 18% of the total), which was nevertheless equivalent to a doubling of "energy" funding at current prices. This upward movement was due to a marked intensification

of efforts in the field of non-nuclear energy and the inclusion of the demonstration activities previously carried out in *Thermie* and other programmes. Energy-related research has a lower share in FP5 (14%). The resources allocated to nuclear fusion and non-nuclear energy research will decrease in real terms while nuclear fission safety research remains stable in absolute terms.

- Changes in environmental research: the relative stability of the importance attached to the environment in Community RTD policy does not adequately reflect the efforts made by the European Union in this high priority area. Not only have Community funds assigned to this priority more than quadrupled during the period from the First to the Fourth Framework Programme, but practically all RTD projects financed through the Framework Programmes now take environmental aspects into consideration. When looking at the Fifth Framework Programme (where environmental research *stricto sensu* decreases slightly in relative terms but increases by 17% in absolute terms), a stronger emphasis on sustainability is visible in the other thematic programmes.
- The Fifth Framework Programme, after the Fourth, significantly reinforces the importance attached to international co-operation activities and activities concerning innovation and SMEs (the shares of these activities have grown by a factor of two or three, respectively, compared to the Third Framework Programme). In particular, this reflects the integration since the Fourth Framework Programme of activities that were previously financed from outside the Framework Programmes for RTD (such as International Scientific Co-operation, COPERNICUS, PECO or SPRINT). With the Fifth Framework Programme, the commitment to allocate at least 10% of thematic programmes to SMEs reflects a growing priority. These two types of activity will be strengthened as well in the thematic programmes.
- "Improving the human research potential and the socio-economic knowledge base" increased its share substantially between the Second and Third Framework Programmes and again between the Fourth and Fifth. With the latter, it reaches 9% of the total against 7% previously. Furthermore, a significant part of training activities is henceforth addressed through "thematic" programmes.

Finally, it should be noted that the share of the EU's own laboratories is decreasing in the recent years. The direct actions of the Joint Research Centre will see their budget decreased by 3 % in real terms between the Fourth and Fifth Framework Programmes. The decrease is particularly significant in the field of nuclear research.

4.3 – The micro-level of "key actions": de-compartmentalising research for solving problems faced by society

Once a number of strategic lines of action have been decided at EU level for five years and need to be translated into specific programmes, how should these be organised? How is EU research support organised?

Research programmes can be divided into a certain number of "building blocks". These can be designed on the basis of technology families (like information and communication technologies, materials technologies, biotechnologies ...), types of activities (project funding, infrastructure funding, training funding ...) or problems. Between the Fourth and the Fifth Framework Programmes, a shift was decided: specific programmes will be focused primarily on problems that European societies have in common.

The group of activities that mobilise resources on those problems is called a "key action". These key actions, after the experience of co-ordinating research related to a common objective through Task Forces (in the context of the Fourth Framework Programme), have become the main building blocks of EU specific programmes (they absorb from 77% to 95% of the budget of each specific programme, the rest being allocated to generic research and activities to support research infrastructures).

In June 1995, the European Commission set up Task Forces on Industry and Research in order to identify research priorities in certain areas of key importance to industry and European society, to ensure better co-ordination of European and national work and resources, and to encourage the emergence of an environment which would be favourable to this research and demonstration work in terms of funding and the regulatory framework (European Commission, 1997). Eight areas received attention (new-generation aircraft, multimedia educational software, the car of tomorrow, vaccines and viral diseases, trains and railway systems of the future, inter-modality in transport, maritime systems of the future and environmental technologies for water). The experience gained from these Task Forces was put to good use under the Fifth Framework Programme in the design and implementation of key actions.

There are 23 such actions, and they are spread over the four thematic programmes. They concern important problems for our societies, such as food and nutrition, the city of tomorrow, ageing of the population and information and communication services for the citizen. They are novel in a number of respects:

- They focus on all manner of bottlenecks (scientific, technological and socio-economic), which hamper the resolution of problems common to all 15 member states.
- As part of a "systemic" approach they will mobilise the resources of the various relevant disciplines, technologies and areas of know-how, i.e. the full range of activities needed to attain their objectives, from basic research through to demonstration. The economic and social sciences should dovetail with the "hard" sciences in respect of specific research tasks relevant to their objectives.
- They will seek to ensure that public and private work done in Europe in the specific area of their remit is conducted along convergent lines. Relations with other fora for European co-operation such as EUREKA or ESA will be strengthened through them.
- Their substance and development will be determined, throughout the period of their implementation, in close consultation with research players and users (through the External Advisory Groups, see Section 5.3).
- Their research objectives will be determined rigorously and precisely but the conditions and methods of implementing them need to be flexible (a mixture of "top-down" and "bottom-up" approaches).

A focus on common economic and social problems, interdisciplinarity, European co-ordination, involvement of users, a pattern of ongoing development and flexible implementation: these are the main features of these actions which exemplify the break between the Fifth Framework Programme and its predecessors.

4.4 – The global and regional dimensions

National research and innovation systems are more and more affected by the trends towards globalisation, and, in Europe, "Europeanisation" and regionalisation. The EU research and innovation policy has integrated these dynamics of change and has developed specific activities to help research actors cope with them. The political objective of strengthening economic and social cohesion in the European Community, integrated in the Treaties by the Single European Act in 1987, has been translated into a rapid expansion of regional and social policy activities funded by the EU Structural Funds (the Regional, Social and Agricultural Guidance Funds) in member states less-favoured or declining regions. Research and innovation-related measures have taken a growing share of structural interventions throughout the 1980s and 1990s. On the other hand, the increasingly global nature of the research process has been taken into account in EU research and innovation policy framework through a great number of co-operation agreements between the EU and third countries and more recently an opening up of the Framework Programme to a full participation of more than 15 non-EU countries. The following sections will present briefly the regional and global dimensions of EU research and innovation policy.

Regional dynamics: "strengthening economic and social cohesion" of Europe in the research field

"Economic and social cohesion" is enshrined in the Treaty on European Union together with EMU and the completion of the Single Market. This expression reflects a political commitment by the EU to bridge the gap between the more advanced and the less-favoured regions of Europe.

The enlargement of the European Community to the less-favoured Southern European member states (Greece, 1981; Spain and Portugal, 1986) has significantly increased the emphasis on regional development issues. A new institution, the Committee of Regions (a consultative body responsible for bringing a local and regional viewpoint on European Union legislation and policy) was created in 1994.[1]

Specific instruments, the Structural Funds and the Cohesion Fund have been redesigned since the end of the 1980s in order to support the catching-up process in the less-favoured regions (LFR) of Southern and peripheral Europe (mainly in Spain, Greece, Portugal and Ireland) and the reconversion of industrial and rural areas affected by economic decline.

The Second European Report on Science and Technology Indicators (European Commission, 1997) has provided an analysis of Europe's "cohesion gap" in the field of science and technology. The cohesion gap is obvious when one compares the funding of research, measured as a percentage of GDP: Sweden allocated the highest percentage (3.04%) of GDP in 1993 to R&D activities, Greece the lowest (0.66%). The gap is even more striking when comparing GERD/GDP figures of (sub-national) regions.

1. It is composed of 222 members who are elected officials from cities and regions across Europe. The Committee was consulted on the European Commission's proposals for FP5.

For example, if the EU average is rated 100, Denmark is rated 318 and the Greek region of Dytiki Makedonia three. In terms of number of patents per million inhabitants, the German *Land* of Baden-Württemberg reaches a score of 481 when the Italian region of Calabria scores one.

In complementarity with the Framework Programmes, which have supported the transnational parternerships between research organisations throughout the Union, the Structural Funds have allocated an increasing amount of funding to RTD and innovation-related activities. The following figure shows that between the periods 1989-1993 and 1994-1999, the percentage of Structural Funds allocated to RTD has doubled (up to more than 5%) in the less-favoured regions (known as "objective 1" regions) and increased significantly in declining regions ("objective 2" regions).

Before the reform of 1988, the Structural Funds financed only relatively few R&D activities: about 190 million ECU were spent on actions of this type, chiefly in the form of investment in research centres and technical schools. In the period 1989-1993, there was a considerable increase in the funding of R&D measures from 190 million ECU to 3920 million. For the 1994-1999 period, it went up to 8520 million ECU, the percentage of R&D measures in the total of the activities supported by the Structural Funds decreasing slightly (from 5.74% to 5.60%).

FIGURE 6.3 – Evolution in the RTD component of structural interventions

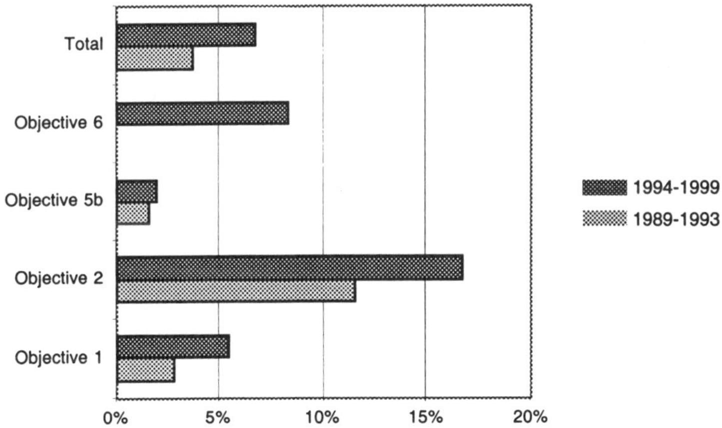

Source: European Commission, Communication of 27 May 1998, COM(98)275, "Reinforcing cohesion and competitiveness through research, technological development and innovation".
Note: "Objective 1" regions are regions whose development is lagging behind; "objective 2" regions are regions seriously affected by industrial decline in need of reconversion; "objective 5b" regions are rural areas in need of development and structural adjustment; "objective 6" regions are those with extremely low population intensity.

At the same time, the types of RTD-related measures financed by the Structural Funds have evolved from the 1989-1993 to the 1994-1999 periods. In some countries and regions, more effort has been put into building human capital, brokerage and demand stimulation after an initial emphasis on the financing of classical RTD infrastructure.

What is the impact of EU policies towards the reduction of its technology gap? The Second European Report on Science and Technology Indicators (1997) has presented an analysis of the evolution of the technology gap. The authors concluded that from 1989-1993, there was some narrowing in the technology gaps. However, the gulf remains significant. One explanation for this could be the "twin peak" theory proposed by some academics, in which a group of the rich and a group of poor regions form two convergence clubs. If this hypothesis holds, you can come top of the second division, but you can never be promoted to the first. The authors find no easy correlation between the progress made by a region and the amount of EU funding received. They conclude that, perhaps, "Europe plays an important conditioning role in supporting cohesion, but its policy can only be effective if it is complemented by member state actions".

Analysing the performance of those "second division" regions which have performed well, two major factors become apparent: "these regions belong to a country with a well-elaborated innovation system", the authors find, "and each of them is geographically located close to absolute leaders". Science and technology, it seems, is subject to the laws of diffusion. The authors point out that statistics are not always comparable. A region which doubles its output of patents from two to four and then four to eight in successive years, is not to be held up as a success against a region which produces a lower percentage growth, but claims many times as many successful applications. There have, however, been some illustrations of progress. "The new German *Länder* are the most prominent example", the authors disclose. "Other cases include: Ireland, and Algarve, Alentejo and Centro regions (in Portugal)".In its communication of May 1998 on the relations between EU structural and research policies, the European Commission stressed that both the economic gap (measured in terms of Gross Domestic Product per inhabitant) and the technology gap (measured in terms of gross expenditure on R&D by GDP and patents per 1000 inhabitants) between the four Cohesion Countries (Ireland, Spain, Portugal and Greece) and the other member states have decreased from 1989 onward.

Yet the technology gap remains significant both in the private and public RTD systems. Furthermore, the situation seems even more divergent when inter-regional differences are examined. Important regional differences exist also in the countries with more efficient RTD and innovation systems. However, even when the weakest regions of the richest countries are considered, their RTD and innovation system is still more robust and demand-oriented than that of the poorest countries.

There are also indications of increased polarisation between North and South in terms of information and communication technologies, as indicated by domestic and commercial usage of the Internet, home ownership of personal computers, web hosting and tele-working schemes. Broadly speaking, it can be said that the scientific and technological systems of less-favoured regions – and especially "Objective 1" regions – continue to be characterised by: (1) overall low RTD intensity; (2) over-representation of the public sector and lower presence of the private sector; (3) primary emphasis on basic research; (4) low levels of technology transfer between the public and the private sectors and within the private sector itself; and (5) poor linkages to international RTD and Innovation networks.

Such qualitative differences suggest that injections of public funds into research activities in less-favoured regions will produce lower economic return than in more developed regions. More emphasis therefore needs to be put on the private sector. Firms need to be engaged more in the research and innovation process and this should be facilitated by public authorities stimulating venture capital, and providing other incentives to encourage activities such as electronic commerce.

Some pilot activities aiming at supporting the design of RTD strategies at regional level have been funded in the 1990s by the Structural Funds. Under Article 10 of the European Regional Development Fund (ERDF) and in collaboration with the Innovation Programme, Regional Innovation Strategies (RIS – Article 10 ERDF) and Regional Innovation and Technology Transfer Strategies and Infrastructures (RITTS – Innovation Programme) have been developed with the following main objectives: to create a clear strategic framework for integrating innovation in the productive structure of the region; to establish networks of co-operation within firms and between firms and the public sector; and to strengthen the RTD and innovation supply in the region.

Funded by Article 6 of the European Social Fund and Article 10 of the ERDF, Regional Information Society Initiatives have been launched to take account of the growing importance of information and communication technologies for regional development, work organisation, human resources, employment and equal opportunities.

BOX 6.2 – A shared responsibility for regional development through research and innovation (2000-2006)

The role of local and regional levels could be to:
- carry out analysis of regional and local needs and potential;
- develop the strategic framework in which research, technology innovation and related policies should be embedded;
- implement an agreed programme for RTD and innovation which incorporates specific priorities and measures, suitable for delivery by appropriate agencies;
- organise a streamlined, focused and inclusive regional partnership which takes responsibility for effective strategic economic development planning, co-ordination of policy instruments and provides the necessary finance to fund the strategy and the actions.

The role of the member states could be to:
- determine national framework conditions which can assist RTD and Innovation efforts;
- shape national RTD and Innovation policies and systems including the distribution of technological installations and programmes to assist usage of these facilities by business in the regions;
- prepare the development plans to be co-financed under the Structural Funds.

The role of the member states and the EU could be to:
- establish the long-term strategic objectives of RTD and Innovation policies, based on a comprehensive understanding of the strengths and weaknesses of the RTD and Innovation system and the economic system of each member state;
- ensure that interventions are coherent and that available instruments are aligned and directed towards compatible goals.

Source: "Reinforcing cohesion and competitiveness through research, technological development and innovation", COM(98)275, European Commission.

Despite all these positive changes, a large part of structural interventions still tends to be directed towards support of the existing scientific (public-oriented) system, especially in the "Objective 1" regions, perpetuating and eventually reinforcing the structural problems besetting the regional innovation system. This issue needs to be examined critically. The European Commission concluded that for the next period of Structural Funds interventions (2000-2006), there is a need to go further, i.e. towards a common strategy, involving local, regional, national and EU initiatives (Box 6.2).

The global dimension: between research policy and external relations imperatives[1]

Scientific research and technology are becoming increasingly internationalised and for many reasons. The information society and a general move towards globalisation mean that researchers may now exchange information and expertise as never before. New centres for research are emerging in various parts of the globe. Issues critical to man's future, namely health, food and the environment, are assuming a worldwide dimension. The great challenges of tomorrow, in terms of space, sustainable energy policies and biotechnological mastery, involve prohibitive research costs that require international action. Links with other European research co-operation initiatives help finally to consolidate a European space of research and innovation. That is why Community research policy, as embodied in the Treaty itself, is fundamentally open to exchange with the rest of the world.

Moreover, EU international scientific and technological co-operation activities take into account the evolution of EU external relations policy. That explains, in particular, why research co-operation with the countries of Central and Eastern Europe, Mediterranean countries and developing countries has been given an increasing importance since the beginning of the 1990s.

In this context, the Fourth Framework Programme (1994-1998) has given international co-operation in RTD its full meaning in two ways:

- Through the innovative launch in 1994 of a specific programme on co-operation with third countries and international organisations. This programme, named "INCO", backs numerous research projects, bringing European and foreign partners together in a variety of fields;
- By making it possible for scientific and technical partners from all European countries, and, in some cases, non-European countries, to participate on a project by project basis (that is, without Community research funding) in various specific Framework Programme initiatives (transport, information technology, biotechnology, environment and climate).

The Fifth Framework Programme (1998-2002) includes a specific programme for "Confirming the international role of Community research". Its objectives are: to promote scientific and technological co-operation internationally; to reinforce Community capacities in the fields of science and

1. For a complete presentation and analysis of "the EU as a world partner", see the Second European Report on Science and Technology Indicators (1997), Part 5.

technology; to support the achievements of scientific excellence within the wider international framework; and to contribute to the implementation of the Community's external policy (including the accession of new EU members from Central and Eastern Europe and Cyprus). With a budget of 475 million Euros (compared with 575 for the INCO programme in the Fourth Framework Programme), it has a dual function: to fund specific actions not addressed by the other programmes and to ensure that the latter implement co-operation activities with third countries in a coherent way.

The programme "Confirming the international role of Community research" will fund the following specific activities:

- Co-operation with third countries: states in the pre-accession phase (promotion of their centres of excellence, measures to facilitate participation in the other programmes of the Fifth Framework Programme); New Independent States and Central and Eastern European countries not in the pre-accession phase (support for their RTD, co-operation in areas of mutual interest); Mediterranean partner countries (support for their RTD capacities, co-operation in areas of mutual interest, measures to facilitate participation in the other programmes); developing countries: a policy dialogue on RTD needs and priorities with groups of countries and regions will be developed together with specific co-operation activities (for example in the areas of sustainable development and health); emerging economies and industrialised countries; exchanges of scientists, workshops and measures to promote partnerships and to enhance mutual access to each other's RTD activities.
- Training of researchers: fellowships to give many researchers from developing countries, Mediterranean and emerging countries the opportunity of collaborating in laboratories in the Community on specific projects of the Framework Programme and young Community researchers the possibility of working in Japan in industrially-oriented laboratories;
- Co-ordination activities: with COST, EUREKA and with international research organisations as well as co-ordination of activities pursued through the other programmes of the Fifth Framework Programme.

Apart from this specific programme described above, the Fifth Framework Programme realises a major step forward in the field of the co-operation between the EU and third countries: 16 non-EU countries will be able to participate in activities alongside the 15 member states. For example, the ten Central and Eastern European Countries which are, together with Cyprus, in a pre-accession phase will contribute financially to an addition to the budget (calculated on the basis of their GDP but with appropriate progressive financing mechanisms) and their research organisations will participate under similar conditions to member states' entities. Similar schemes have been devised for Norway, Lichtenstein, Iceland and Israel and subsequently will be for Switzerland. In other words, the enlargement of the EU is already effective in the field of research policy and 31 countries are in a position to benefit from research co-operation supported by the EU budget.

In addition to the RTD association agreements quoted above (involving 16 non-EU countries in the implementation of the Fifth Framework Programme), it should be noted that the EU has signed or will sign soon a number of science and technology co-operation agreements with non-EU industrialised countries (Canada, Australia, the US) as well as emerging or transition economies (Argentina, Brazil, China, Russia and South Africa).

5. THE POLICY PROCESS AT EU LEVEL: THE EXAMPLE OF THE FIFTH FRAMEWORK PROGRAMME

The specific – transnational – character of EU research and innovation policy can be appraised when looking at the policy process according to which decisions are prepared and made. Specific consultation processes and criteria are used to define its priorities for a five-year period (Section 5.1). The decision-making process starts when the European Commission introduces its proposals and it involves two complementary legislators: the Council of Ministers, representing national governments, and the European Parliament elected directly by European citizens (Section 5.2). Moreover, the European Commission has set up various consultative bodies to help it prepare and implement the decisions in an open and flexible way (Section 5.3).

5.1 – How are the priorities of RTD policy defined? A broad consultation process

Against a background of budgetary stringency, there is general acceptance of the need to make difficult choices in earmarking public funds for a limited number of scientific and technological priorities. In the industrialised countries a variety of approaches have been developed to this end over the last 50 years and more. Recent developments point to a common feature of these exercises: they are both interactive and systemic.

At European level, even though a vast integrated exercise in technology foresight[1] is not yet on the agenda, the results of national forecasting initiatives have been analysed and taken on board. The Commission, in its proposals for the Fifth Framework Programme, started with a similar approach, initiating a vast debate with:

- member states and the European Parliament;
- bodies representing scientific and industrial circles at national and European level;
- the Industrial Research and Development Advisory Committee (IRDAC);
- the European Science and Technology Assembly (ESTA);
- numerous companies and public and private research bodies.

It received a large number of documents from players in the European system of research and innovation. It also learned from the debate on the 1995 Green Paper on Innovation, which yielded many interesting suggestions for Community research activity. In identifying priorities for the Fifth Framework Programme the Commission drew on internal studies and analyses, including the reports of the Task Forces on Research and Industry.

The features of the European machinery for setting scientific and technological priorities may thus be summed up as follows: it is a systematic process, based on both analysis and forecasting and on broad consultation of the players and decision-makers. The key elements to be taken into account at European level in the selection of scientific and technological priorities are very specific: they must help to identify the "added-value" of Community action in relation to action by member states or other forums of international co-operation. In other words, the programmes covered by the Fifth Framework Programme adopt only priorities which satisfy three categories of criteria (cf. Box 6.3): criteria related to social objectives; criteria related to economic development and scientific and technological prospects; and criteria related to the Community "added-value" of an activity.

The elements of economic and social assessment set out below apply to each of the four thematic priorities of the Fifth Framework Programme.

1. Cf. Technology Foresight: Perspectives for European and international co-operation, PREST – University of Twente – ANEP, Final Report, April 1996; this study, discussed at a European RSEF/ETAN Workshop in April 1996, calculated that a large "UK-style" technology foresight programme, involving 40 000 people in Europe, might cost up to 72 million ECU.

BOX 6.3 – Criteria for selecting the themes and objectives of Community activities

"... in pursuit of a cost-benefit approach dictated by concern for optimum allocation of European public funding and in accordance with the subsidiarity principle, themes for the Fifth Framework Programme and the related objectives are selected on the basis that the Community shall take action only if and in so far as the objectives cannot be sufficiently achieved by the Member States.
In application of the foregoing principles, the Framework Programme shall be defined on the basis of a set of common criteria, divided into three categories:

Criteria related to the Community 'value added' and the subsidiarity principle:

- need to establish a 'critical mass' in human and financial terms, in particular through the combination of the complementary expertise and resources available in the various Member States;
- significant contribution to the implementation of one or more Community policies;
- addressing of problems arising at Community level, or questions relating to aspects of standardisation, or questions connected with the development of the European area, so as to select only objectives which are more efficiently pursued at the Community level by means of research activities conducted at that level.

Criteria related to social objectives:

- improving the employment situation;
- promoting the quality of life and health;
- preserving the environment, in order to further major social objectives of the Community reflecting the expectations and concerns of its citizens.

Criteria related to economic development and scientific and technological prospects:

- areas which are expanding and create good growth prospects;
- areas in which Community businesses can and must become more competitive;
- areas in which prospects of significant scientific and technological progress are opening up, offering possibilities for dissemination and exploitation of results in the medium or long term, in order to contribute to the harmonious and sustainable development of the Community as a whole."

Source: Decision No 182/1999/EC of the European Parliament and of the Council of 22 December 1998 concerning the Fifth Framework Programme of the European Community for RTD&D (1998 to 2002).

5.2 – How are decisions made? The "co-decision" procedure

The decision-making process at European level is based on the interplay of Community institutions, i.e. the European Commission, the Council of Ministers and the European Parliament. The European Commission, which possesses executive powers, is the only one of these institutions with the "right of initiative". This means that, for any decisions to be taken at the level of the European Community, there is a need for a proposal to be put on the table of the Ministers and the European Parliament (EP) by the European

Commission. The way the Council of Ministers and the EP may decide on a Commission's proposal is based on the Treaties. "Co-decision" means that the Council and the EP decide jointly on matters of EU interest. The Economic and Social Committee gives advice on the Commission's proposals.

The decision-making procedure followed in order to decide on the efforts and ways for developing a common RTD policy for the years 1998-2002, i.e. on the Fifth Framework Programme, can be described as follows. The "pre-decision" phase took roughly one year:

- In April 1996, Commissioner Edith Cresson (responsible for research, education, training, youth and innovation) made a first consultation of the Research Ministers of the member states on what could be the priorities for the programme.
- in July of the same year, the European Commission published a communication, titled "Inventing tomorrow", spelling out its approach concerning the Fifth Framework Programme and its first ideas on the overall structure this programme could have.
- In order to stimulate the debate further (including with the actors, decision-makers and users of research in Europe), it circulated, between the end of 1996 and beginning of 1997, two "working documents" describing respectively the structure and key elements of the future Framework Programme and its science and technology objectives.

The formal decision-making process started when the European Commission diffused, in April 1997, its legal proposals for the Fifth Framework Programme (with two complementary components: a FP based on the EC Treaty and a FP based on the EURATOM Treaty). The European Commission's proposals of 9 April 1997 took account of the assessment reports for the five previous years of Community activity (Davignon report, see Section 3.3.above).

The first reading by the European Parliament took place at the end of 1997 and the Council of Ministers reached a common position in February 1998. After a second reading by the EP in June 1998, disagreements between the Parliament and the Council of Ministers (on the total budget to be allocated to the programme and elements of its contents) led to the setting-up of a Conciliation Committee, which, after four meetings, reached agreement in November 1998. The final adoption of the Fifth Framework Programme took place in the Council of Ministers on 22 December 1998.

A similar decision-making process was necessary for decisions on the specific programmes implementing FP5 at a more detailed level. It started when the Commission presented its proposals on these programmes in June 1998 (after a common position was reached by the Council) and ended with a decision by the Council in January 1999.

The decision-making process of the Fifth Framework Programme took place under the rules of the Maastricht Treaty, which stipulate that, in this co-decision procedure, member states need to decide unanimously on the Framework Programme (which represents the strategic decision, followed by implementing decisions on the specific programmes). The Amsterdam Treaty, by replacing the unanimity clause by a Council decision taken by a qualified majority (a majority of member states votes weighted according to the size of each State) will certainly speed up the decision-making process and allow for more clear-cut choices to be made on priorities.

5.3 – Strengthening participative governance: the establishment of new advisory structures

In environments marked by high uncertainty and complexity, it is more and more recognised that any kind of action is driven by bounded rationality. This has influenced the way governments have reflected upon their role and their need to permanently seek advice from major stakeholders in society in order to define and implement public activities with maximum efficacy and transparency.[1] The European Commission has, since the 1980s, recognised these trends in governance models by instituting various advisory bodies, helping it to better prepare and better implement the decisions made at EU level. The European executive needs to draw upon three channels of partnerships and legitimacy:

- the channel of member states, i.e. of elected national governments;
- the channel of the European Parliament directly elected by the citizens within the Union;
- the channel of "civil society", i.e. the various stakeholders which are interested in being involved in EU decision-making which directly or indirectly influence their activities.

In the field of research policy, the Commission works closely with national research administrations at two levels:

- at strategic level, the Committee for Scientific and Technological Research (CREST) advises both the Council of Ministers and the

1. For an excellent overview of political theory developments concerning these issues, see De Munck (1999).

Commission on all important policy issues or decisions (for example on the Commission's proposals for framework programmes);
- at the level of specific programme implementation, the Commission is assisted by programme management committees where member states' officials are represented (these committees are chaired by a Commission official).

Secondly, the Commission informs regularly the special European Parliament Committee on the development of EU research activities. Up until April 1999, the relevant Committee was the one responsible for Energy, Research and Technology. Since then, research policy issues have been handled by the Industry, Foreign Trade, Research and Energy Committee.

Thirdly, two major stakeholders (industrialists and researchers) in the research field were represented in distinct advisory bodies: the Industrial R&D Advisory Committee (IRDAC) and the European Science and Technology Assembly (ESTA).

Two recent developments are worth mentioning for what concerns relations between the Commission and research actors and users: the creation of External Advisory Groups (EAGs) for accompanying the development of the key actions and the setting up of the European Research Forum.

In November 1998, the Commission appointed 278 experts to provide independent advice on the implementation of the Fifth Framework Programme for research and technological development. Selected from nearly 5000 candidates from Member and associated states, they will serve on 17 advisory groups. Each EAG will hold regular meetings to review the draft work programmes and provide the Commission with advice on the evolution of the research work to be carried out under the key actions.

In addition, the European Commission decided in October 1998 to create a European Research Forum. On the grounds of rationalisation and in order to increase the synergy between industry and research, it thought it was appropriate to replace the two existing committees, namely IRDAC, and ESTA, by a single body entitled the European Research Forum, comprising a maximum of 60 members appointed by the Commission and distributed more or less evenly between two "chambers" working in close co-ordination: an academic and scientific chamber and an industrial, services and users chamber.

The missions of the Forum are of a more general nature than the External Advisory Groups as far as they concern the study of strategic issues relative to the formulation and implementation of EU RTD and innovation policy as well as issues related to technology policy of "European interest in general".

In order to maintain a link between this general consultative body and the more specialised Advisory Groups, the chairmen of these latter will figure among the members of the Forum, in operation from 2000.

In conclusion, it can be said that by setting up the External Advisory Committees attached to the key actions of the Fifth Framework Programme, the EU is moving further in the direction of a "stakeholders' democracy" in the field of research policy.

CONCLUSIONS

Research and innovation policy is a key dimension of the European integration process and of the EU strategy for growth and job creation. Influenced by the general evolution of the EU, in particular by the shift from a mainly economic construction to a political one, it has also contributed to it. Institutional innovations, such as, previously, the Framework Programmes, the partnerships and networks leading to a new form of transnational co-operation and today, systemic innovation policy and problem-oriented key actions, demonstrate that Europeans have learned how to share their destiny in a field like research and innovation, where lies an important element of their common future. Tomorrow's challenges like EMU, the next enlargement to 11 possible new member states and the evolution of the Common Foreign and Security Policy (CFSP) may require appropriate responses from the EU research and innovation policy (How will it contribute to reduce the vulnerability of European economies to possible asymmetric shocks? How to avoid the enlargement making the problems underlying the notion of a "European paradox" more serious? What EU activities might be required to strengthen the industrial and technological base of the European defence industries?). The question remains open to see if Europeans are ready to go a step forward in the direction of a more coherent European system of research and innovation, where they not only co-operate on specific activities at EU level but they also better co-ordinate their national policies and programmes between themselves and with the Union. Inventing the means and methods of such a new policy breakthrough will probably be one of the key tasks for the next decade, especially if the European Union really wants to become the key player in economic and political terms that it deserves to be, given its past and current contribution to the world's scientific, technological, socio-economic, political and environmental progress.

ANNEX

TABLE 6.1 – The co-evolution of EU research and innovation policy with the broader European integration process (1952-2002)

Major stages of the European integration process	**EU research and innovation policy: the legal basis**
The first Treaties (the European Coal and Steel Community, 1952; the European Atomic Energy Community, called Euratom, 1957; the European Economic Community, 1957).	Specific provisions for coal and steel research, for nuclear energy and agricultural research. Article 235 of the EEC Treaty allows for the launching of new initiatives.
The Single European Act (1987) speeds up the establishment of a Single European Market and strengthens the accompanying structural (regional, social) policies.	A new Title for RTD is introduced in the EEC Treaty describing the objectives, implementation instruments and decision-making process.
– The Maastricht Treaty (1993) pushes for Economic and Monetary Union, and introduces two new "pillars" (internal affairs and justice, common foreign and security policy) in the architecture of what become the European Union. – White Paper of the European Commission on "Competitiveness, Growth, Employment" which pleads for a new model of development for Europe (1993).	– The objectives of the EU research policy is broadened to the support of other EU policies. – The Framework Programme is adopted by codecision between the Council and the European Parliament; unanimous decision is still required in the Council.
– The Amsterdam Treaty (1997) moves the EU a new step forward towards political Union; employment becomes a priority; parts of the "second pillar" (justice and home affairs) is integrated into the EC Treaty; common foreign and security policy is strengthened. – The European Commission document "Agenda 2000" (1997) becomes the basis for new decisions on the EU financial perspectives 2000-2006 (agreement between the Council and European Parliament reached in May 1999).	Unanimous decision in the Council on the Framework Programme is replaced by qualified majority decision.

EU research and innovation policy: the main policy developments	**EU research and innovation policy: the main focus**
– Establishment and operation of the Joint Research Centre in the context of Euratom. – Incremental diversification in other fields by EEC programmes decided on the basis of Article 235. – Creation of the Committee for Scientific and Technical Research (CREST). – First Framework Programme (1984-1987).	– Initial focus on (nuclear) energy research. – Diversification into non-nuclear environmental, materials and biomedical research. – Development of new type (shared-cost) programmes in the field of IT (ESPRIT), telecom (RACE) and industrial technologies (BRITE) research.
– Second (1987-1991) and Third (1990-1994) Framework Programmes. – Creation of new advisory structures representing industry (IRDAC) and scientists (CODEST).	– Generalisation of the shared-cost model of programmes. – Development of new horizontal activities in the field of technology transfer, international science and technology co-operation, technology assessment, mobility of researchers.
– Fourth Framework Programme (1994-1998). – European Commission's communication on "Co-ordination through co-operation" (1994) pleading for a reinforced co-operation between national RTD policies and programmes.	– Increasing importance of socio-economic research horizontally. A specific programme is also allocated to it for the first time. – Experimentation of the "Research-Industry Task Forces" co-ordinating research activities financed by various programmes around a common objective (e.g. the car of tomorrow).
– European Commission's Green Paper on Innovation (1995) followed by the Innovation Action Plan (1996) represents a first attempt to have a comprehensive research and innovation policy framework. – EC's White Paper on Education and Training (1995) focuses on new requirements for learning in a knowledge-based society. – The Fifth Framework Programme (1998-2002) adopts a systemic and problem-solving approach to EU Research and innovation policy. – The EC communication on "reinforcing cohesion and competitiveness through research, technological development and innovation" (1998) emphasise the need for new type innovation measures in the next generation of EU structural programmes.	– FP5 introduces a new way of organising EU research activities: the "key action". – The architecture of the FP is streamlined: four "thematic" programmes and three "horizontal" programmes. – Increased linkages between RTD and innovation policy and other EU policies. – Opening up of the FP to 31 countries (EU 15, EFTA, Accession countries and Israel). – More and more provision of services (such as IPR helpdesk) in addition to funding of research co-operation. – Increasing focus on SMEs through specific measures.

REFERENCES

Amable, B., R. Barré and R. Boyer, 1997, *Les systèmes d'innovation à l'ère de la globalisation*, Paris: Economica.

ANEP, PREST and University of Twente, 1996, *Technology Foresight: Perspectives for European and International Co-operation*, Final Report for the European Commission.

Caracostas, P. and U. Muldur, 1997, *Society, the Endless Frontier*, Luxembourg: Office des publications officielles des communautés européennes (OPOCE).

Caracostas, P. and L. Soete, 1997, the Building of Cross-Border Institutions in Europe: Towards a European System of Innovation?, in C. Edquist (ed.), op. cit.

Davignon Panel, 1997, *Five-Year Assessment of the European Community RTD Framework Programmes*, an independent panel chaired by Viscount E. Davignon, Luxembourg: OPOCE.

De Munck, J., 1999, *L'institution sociale de l'esprit*, Paris: PUF.

Dodgson and Bessant, 1996, *Effective Innovation Policy: A New Approach*, London: ITP.

Edquist, C. (ed.), 1997, *Systems of Innovation; Technologies, Institutions and Organisations*, London: Pinter.

Edquist, C. and B. Johnson, 1997, Institutions and Organisations in Systems of Innovation, in C. Edquist (ed.), op. cit.

European Commission, 1994, *First European report on science and technology indicators*, Luxembourg: OPOCE.

European Commission, 1995, *The Demographic Situation in the European Union*, Luxembourg: OPOCE.

European Commission, 1995a, *The Green Paper on Innovation*, Luxembourg: OPOCE.

European Commission, 1995b, *The White Paper on Education and Training, Towards the Learning Society*, Luxembourg: OPOCE.

European Commission, 1996, *The Green Paper on Education, Training and Research: Obstacles to Transnational Mobility*, Luxembourg: OPOCE.

European Commission, 1997a, *Second European Report on Science and Technology Indicators*, Luxembourg: OPOCE.

European Commission, 1997b, *The Single Market Review, Background Information for the Report to the Council and European Parliament*, Bruxelles.

Eurostat, 1994, *Report on Demographic Patterns in Europe*, Luxembourg: OPOCE.

Guzetti, L., 1995, *A Brief History of European Union Research Policy*, Luxembourg: OPOCE.

Larédo, P., 1995, Structural Effects of European Research Programmes, *Scientometrics*, 34, 3.

North, D., 1990, *Institutions, Institutional Change and Economic Performance*, Cambridge: Cambridge University Press.

United Nations, 1994, *The Sex and Age Distribution of the World Populations*.

World Bank, 1994, *Advertising the Old Age Crisis*, Oxford: Oxford University Press.

CHAPTER 7

THE GERMAN INNOVATION SYSTEM

FRIEDER MEYER-KRAHMER

1. BASIC ECONOMIC FEATURES OF THE GERMAN INNOVATION SYSTEM

Germany as a country with no natural resources (except coal) is traditionally driven by knowledge creation and technologies. It is characterised by a broad science and technology (S&T) infrastructure, a well-educated workforce, an export-oriented industry (large as well as small and medium-sized firms) and an active civilian government technology policy towards research and diffusion.

The contribution of R&D-intensive branches of industry to domestic product is much higher in Germany (12.2%) and Japan (11.5%) than in Great Britain, the US and France (7.7 to 8.5%). Germany is the leader in high-tech, ahead of Japan, whilst its significance in France and Great Britain as well as in the US and Italy is relatively slight. In Germany, even leading-edge technology contributes almost as much to the value-added of the entire economy as in Japan, and approximately as much as in the US (see NIW/DIW/ISI/ZEW, 1998). However, the 1990s marked a change, inasmuch as the research-intensive sector worldwide tended to shed jobs even faster than the other branches of industry. In West Germany and Japan, in particular, which were able to increase employment the longest, the loss of jobs in this phase took place fastest. In West Germany, all the jobs in technology-oriented industrial areas created in the employment upswing since the mid-1980s vanished within a mere three years. The growth and employment centres have shifted to the US and Great Britain in the 1990s, for only there (since 1992) has there been a renewed upwards trend in the research-intensive sector – with a booming domestic economic situation, favoured also not least by the proximity of strongly growing markets with high technology needs in South East Asia, with which the exchange of goods and technologies has markedly been intensified.

From 1980 to the mid-1990s, when dividing industrial sectors into three different R&D intensity categories, increases in the numbers of persons employed in Germany were registered only in branches of leading-edge technology and higher-quality technologies. The non-R&D-intensive industries have cut back on employees. For all categories, however, it is true that jobs have been lost since the great recession of 1992. To sum up, this means that the R&D-intensive industries are the potential winners of the structural change, but that they are also subject to fluctuations in the economic cycle, and for this reason they cannot be expected to exercise such positive effects on employment that they alone can cure mass unemployment. Growth of employment can be observed mostly with regard to high-tech start-ups and innovative services.

Not only ever increasing shares of the value-added, but also innovation potentials are shifting from industry to the service sectors. At the same time, an ever increasing number of employees within industry are carrying out (more highly skilled) service functions, whilst production intensity continues to decrease. In the future, there is little expectation that employment problems in Germany, as well as in many highly developed industrialised countries, can be solved by a rapid expansion of R&D-intensive industrial branches. The economic structure in the industrialised countries is shifting increasingly in favour of the service sector, which is a significant user of new technologies. This is also the case in Germany, which has been able to keep pace with the sectoral structural change. Above all, knowledge-intensive service firms are on the advance: more than a third of all wage and salary earners are employed in this sector. Success is increasingly dependent on the use of new technologies and of innovations. This creates new markets for technology producers (especially information and communication technologies, infrastructure facilities in the traffic and communications area, and medical technology). Technical change is oriented increasingly towards the needs of the service sector. In parallel, the significance of research and innovation activities is growing continually in the service sector: in the US a third, and in Great Britain a quarter of the technical knowledge of service firms is already based on their own R&D efforts. In Germany, company-oriented services are expanding fastest in those areas where there is a corresponding demand from innovative industries.

An analysis of the technological profile of countries and their deviation from the world average (Gehrke and Grupp, 1994) shows that smaller countries are very specialised, whereas larger countries display a spread in their profiles. The theoretical approaches of foreign trade theory and business administration, which regard the formation of own specialisation patterns as a worthwhile strategy for attaining comparative advantages, are confirmed here. A deviation can be determined for the EU, where, as a result of the European domestic market, the specialisation patterns between the individual countries are getting closer. When the industry branch specialisation in Germany is considered (Figure 7.1), it can be seen that particular specialisation advantages exist in the clusters surrounding vehicle construction, mechanical engineering and the chemical industry, whereas the sectors of the information technology (IT) industry are characterised by specialisation disadvantages across the whole scene[1] (see below).

The strengths and weaknesses of the German innovation system are summarized in an extremely simplified fashion in Table 7.1 (see also NIW/DIW/ISI/ZEW, 1998; Krull and Meyer-Krahmer, 1996; BMBF, 1999). One of the strengths of the German innovation system lies in advanced technology branches, which were already identified in the branch specialisation (automobile and mechanical engineering, chemical industry). A barrier can be seen primarily in the failure to link up with the leading-edge technologies. The most important areas of leading-edge technology in the next ten years will be information technology, biotechnology and new materials. Due to the presumptive path dependency of innovation systems, it is of strategic significance to link the top technologies to the existing competences in the German clusters. The failure to link up with the top technologies goes hand in hand with a blockade of product and service strategies, as companies are still very much oriented towards their traditional concepts and product lines. The sectors find it difficult to integrate new products and technologies at the fringes. This makes the increasingly important cross-border and cross-sector activities more difficult. Nevertheless, a strong(er) specialisation in highly productive, company-associated services can be discerned.

Closely connected with this is the opening up and exploitation of new markets. Germany offers a very disparate picture in this respect: while markets

1. In addition, we refer to the comprehensive analyses of NIW/DIW/ISI/ZEW (1998) on the technological performance of Germany.

FIGURE 7.1 – Germany's technology and trade portfolio in R&D-intensive goods

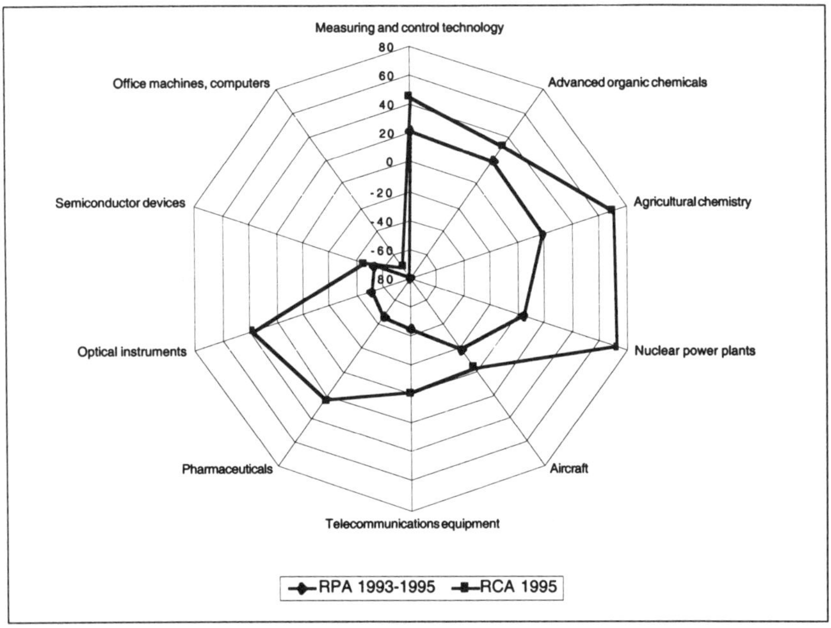

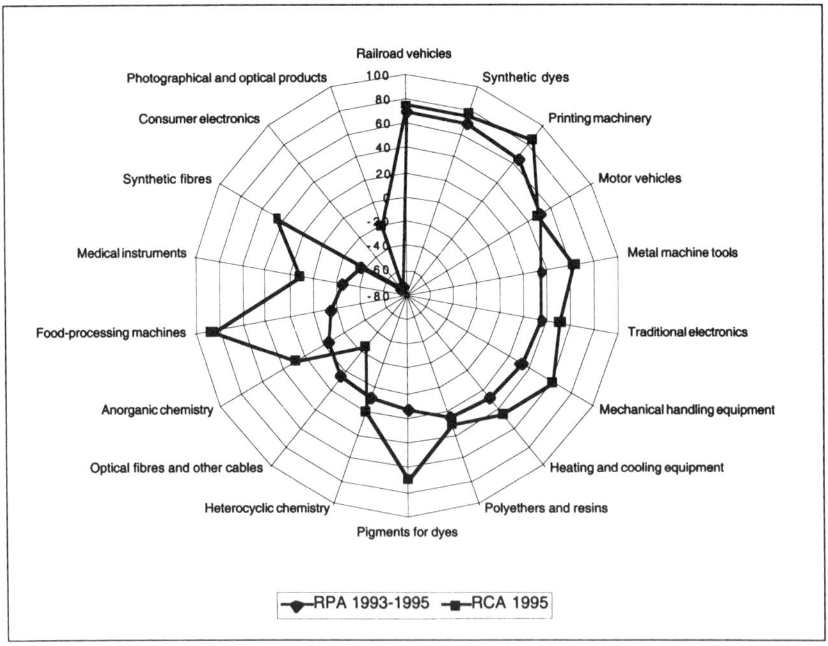

Sources: EPO, FhG-ISI calculations.
Notes: Relative Patent Activity (RPA): a positive value indicates that the share of patents in this field is larger than for total patents.
Revealed Comparative Advantage (RCA): a positive value indicates that the export–import ratio is higher for the product group than for total manufactured goods.

TABLE 7.1 – Elements of the German innovation system

Strengths	Weaknesses
High-level technology	Inadequate link-up to leading-edge technology and to product/service strategy
Existing markets	New markets
Decentralised research system	Fragmented structure
High degree of internationalisation in R&D	Domestic orientation of technology policy
Qualified staff and workers	Drop in incentives to invest in training and education
Adoption of long-term orientation (entreprises, science, policymakers)	Inadequate incentive mechanisms, limitation to traditional fields

Source: ISI.

for environment-friendly technical solutions have been developed dynamically (not least because of the demand created by publicly-funded measures) and the environmental industry occupies a leading position, numerous new market developments are – especially in the area of communication and information technology – lagging far behind, when compared with the dynamism in the US, for example. There is no doubt that job creation opportunities have been neglected in these cases.

The research system in Germany is highly differentiated and decentralised (see section 2). Internationally, this is seen as a characteristic advantage of the German innovation system (Malerba, 1994), which should not be destroyed by extensive priority setting. Such a system, however, must be closely networked and exhibit sufficient flexibility and dynamic. This is not the case at present. There is a strong need to improve the system towards a competitive, flexible and high-quality public research infrastructure facilitating the practice of modern modes of knowledge production and offering excellent, user-oriented academic teaching.

The high degree of internationalisation of R&D is a plus point for Germany as a location. For the US, Germany is the most important foreign research location and in the EU countries, only in England do the Japanese conduct more R&D than in Germany. It is a generally perceived phenomenon in many developed industrial states that the globalisation of the markets and the internationalisation of R&D have been manifested relatively little in the national technology policies. The smaller countries have proved in the last few years to be more open than the larger countries, which, because of their inertia and their greater technical-scientific power, still have to embark on the necessary adaptation processes. In Germany, too, the national orientation predominates.

The availability of qualified personnel is an essential precondition for the strength under point (2) of Table 7.1. The decreasing willingness of companies to invest in training/education is a dangerous and general barrier. Fundamental suggestions for reform are needed here. Classical strengths (good infrastructure, human resources, education system, research) are still effective, but because of the increasing mobility of research and human resources, the acquisition of knowledge, and an efficient education system, as well as "lead" markets will become more significant. The institutional system of industrial relations (co-determination, autonomous wage bargaining) and trust and co-operation between management, banks and the participating workforce, lead to relatively small frictions when introducing innovations. However, there are still not sufficient firms prepared to risk start-ups or banks prepared to play a more offensive role, or private investors to finance innovation.

With regard to the long-term orientation of science, politics and industry, we can determine that this is drastically decreasing, compared to the 1970s and 1980s. A development of this kind is, as we see it here, the last of the fundamental barriers in the German innovation system. As today's product strategies and tomorrow's markets require learning processes over several years, it is of crucial importance to uphold farseeing strategies. This is especially significant for the decoupling, which is necessary in the long term, of resources consumption and growth, as well as for the target of achieving a sustainable economy.

2. PUBLIC AND PRIVATE RESEARCH LANDSCAPE

In Germany, S&T planning, policy-making and funding take place on such a wide variety of different levels, and there are so many institutional structures, that it is difficult to avoid oversimplification when one tries to draw a coherent map of the present system. Not only do we have a wide variety of institutional actors, but within each institution we also find quite diversified structures funded by a multiplicity of sources and consisting of many subsystems (Krull, 1991b, Keck, 1993, Schmoch, 1996, Krull and Meyer-Krahmer, 1996). All of these have their specific functions concerning the objectives, the resources and the amount of freedom granted to the respective scientific and technical staff in conducting research.

When looking at the total R&D budget (see Figure 7.2), one of the most striking features is that in 1997 more than two-thirds of R&D activities were carried out in industrial laboratories. Throughout the 1980s industry's share of R&D conducted steadily increased, and in 1990 it reached a share of more than 70%. Since then, according to the data available, industry's contribution has declined over the last years to around 67%. In contrast to France, private industry is, in this respect, by far the most important actor in the German innovation system.

Due to the federal structure of the German state and the division of power anchored in the constitution (which assigns the individual state (*Land*) the task and authority of running the higher education institutions), a marked emphasis on the freedom and autonomy of science, and a market economy which fosters the initiatives of industry and private foundations, there is only limited room for centralised decision-making. However, the Federal Ministry for Education and Research (*Bundesministerium für Bildung und Forschung* – BMBF)[1] does take responsibility for the general principles and the legal framework governing the publicly financed areas of S&T. Figure 7.3 gives an overview of the actors performing R&D in the German innovation system.

2.1 – Industry and its confederation

In 1993 there was a total of about 11500 industrial businesses conducting research in Germany, of which some 9200 were based in the old *Länder* and 2300 in the new *Länder* (including East Berlin). About 90% of these research-performing companies employed fewer than 500 people each and thus fell into the category of small and medium-sized enterprises (SMEs).

---

1. The responsibility for technology has been moved to the newly-named Federal Ministry of Economics and Technology (*Bundesministerium für Wirtschaft und Technologie* – BMWi), since October 1998 responsible for near market research and technology support.

FIGURE 7.2 – The German research budget (1997) – performance of R&D

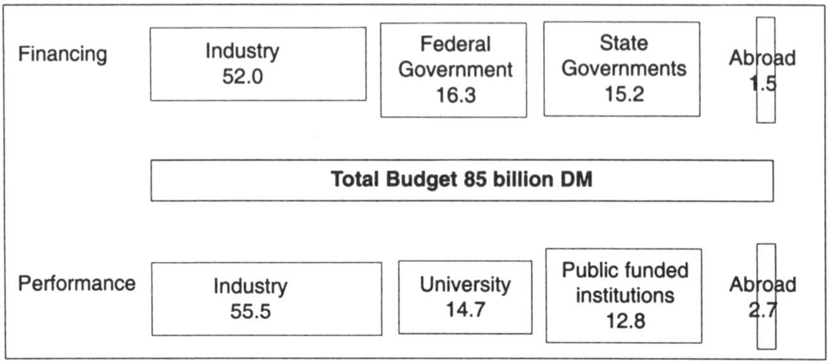

Source: BMBF.

FIGURE 7.3 – The German research landscape (1997)

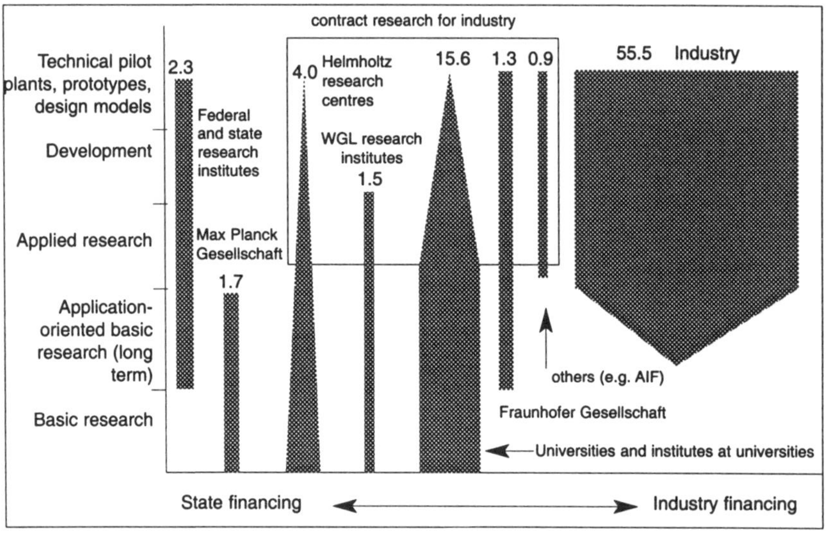

Source: BMBF. Design: IS.

However, a considerable number of companies are engaged in research on a regular, albeit discontinuous basis (i.e. a completed R&D project is not immediately followed by the next one) therefore, it is assumed that the research potential of the business enterprise sector is much greater. Estimates made by ISI suggest a total of about 25 000 companies.

Large companies account for 85% of total intramural business enterprise expenditure on R&D, followed by SMEs with some 14% (their share is rising). Looking at the total expenditure on R&D in Germany, one can see how important the business enterprise sector is for R&D. With shares of 62% (financing) and 67% (performance) in gross domestic expenditure on R&D in 1997 this sector is, quantitatively, the most significant R&D player in Germany. A comparison with the other G7 states shows that in terms of performing R&D this also applies to all other major industrialised countries. In the early 1980s intramural R&D expenditure in the business enterprise sector had double-digit rates of increase, which were considerably higher than those of government expenditure. This growth slowed down considerably during the rest of the 1980s, and the first half of the 1990s was characterised by a stagnation of this indicator. A similar slackening dynamism in R&D expenditure was also identifiable in the other major industrialised nations. With a probable increase of the business sector's intramural R&D expenditures of 5% between 1995 and 1997, this is the first growth in real terms since the 1980s. Financial R&D resources of the business enterprise sector comprise both intramural and extramural R&D expenditure. The total volume of the latter of about DM6.2 billion (1995) is low, compared with the former. However, their development in the period under review shows that their relative importance was increasing continuously.

R&D in the business enterprise sector is mainly performed in companies, but one must also consider the institutions for co-operative research and experimental development. These mainly include institutes set up by research groups formed by medium-sized companies, which make up the Confederation of Industrial Research Associations (AiF). In 1996 there were 109 such research associations with 57 institutes of their own and a total budget of around DM620 million. Co-operative research within the framework of the Federation of Industrial Research Associations (AiF) is an instrument to perform projects which surpass the capacity of single small or medium-sized enterprises (an extensive analysis of AiF can be found in Schmoch 1996). Those companies directly involved in the definition and supervision of projects benefit primarily from the results of co-operative research, but also technology transfer to other member companies is undertaken. A major focus of co-operative research is its limitation to pre-competitive problems, since several companies in the same subsector of industry have to co-operate. Nevertheless, in some industries the ratio of co-operative research to their total R&D activity is quite high. However, intensive use of co-operative research primarily concerns less research-intensive branches, whereas in research-intensive industries such as chemistry, electrical engineering and aeronautics the role of co-operative research is negligible.

2.2 – *The universities and Fachhochschulen*

The backbone of the German public S&T system (which accounts for the largest share of the publicly-financed part of research activities) is the universities. All in all, there are 84 of them. They are to a large extent financed by the state governments and have a threefold mission: educating students, training doctoral students and researchers, and carrying out research. The unity of research and teaching is an important characteristic of German universities. The training of students both provides the universities with a new generation of young scholars, and constantly gives fresh impetus to university research. By training students and researchers, the universities contribute to all non-university research activities and act as a stronghold of basic research.[1]

Even though there are differences from university to university, it can rightly be said that in no other scientific institution in Germany is the spectrum of different research activities so broad. Such diversity means that research at universities – particularly in the humanities – is pursued in relatively small units. Research projects involving just one or a few scientists are frequent and large-scale projects in S&T are seldom found at universities. This division into numerous small units makes it easier for the universities to cover a wide range of research in a variety of individual projects. On the other hand, this creates problems as far as their ability to contribute to the solution of complex issues entailing high expenditure over a long period is concerned. As the requirements of research (particularly the need for expensive equipment in the sciences, biomedical research and engineering) are constantly mounting, the concept of differentiation and specialisation is increasingly making progress. Now there is a remarkable consensus that concentration of efforts must be achieved among the universities.

In addition to the research carried out at universities, in recent years more than 220 *Fachhochschulen* (a specifically German type of higher education institution, especially in technical, managerial and social subject areas) have played an increasingly important role in training students as well as in S&T, especially in technology transfer at local and regional levels, in qualified consultancy for industry, in further training as well as in creating a link between regional requirements and applied research. They have become important contact points and providers of information, especially for small and medium-

1. An extensive description can be found in Krull and Meyer-Krahmer (1996).

sized enterprises. With their specific role in application-oriented R&D, the *Fachhochschulen* can complement the university research system. Research and development activities at *Fachhochschulen* largely depend on external funds and technology transfer revenues.

2.3 – The Max Planck Society and its Institutes

The *Max-Planck-Gesellschaft zur Förderung der Wissenschaften* (MPG) is a non-profit research organisation founded in 1948 as a successor to the *Kaiser-Wilhelm-Gesellschaft*, which was originally established in 1911. Institutes, research units and working groups run by the MPG are mainly devoted to basic research, to a large extent in the natural sciences and, to a smaller extent (about 14%), in the humanities. In contrast to the higher education system, which must cover all academic disciplines and employs some 102 000 academics (among them some 30 000 professors), the MPG can concentrate its funds and energy on selected key areas of basic research. In 1997, the MPG employed some 11 175 people in its research establishments (including 3015 scientists). In addition there are some 5700 fellows and guest scientists – almost 40% of them from abroad – who work at the 70 Max Planck institutes each year.

In its support of basic research, the MPG is highly selective. It does not primarily attempt to close national gaps in research, but tries to give strong, concentrated support to a few carefully selected fields. In this respect, the MPG proceeds from the following principles:

- The MPG takes efficient, widespread university research for granted, and pursues basic research in order to complement the research efforts of the universities (the principle of subsidiarity – which also applies to other areas of non-university research).
- The MPG supports basic research in particularly promising fields requiring interdisciplinary approaches, a great deal of institutional flexibility and international orientation. It tries quickly to embrace newly developing fields outside established disciplines (the principle of emerging fields).
- In order to fulfil its function, the MPG requires outstanding scientists who are particularly qualified for the position as head of their respective departments or institutes. For every new initiative, it is crucial to find researchers whose past achievements provide proof of their outstanding innovative capabilities and who can be expected to use the freedom and the research opportunities offered by the

position as a scientific member and director of a Max Planck institute to make exceptional progress (the principle of the eminent scientist or 'Harnack principle').

2.4 – The Hermann von Helmholtz Association of National Research Centres

The *Hermann von Helmholtz-Gemeinschaft deutscher Forschungszentren* (HGF – formerly, the *Arbeitsgemeinschaft deutscher Großforschungs – einrichtungen*) comprises 16 national research centres, which in 1997 had a combined budget of DM4.2 billion, 90% provided by the Federal Government and the rest by the Government of the state in which the respective centre is located, and some 25 000 personnel (including some 10 000 scientists). The national research centres cover a wide spectrum of R&D, ranging from basic research (in high energy physics, biomedical research, the marine sciences and so on) through mission-oriented or strategic research (aerospace, information technology, biotechnology and so on) to development of industrial technology (in areas like microelectronics, nuclear safety and super computers). These centres also operate big machines (for example neutron sources, accelerators, colliders, super computers, wind tunnels), which to a large extent are used by university researchers.

The national research centres are legally independent bodies, usually taking the form of a limited company or a foundation, with a reasonable amount of autonomy in relation to determining their research priorities. However, Federal Government authorities provide general guidance (*Globalsteuerung*) and use supervisory boards for laying down the overall objectives of research and the extent of core funding. It is mainly the BMBF which co-ordinates the work of the national research centres and, through its priority programmes, influences the process of priority-setting within each of the centres.

2.5 – The Fraunhofer Society (FhG)

A special role has been assigned to contract research, through the establishment of the Fraunhofer Society and its institutes. The *Fraunhofer-Gesellschaft zur Förderung der angewandten Forschung* (FhG) operates a network of 47 institutes with some 8000 personnel, a third of them scientists, and a 1997 budget of DM1.3 billion. Its primary task is to carry out a broad range of strategic and applied research for government and industry. The institutes perform contract research and offer information and services related to new technologies, products and processes.

The Fraunhofer institutes receive public core funding (90% from the Federal Government and the rest from the government of the state in which the institute is located), but the actual budget depends on the amount of money earned by the institute through contract research. If the institutes meet the targets agreed for contract income (usually a fixed proportion of the overall budget, varying between 70% and 80%) they enjoy considerable autonomy. Since 1973, when it was decided that the basic funding of the FhG should be financed jointly by the Federal and the State Governments, this has been the fastest growing sector of non-university research. Over the 15-year period 1975-1990, on average, one institute a year was set up in West Germany after consideration of detailed proposals on needs and opportunities. As a result of the restructuring of the East German research institutions, in 1990-1991 the FhG set up 15 new research institutes and R&D groups in East Germany.

The most important criterion for setting up a new institute or an R&D group is whether the demand for contract research in a given field (for example microelectronics, information technology, production engineering, advanced materials, environmental research, innovation studies) will support an institute in two or three years' time. Fraunhofer Institutes have become important catalysts, above and beyond their own activities, in the process of transfer of technology from basic research institutions to industry. As small and medium-sized enterprises, in particular, benefit from co-operation with Fraunhofer Institutes (by increasing their rate of innovation), they are increasingly being seen as an important part of the publicly-financed R&D infrastructure for regional development. Therefore, the FhG is expected to contribute to the innovative strength of German industry:

- it should help to identify innovation potential at an early stage by closely watching and fostering technological developments;
- it should meet the demand of industry for specific long-term contract research and technical assistance;
- it should adapt its tools and abilities to meet the requirements of the research market, and perform research work for public-sector clients.

2.6 – *The Gottfried-Wilhelm-Leibniz Association of Research Institutions*

The *Gottfried-Wilhelm-Leibniz-Gemeinschaft der Forschungseinrichtungen* (formerly *Blaue Liste* institutes) comprises 82 research institutes, which differ very much in terms of function, size, location and legal structure. Apart from a number of institutes in the humanities, most of the Blue List Institutes are either mission-oriented research institutes in the sciences and the humanities, or service institutions (such as the Information Centre for Chemistry in Berlin, the Library of Technical Information at Hanover, and the Central Medical Library at Cologne). These institutes have to meet the following requirements: the annual budget must exceed DM2.5 million (or DM1.8 million, if the institute is mainly operating as a service institution for other researchers), the institute's work must be of importance above regional level, and should be in the interest of the Federal Republic of Germany as a whole. In 1997, the 82 institutes had a total budget of DM1.7 billion and some 12 000 personnel. Roughly 50% of their budget is usually provided by the Federal Government, and the rest comes from either the government of the state in which the institute is located or from all of the states (in the case of a service institution).

After the huge expansion of the Blue List sector in 1992 (some 34 new institutes in the new *Länder* were added to the previous list), the Science Council prepared recommendations for the restructuring of the Blue List and, as a consequence of these recommendations, evaluated the performance of all of the Blue List institutes, including a statement concerning the fulfilment of the above-mentioned criteria and a recommendation on whether joint funding of the institutes should be continued or not.

2.7 – *Private foundations*

All in all, there are some 900 private foundations offering support for higher education and research in the order of DM500 million per year. Most of these foundations are quite small, with a capital stock of less than DM1 million and revenues which only allow for small-scale activities such as a few scholarships or a prize for specific achievements. However, they add a pluralistic note to the German S&T system, and often encourage researchers to pursue non-conformist, high-risk projects which otherwise might not get funded. Some of the major foundations such as the *Robert-Bosch-Stiftung*, the *Thyssen-Stiftung*, and the *Volkswagen-Stiftung* (which is by far the biggest foundation offering support for higher education and research in the order of DM100 to 150 million per year) as well as the *Humboldt-Stiftung* and the *Stifterverband für die Deutsche Wissenschaft* are strongly committed to international co-operation in S&T, and offer a wide variety of funding opportunities for foreign scholars.

2.8 – Military research

The Federal Government's position on military research is that if Germany's Federal Armed Forces are to perform their mission credibly and play their role in the Alliance effectively, they have to be equipped with state-of-the-art technology. Defence research and technology use the results of civil research in order to tackle specific defence issues. The funding for such research, which is also within the framework of international co-operation, is provided by the Federal Ministry of Defence. All in all, military R&D with an overall budget of around DM3.0 billion (1997) plays a rather minor role in Germany.

Other actors in the German system of innovation, such as the Federal and *Länder* research establishments, play specific roles in providing research results to be used for policy making or administration and are discussed in Section 3.

3. POLICY PRIORITY SETTING AND GOVERNANCE

Germany has a decentralised and diversified research system. Against this background, it is not surprising that German R&D policy relating to industry over the last decade was described as "diffusion oriented" rather than "mission oriented" (cf. Ergas 1987). A strategic industrial policy with definite missions, such as that pursued in France, can hardly be implemented in a decentralised research system. Since the 1950s, German R&D policy – as in other Western European industrialised countries – has continually extended the scope of its activities. In the 1950s, the core area of R&D support comprised the (basic) university research financed by the *Länder* and the special area research carried out in federal and state governments' research establishments. Large technology programmes were added from the mid-1950s onwards, showing a marked orientation towards goals similar to those of the US, mainly in the fields of nuclear technology, aerospace and data processing, and later microelectronics.

From the beginning of the 1970s, in order to create and support the conditions necessary for the export of technology-intensive goods, public funding flowed into research projects of industry and institutes for applied research to promote cross-sectoral technologies (for example materials), key technologies (for example microelectronics) and technological systems (for example transport systems). During the 1970s, the reform policies of the government of the Social Democrats complemented these policies with research activities in the areas of environment, public health and working conditions. Increasingly, the diffusion of technologies came into the focus of policy. This includes the support of R&D activities, the strengthening of the infrastructure for the support of technology transfer from the science system into industry, as well as the support of SMEs and less-developed regions. In the 1980s increased internationalisation and the technology race within the "Triad" enforced the support of exploitation of the research system and the creation of new firms. In the 1990s the reunification of Germany and new modes of priority actions dominated the policy agenda.

TABLE 7.2 – Instruments of national technology policy

In a narrow sense	In a broader sense
1. Institutional promotion: - national research centres; - Fraunhofer Society; - Max Planck Society; - higher education institutes; - other institutions. 2. Financial incentives: - indirect promotion; - indirect-specific promotion; - R&D projects; - risk capital. 3. Other infrastructure plus technology transfer via: - technology centres; - information and consulting; - demonstration centres; - co-operation, networks and change of personnel (e.g. change of jobs).	4. Public demand: Use of public institutions to increase market pull for "desirable" technical developments, e.g. environment-friendly consumer goods (recylcing paper, "eco" car). 5. Issues concerning company strategy: - orientation knowledge and long-term visions; - technology assessment; - technology council; - to increase awareness of the importance of innovations. 6. Education and training: Early arrangement of training opportunities for potential needs. 7. Regulatory policy measures: - competition policies; - legal framework; - influence on private demand.

Source: ISI (Meyer-Krahmer and Kuntze, 1992).

Since the 1980s the federal states have increasingly promoted R&D and innovation within regional industry. This can be illustrated by support for SMEs. Within the federal constitutional framework, state governments have the general responsibility for supporting SMEs. Financially stronger states, such as Baden-Württemberg or North Rhine Westphalia, have developed a fairly systematic and substantial set of innovation support programmes, whereas less-developed states, such as the coastal states, have not been able to do so. It should be noted that the state ministries for education provide limited funds to universities for the running of industry liaison offices.

In the meantime, public technology and innovation policy in the Federal Republic of Germany has become one of the most important arenas for tackling economic structural change, modernising the national economy and, increasingly, meeting challenges to society. Tighter public budgets, the control requirements of Parliament, the Commission of the European Communities and the need for policy administrators to learn by using the new policy instruments, have led to a strong upswing in the evaluation of technology policy programmes and instruments in Germany. The main aims of federal technology policy in Germany are to:

- continually broaden knowledge about mankind and its natural environment, and to offer new orientations;
- create conditions that will strengthen the husbanding of resources and the environment;
- improve living and working conditions;
- increase economic performance and competitiveness.

The most important approaches and instruments of current German technology policy include (see Table 7.2):

- regulatory intervention (such as regulatory policy, technical norms and protective regulations);
- financial incentives (promotion of research and development and other innovative projects within enterprises by direct support or tax measures, promotion of co-operation of enterprises with public or semi-public research institutions);
- promotion of public demand for innovative products (procurement by the public sector);
- building up the public and semi-public infrastructure (such as provision of research and development institutions, technology transfer institutions, advisory institutions);
- support for the formation of consensus on technical developments in the economy and society.

FIGURE 7.4 – Federal R&D expenditure by funding area (1997)

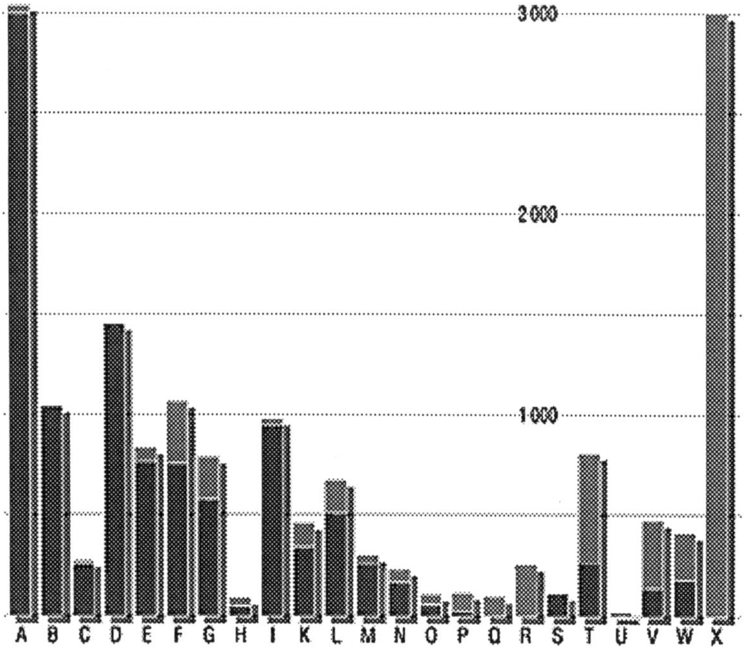

Funding area

A University supporting organisations (construction and special programmes)
B Large-scale equipment for basic research
C Marine and polar research and technology
D Space research and technology
E Energy research and technology
F Environmental and climate research
G Health research and development
H R&D on working conditions
I Information technology (including multimedia and production engineering)
K Biotechnology
L Materials research; physical and chemical technologies
M Aeronautical research
N Research and technology for ground transport (including traffic safety)
O Geosciences and raw materials supplies
P Regional planning and urban development, building research
Q R&D for the food sector
R R&D in agriculture, forestry and fisheries
S Educational research
T Innovation and improved basic conditions
U Specialised information
V Humanities, economic and social sciences
W Other activities not assigned to other sectors
X Defence research and technology

Unit: Million DM.
Source: BMBF, Facts and Figures 1998.
Note: In black BMBF budget, in grey, other Ministries.

The Federal Government's R&D activities can be subdivided into specific subject-related funding areas and priorities, based on the Federal Government's R&D planning system, irrespective of which federal department finances the activities, and irrespective of whether the funds involved are tied to specific institutions or projects or whether the activities are contributions to international research programmes. This structure is illustrated by Figure 7.4, which provides an overview of the Federal Government's R&D spending (budgeted for 1997) in the various funding areas.

Departmental research is aimed at obtaining scientific findings which are directly related to the fields of activity of a given ministry or department. Such findings are used as a basis for decision-making to ensure proper execution of departmental functions. If the general status of knowledge is not sufficient for this purpose, the necessary research activities will be carried out, primarily by federal institutions. In the area of responsibility of the Federal Ministry of Transport, for instance, the technical rules for the expansion of the federal networks of roads and waterways are continuously being optimised. The Federal Institute of Physics and Metrology, which is under the responsibility of the Federal Ministry of Economics, carries out a considerable amount of research aimed at continuously optimising the status of public metrology and testing, in response to increasingly demanding requirements. If research is to be conducted by external institutions, precise terms of reference are set out, and contracts for the corresponding research projects are invariably awarded on the basis of public tenders. It goes without saying that there is overlap between the R&D activities of the various ministries. In the interest of research efficiency and in order to avoid duplication, it is therefore necessary to co-ordinate the research efforts made by the various ministries involved. The Federal Government's Co-ordination Concept has been developed for this reason. Its primary objective is to co-ordinate the contents of research activities, which also includes the development of co-operative research programmes for several ministries (government programmes). Such programmes are currently in operation in several fields such as health research, work and technology, and aeronautical research.

Responsibilities in the area of research promotion are divided between the Federal Government and the state governments according to the federal system, as determined by the Basic Law. According to Article 30 of the Basic Law, the exercise of state powers and the fulfilling of state functions – including those relating to the state promotion of research – are a matter for the state governments, unless otherwise stipulated or allowed for by the Basic Law itself. According to the distribution of powers laid down in the Basic Law, the Federal Government is responsible, among other things, for the legislation relating to the promotion of scientific research (Article 74 No 13 of the Basic Law); it has not, however, availed itself of its power to issue a general research promotion law. The power of the Federal Government to issue framework regulations concerning the general principles of the university system (Article 75 No 1a of the Basic Law) is also relevant to university research. Most of the R&D expenditure of the states goes to the universities (around two-thirds).

In addition, intermediate bodies like the Federal State Commission for Educational Planning and Research Funding (*Bund-Länder-Kommission für Bildungsplanung und Forschungsförderung* – BLK) and the Conference of Ministers of Education and Science (*Kultusministerkonferenz* – KMK) take on responsibility for co-ordinating higher education and research policy matters across the Federal Republic.

Usually, the co-operation of the federal and state authorities is quite close, and the consensus mechanism for decision-making in the intermediate bodies paves the way for more and more jointly funded institutions and programmes. However, it should be noted that the federal as well as the state governments also run their own laboratories, especially in areas where the implementation of regulations and standards needs to be supervised, for instance in health and environmental research, but also in materials and other industrial technologies.

3.1 – The Science Council

The Science Council (*Wissenschaftsrat*) is a science policy advisory body set up in 1957 to advise the German federal and state governments on all matters of higher education and research policy. Its main function is to prepare reports and recommendations on the structural development of German universities, *Fachhochschulen* (Technical Colleges), and research institutes, and on important aspects of higher education in general. Reports and recommendations are prepared in committees and working parties made up both of academics and of representatives of federal and state governments. The Science Council has an explicit mandate to consider the quantitative and financial implications of its proposals as well as problems related to their implementation. Prominent activities of the Science Council in the 1990s were advice on the restructuring of the former East German innovation system after German unification, advice on the development of higher education in the 1990s, on the establishment of centres of excellence in selected research fields, to forecast the availability of, and demand for, research personnel in universities, *Fachhochschulen* and society at large, or to evaluate research institutes. The Science Council regularly publishes reports and statistical material on various aspects of the higher education and research system, which can be obtained from its headquarters in Cologne.

3.2 – The German Research Association

The German Research Association (*Deutsche Forschungsgemeinschaft* – DFG) serves all fields of science, engineering and the humanities by financing research projects and offering scholarships. It performs this task as an autonomous self-governing body within the German research system. Scientists and scholars choose their own subjects for research, and through electing their own peer reviewers, contribute their influence to decisions on resource allocation in universities and research institutes. It receives close to 100% of its funds from public sources. Roughly, 60% of a total of around DM2.1 billion come from the federal government, chiefly the Ministry for Education and Research (BMBF), and 40% from the 16 state governments.

The DFG spends approximately half of its funds on individual, unsolicited investigator-initiated projects where proposals are usually reviewed by mail (*Normalverfahren*). Almost 30% of its funds are spent on some 200 research centres (*Sonderforschungsbereiche*), which are reviewed by site-visits at three year intervals. In this case, proposals must come from universities (as institutions), but are otherwise unsolicited. Nearly 20% of its funds are spent on solicited, individual investigator-initiated projects in co-operative research programmes usually called priority programmes (*Schwerpunktprogramme*) for which panel review following proposal discussion with the applicants is normally used. All funding decisions on more than 12 000 research projects per year rely on peer review.

3.3 – Re-evaluation of the role of public research

When in October 1990, after more than 40 years of separation, the two parts of Germany were legally unified, there was great enthusiasm for jointly rebuilding the nation, and there were even higher expectations concerning the improvements that could be made over the next five or six years. Like many other areas of the political, economic and social systems developed after World War II in the Federal Republic of Germany (FRG) and the German Democratic Republic (GDR), science and technology (S&T) did take different courses.

While the West German S&T system was established along federal lines, with primarily decentralised ways of decision-making based on freedom in teaching and research, and on the autonomy of higher education institutions and science organisations funded by many sources, the GDR structured its S&T system according to the Russian mode.

In comparison with the GDR's S&T system, the federal, democratic structures of S&T support in West Germany appeared to present virtually no problems. Therefore, it seemed that those East German policymakers and politicians who won the first free elections in spring 1990 wanted to introduce an S&T system similar to the West German one. Nevertheless, several reports revealed many unsolved problems, deficiencies, organisational shortcomings and inherent contradictions in certain sectors of the S&T system in West Germany. It seemed only logical that the West German Science Council, when asked by the East German government to conduct an evaluation of the research institutes and to prepare recommendations for the restructuring of the entire S&T system, wanted to use its previous analyses of shortcomings in the West as a basis for the development of new concepts which could be valid for both parts of the country.

While the deficiencies of the West German system were well known, basically the West German system was adopted in East Germany, with unification, and none of the indigenous elements survived.

In the 1990s the German Federal Government started a series of different exercises of re-evaluation of the role of public research. Among them was the setting-up of a non-permanent commission on the future of basic research and larger laboratory facilities. Although reluctant to perform a thorough research forecast the commission gave, however, very concise and guiding recommendations, most of which have had decisive influence on governmental science policy decisions. The departments of the Federal Ministry for Education and Research have continued to refer to the advice of the commission. For example, action was taken to push Germany to the front of bioscience research by the end of the 1990s. Another example of this commission still casting a long shadow is the National Health Council updating the relevant subject of medical research.

The roles which the large national multidisciplinary research centres at Karlsruhe and Jülich should play, had been discussed for years, especially that part of their work that was intended to be directed towards innovation goals in important technological fields. Therefore, industry was asked by the BMBF to examine the technological power and capacities of these two research centres and to recommend how to improve partnership and co-operation in a long-term planning process. The committee known as the 'Weule Commission' called for at least a doubling of the proportion of innovation-oriented research, as being of possible importance for economic competition.

Unification of the two German systems and the previously started debate on the functions of public research, in the end, led to a common vote (in 1996) of the conference of the Federal Chancellor and the state Prime Ministers to undertake an evaluation of major parts of the German innovation system. Those organisations that are jointly funded by the federal and state bodies – the Max Planck Society, the Leibniz Institutes ("Blue List"), the German Research Association, and the Fraunhofer Society – should be assessed in view of their role and position in the national system of innovation (hence "system evaluation"). The following general conclusions emerge from this evaluation of the German public R&D system:

- there is a need for public sector R&D institutions to respond faster to changing competitive environments, which requires more systematic strategic planning;
- more interdisciplinarity is needed to cope with questions of large impact in systems of growing complexity;
- for a faster adaptation of research results as well as for a clearer understanding of results needed, more thorough links in the chain of basic research–higher education research, as well as extra-university research–industry is needed;
- as a means to achieve these goals, the introduction or intensification of internal competition in the organisations of public research is being discussed; and, finally,
- a restructuring of the salary system is being proposed to allow for more performance orientation.

The increasing technological change and the globalisation of the markets, as well as the special situation after the reunification of Germany with its severe budget restraints made the responsible persons at the BMFT change their minds (Martin, 1995). Longer-term perspectives and strategies to make better use of the limited resources were sought. The selection of support and more goal-oriented prioritisation of certain technologies seemed to be necessary. On the other hand, the state had to be careful not to intervene too much in the market and its self-regulating forces, or in the self-organised science system.

In the beginning of the 1990s, the necessity of concentrating their resources made all parties more interested in foresight and therefore some long-term prospective studies were commissioned in 1991, in order to get some early indications of the most promising developments in science and technology.

After years of considerable investments in basic science facilities and also in large technological projects, the German Federal Government started once more to reflect upon the role of the national laboratories, the question of how to focus limited resources on the most promising fields of basic research, and how to optimise the interaction between science and industry during the innovation process (see Cuhls, Uhlhorn and Grupp, 1996, p. 76). In 1994, the German Science Council recommended a foresight process guided by the science community itself and with the ambition of becoming recognised in a partnership with industry and government.

In 1990 the German Federal Government started a series of different foresight exercises. From the beginnning, it was intended to involve most of those persons, who supposedly should have had deeper insights into technological trends emerging out of new research results. As a subsequent objective, new ideas should be transported to a wider public to initiate creative discussions. As a major step forward towards a national policy, a strategic panel was established where all three parties involved – science, industry and government – could exchange arguments and clarify positions with regard to R&D priorities. In spring 1995, Chancellor Kohl set up a Council on Research, Technology and Innovation, which finished the first round of deliberation with recommendations on action in the sector of the information society. In 1996, work on bioscience began (the Council was abolished when the new Government was elected in 1998). Reports of the Council have covered the topics of the information society, biotechnology and competences for global competitiveness. Parallel to this, the Parliament has set up a commission of inquiry (*Enquete Kommission*), therefore guaranteeing an independent view of the legislative body.

4. PRIORITY PROGRAMMES AND THE SUPPORT TO SME

In its 1996 report on research, the Federal Government outlined its perspectives on research and technology policy under the heading "Making the future possible", and assigned the following central objectives to it:

- promotion of high technologies as drivers of innovation;
- innovation-oriented research policy;
- cultural vitality and performance;
- research to provide for and shape the future;

– safeguarding and improving scientific excellence;
– strengthening and networking the research system;
– developing the research system in Germany's new *Länder* (states);
– internationality and international co-operation.

From these objectives the following "visions for a future-oriented research and technology policy" were derived:

– high technology for manufacturing and services – opportunities for more employment;
– number one in Europe – opportunities for Germany offered by biosciences and biotechnology;
– growth through knowledge: the future of the information society;
– energy and the environment;
– mobility – decoupling growth and economic resource consumption;
– competitiveness through international co-operation.

New technological developments are increasingly produced at the boundaries of traditional disciplines and entrepreneurial core competences. Often they are so complex that a single business enterprise cannot provide all the know-how resources required to cope with the tasks in hand. In particular, in those cases where companies need to combine highly different technology areas, at the same time involving public research institutions and SMEs (for example microsystems, and the versatile use of new materials), high risks and communication problems obstruct the further development of technologies and their applications.

4.1 – The support to "lead projects"

Acting as a facilitator, the Federal Government supports the process of innovation by funding such projects. The joint participation of several companies and research institutions in research networks helps to make more efficient use of scarce research capacities, to accelerate the technology transfer between science and industry, to produce synergy effects and to reduce the selectivity of funding.

In view of the increasing globalisation of the international markets, a clear-cut competence profile is increasingly required. A performance-oriented research landscape and the development of competence centres, which enjoy worldwide recognition, the Federal Ministry for Education and Research (BMBF) sees therein the best guarantees for progress and economic success. Together with the central organisations for industry and for science, the BMBF is backing, above all, a stronger orientation of German research and development towards innovation. "Lead projects", as a new element in federal research promotion, should fulfil an important "pace-maker" function in this plan.

In technologically decisive strategic research fields, these lead projects should ensure that future-oriented, pioneering ideas are transformed more quickly into marketable products, processes and services. In order to reach the targeted goals, enterprises and universities, research institutions and users are combined to collaborate closely in networks, across sectors and disciplines. Industry and end-users are directly integrated in the research and transfer process.

There are public calls for proposals for lead projects by the BMBF, in the form of competitions for ideas. Within a given subject framework, the participants in the open competitions autonomously develop proposals for programmes of work. Independent juries of experts choose the best ideas and most promising suggested solutions in a two-step selection process. In the first step, up to 15 ideas per subject field are selected. The winners then have the possibility to elaborate their concept in detail. In this phase, further partners, in particular small and medium-sized enterprises, can apply to the winners to join their projects. Finally, the jury chooses up to five lead projects per subject field. The BMBF provides funding for the elaboration of the award-winning concepts and their later realisation. It is expected that the participants also contribute their own resources and bundle them in the project with the BMBF funding.

The subjects for the lead projects are laid down by the BMBF, in co-operation with top representatives from science and industry. In February 1997, the Ministry gave the starting signal for the first four concept competitions, and in December 1997 for a further three. Competitions were held, or are being held, in the following subject fields of special relevance to society:

- innovative products on the basis of new technologies, as well as the corresponding production processes;
- utilisation of knowledge accessible worldwide for education and further education, as well as for innovation processes;
- diagnosis and therapy by means of molecular medicine;
- mobility in conurbations;
- energy production and storage for decentralised and mobile utilisation;
- man–machine interaction in the knowledge society;
- nutrition – modern methods of food production.

4.2 – The support of small and medium-sized enterprises

The support of small and medium-sized enterprises is one focus of the Federal Government's programme of funding R&D in the business enterprise sector. The funds appropriated for this purpose in 1995 amounted to about DM1.2 billion, of which DM695 million were accounted for by indirect measures (non-technology specific). It should also be taken into account that many *Länder* governments consider funding R&D in SMEs an important component of supporting the business enterprise sector on a regional basis, for which they appropriate special funds. Finally, it should be pointed out that through the *Kreditanstalt für Wiederaufbau* (KfW; Reconstruction Loan Corporation) the Federal Government allocates more than DM1 billion every year under innovation loan programmes to financing innovations in SMEs. Since funds from the "European Recovery Programme"[1] are used for this purpose, these programmes have only relatively little impact on the federal budget.

Many new and small companies are confronted with a very specific problem: in the years prior to market launch, they do not have the necessary funds to finance high development expenditure. Hence, they depend on equity investments to implement expensive and high-risk innovation projects (see Koschatzky, 1997). The Federal Government has also responded to these needs by means of the following programmes:

- 'Direct-investment capital for young technology-based companies' (BJTU) is a pilot project, which created incentives for making equity investments in such companies. Between August 1989 and early 1995, a total of approximately DM314 million was mobilised for 336 companies.
- 'Direct-investment capital for small technology-based companies' (BTU) is a follow-up programme which provides two services: refinancing by means of the KfW, as well as co-investments by the *Deutsche Ausgleichsbank* (German Equalisation Bank) (see Kulicke, Bross and Gundrum 1997). The total equity investment volume envisaged amounts to DM900 million; in 1995, over DM120 million was mobilised.

In Germany's new *Länder*, these support programmes are complemented by a pilot project for 'New technology-based firms' (NTBFs) which offers not

1. The European Recovery Programme (ERP) is the German successor of the post-World War II Marshall plan initiative.

only funding but also consultancy services (see Pleschak and Werner, 1998). Until the end of 1995, the project had helped to establish 280 new companies and the total funding volume amounted to DM195.7 million.

5. TECHNOLOGICAL DIFFUSION AND RESEARCH LINKS

5.1 – Inter-sectoral technology flow: links between producers and users

Discussions about the importance of national innovation systems in view of the internationalisation of research and development and the globalisation of markets and corporate associations have also enlivened the debate on the characteristics of an efficient innovation system. Particular attention has been drawn in the literature to the vertical interdependences between suppliers, manufacturers and customers (von Hippel, 1988, Porter, 1990), interdependences between technologies (Kline and Rosenberg, 1986 "chainlinked technologies") and network externalities in the application of technology (see David, 1987, Arthur, 1988). In addition, emphasis has been placed on the necessity of close links between basic, applied and industrial R&D (Lundvall, 1992, Nelson, 1993, Mowery, 1992). From this viewpoint, the fundamental characteristics of modern innovation systems are the intersections of knowledge flows and the significance of spillover effects (see Amable and Boyer, 1992 and the references cited there). Spillover effects as (public) information are frequently referred to in the new theory of economic growth as a major source of endogenously generated growth. The following kinds of knowledge flows can be distinguished:

- Technological interdependences via flow of goods and services (interlinking via purchased materials and services and capital goods); in Germany empirical studies were carried out using input–output analyses (technology flow matrix) by DIW (Meyer-Krahmer and Wessels, 1989), IFO (1989) and Schumacher et al. (1995). All the formal transactions of knowledge can be subsumed here.
- Intra- and inter-sectoral flows of information (an overview can be found in Harhoff, 1991, Harhoff and König, 1993). All informal transactions can be subsumed here.

As has been shown by the analysis of the inter-sectoral linkages between the producers and recipients of technology (Meyer-Krahmer, 1992, Schumacher et al., 1995), roughly half of the accumulated know-how represented by the R&D capital stock of the Federal Republic of Germany is channelled, either directly or indirectly, into exports. For major sectors of the economy, there is a significant relationship between imports and exports on the one hand and the application of R&D on the other, although with varying degrees of intensity. About one-third of the imported know-how ultimately flows back into exports, i.e., relatively R&D-intensive foreign products are included in exports. German industry therefore has a specific strength in using high tech from abroad to be successful in foreign markets. One of the major lessons for science and technology policy is the need for open channels of transmission of technology from abroad. The second lesson is to strengthen the internal absorptive capacity of firms and research institutes as an important prerequisite of an efficient knowledge flow.

Besides the large R&D-intensive industrial sectors such as electrical engineering, chemicals, mechanical engineering and the automotive industry, significant industrial activities exist, for which know-how obtained from other sectors has greater importance than internal R&D activities. As far as the main technological flows are concerned, it can be noted that the most significant channel of diffusion in Germany predominantly follows the characteristics of the manufacturing processes (in contrast with the case of Italy in which diffusion follows the flow of materials). This underlines the importance of a diffusion-oriented technology policy in contrast to a mission-oriented technology policy, which is the dominant policy type in many OECD countries.

5.2 – Challenges to linkages within the research system: the increasing role of science-based technologies and interdisciplinarity

For several decades it has been taken for granted that publicly-financed research, in the long run at least, makes important contributions to industrial growth and competitiveness, but the request for immediate industrial exploitation of science has recently been dominating the debate in Germany. The increased pace of scientific progress and technical innovation, the increasing globalisation of the economy, and the current economic crisis have led to a public debate on the contribution science should make to industrial growth and innovations and, finally, to transforming them into marketable products. Some of the leading representatives of German industry have already argued that one should seriously consider giving up basic research in those areas where only Japanese companies can make use of the results (including large sectors of microelectronics and information technology).

Over and above this request for streamlining or closing down entire research areas, other industrial spokespersons have called for a new set of paradigms when it comes to assessing the performance of publicly-financed research institutions, such as by evaluating their actual or future use for industry (see Krull, 1994b). Demanding that in publicly-financed laboratories, researchers become more "industrially relevant" is just one side of the coin. The other is a severe attack on the principles of self-governance and autonomy, which so far have enabled scientists to pursue qualitative research in relative freedom and with a clear-cut orientation towards scientific excellence, rather than mixing various kinds of objectives and responsibilities.

It may be concluded on the basis of recent studies of emerging technologies in the US, Japan and Germany that the characteristics of technology at the beginning of the 21st century show a series of changes: drastically increasing costs of innovation; the growing significance of interdisciplinarity and the dynamism of overlapping technology areas; an increasingly close relationship between basic research and industrial application, as well as a tighter meshing of research and demand. The growing importance of multi- and interdisciplinarity is a reflection of the fact that a separation of technologies becomes more and more difficult and the overlapping areas are often highly dynamic. These phenomena have several consequences for systems of innovation. Even if not totally new, these phenomena are indications of a long-term process of structural change in knowledge production and diffusion. To an extent they may be new paradigms, with empirical evidence based on recent ISI studies, especially on the excellent studies of Grupp (1993), Grupp and Schmoch (1992), Schmoch et al. (1996a, 1996b). An investigation by ISI (Grupp and Schmoch 1992) on science-based technologies showed that technology can be clearly divided into science-based areas and less science-based ones. The science base is most distinct in the area of genetic engineering and pharmaceuticals, as well as in laser technology. A series of technology areas follow in order of sequence, albeit with a significant gap, which can, more or less, all be grouped under information technology: telecommunications, microelectronic parts, information storage and data processing, photo-telegraphy as well as measurement and sensor technology. These are followed by: optics, surface technology, organic and (only just) inorganic chemistry, as well as food technology. The authors show that the science connection diverges strongly between technological sectors but to a lesser extent between countries. This seems to be an inherent feature of technology. Internationally, high rates of growth in patenting are taking place in the science-intensive technology areas.

Internationally, the technological sub-sectors with the most dynamic development in the 1980s are more dependent on science than the others. In all probability this trend was expected to continue or intensify in the 1990s.

As Grupp and Schmoch (1992) show, in medicine and closely related areas (genetic engineering and pharmacy) and in organic chemistry, university research activities are above the anticipated value based on the science bond. In contrast, they are below average in the entire area of microelectronics and information technology research (electronics, telecommunications, information storage, data processing, photo-telegraphy). Involvement in the area of laser technology is average, as is the case in the areas of measurement technology, surface and food technology. It is clear from this that universities really do take up their intended role of protagonist in scientific technology, although the link between science and technology is much stronger in chemistry and medicine than in electrical engineering or information technology. The "microelectronics deficit" which has been much discussed recently in Germany also extends to the universities.

According to Grupp and Schmoch, the electrical industry in Germany does not seem to have succeeded as well as the chemical industry in stimulating or commissioning relevant research activities at universities. The argument of the closeness of the relationship is not only relevant for research aspects but also for the recruitment of qualified personnel as well as, in the reverse direction, the introduction of the most up-to-date insights from industrial research into academic curricula. This works more distinctly in chemistry than in electrical engineering. Based on an analysis of citations, Lieberman (1978) refers to the importance of informal networks. This shows that successful or deficient linkages between industry and universities occur over a whole range of dimensions and can be measured by a corresponding multitude of indicators.

The first consequence of the paradigms for science and technology policy underlines the relevance of a well-established area: the promotion of university/industry relationships. In this area, a long tradition of different institutional settings, financial incentives and other tools of technology transfer are used. Experience of these approaches is summarised in Table 7.3, which lists the main aims, selected effects and problems of different instruments for improving the links between universities, research institutions and industry. The instruments and institutions themselves are described extensively in Meyer-Krahmer (1990).

Institutional arrangements, such as contract research institutes or divisions, are an efficient way to link universities or public research institutes to industrial needs. A higher degree of user-orientation can be expected. The main disadvantage of this approach is the danger of the reorientation of university research from basic research to short-term, applied research. It is also necessary to avoid the danger of the educational tasks of universities being restricted by this approach. European experiences show that the establishment of applied R&D institutes as 'know-how pools' may be useful with multifunctional tasks (contract R&D, demonstrating, counselling, testing, training) in strategic technology fields. The institutions should be part of a university or closely linked to it. Co-operative research institutes solve sector-specific problems, which do not meet specific needs. In the Federal Republic of Germany, a high participation of small and medium-sized enterprises in such co-operations was observed. Recent approaches promoting national or international R&D networks are increasingly important for international competitiveness. An important precondition is a well-developed private and public R&D infrastructure.

Further discussion of the instruments displayed in Table 7.3, based on (mostly) German experiences show that tax concessions or general subsidies for extramural R&D strengthen existing internal or extramural R&D activities in industry, but have only limited effects on initiating R&D activities or corporations and start-ups. Only the reinforcement of existing R&D activities can be expected from this type of financial incentive. R&D grants for selected co-operation projects, which are combined with national strategic economic planning, are an efficient way to establish R&D capabilities and co-operation in strategic technology fields. The main problems are diverging interests of the research and business systems and ensuring flexibility for revising strategic technology planning under changed conditions. Co-operation also requires technical expertise on the industrial side. Usually extramural R&D must be accompanied by internal R&D capacity within industry.

Transfer units and innovation consultancy seem to be appropriate technology transfer instruments to activate small businesses and to adapt firms' behaviour, planning and attitude to new technology. However, problems relating particularly to the acceptance of newly established agencies may be encountered. These problems could be avoided if existing institutions with high industrial acceptance were used (hosting such new activities) instead of establishing totally new agencies. Transfer of individuals is one of the most efficient ways of technology transfer. Increasing the mobility of R&D personnel between industries, universities, research institutes and government is an important aim in many European countries. Existing barriers are often

TABLE 7.3 – Instruments to improve the links between universities, research institutes and industries

Instruments	Aims and effects	Selected problems
	Institutional arrangements	
Contract research institutes	High degree of user-orientation	Time horizon of research may be too short-term
Co-operative research institutes	High participation of SMEs	For sector specific R&D
Network approach	Establishing effective national and international R&D networks	Precondition: a well-developed private and public R&D base
	Financial incentives	
Tax reduction or subsidies for extramural R&D	Reinforcing existing internal or extramural R&D	Only small effects on initiating R&D co-operation
Subsidies for selected R&D co-operation projects	Establishing strategic technology fields	Diverging interests of research and business systems
	Technology transfer by	
Transfer units, Innovation consultancies	High participation of SMEs; initiating R&D co-operation and start-ups	Low acceptance of newly established agencies by industry and host institutions
Exchange of persons	Increasing mobility of R&D personnel	Different career structure in university and firms

Source: ISI (Meyer-Krahmer 1990).

a result of different organisational and carrier structures in the public and private sectors.

The instruments described above are mainly elements of a technology-push approach. It is well known within innovation research that demand is also an important determinant of industrial innovation and technology development. A concentration of R&D capabilities is necessary in those areas for which comparative advantages and/or high national or international demand can be expected. A national science and technology policy should not only be restricted to its own national R&D activities. Countries, acting as intelligent imitators, often use cheaper and more effective ways of technological development and industrial innovation, rather than following the first innovator strategy with high learning costs. This is true not only for small countries, but also for most countries which spend a significant proportion of their net income on R&D. Up to now, the appropriate mix of "first innovator" and "intelligent fast imitator" strategies is a very controversial issue in the S&T policy debate.

6. INTERNATIONALISATION OF R&D

Internationalisation of R&D can be classified into at least three processes (Archibugi and Michie, 1995): the international exploitation of technologies developed on a national level; the international technological collaboration of partners in more than one country for the development of know-how and innovations, whereby each partner retains his own institutional identity and ownership relationships remain unaltered; the international generation of technologies is carried out by multinational enterprises, which develop R&D strategies to create innovations across borders by building up research networks.

German corporations have also been conducting R&D abroad at a fast-growing rate as they internationalise (this section follows NIW/DIW/ISI/ZEW, 1998, p. 46). This process has not, however, uncoupled itself from the internationalisation of production. R&D spending abroad reached an estimated DM10 billion in 1995, representing 17% of all corporate R&D spending in Germany that year. The chemical and pharmaceutical industries accounted for nearly two-thirds of this amount (Table 7.4). The share of Germany's foreign R&D capability located in the US, the most important foreign research location for German firms, is quite large in relation to Germany's direct investment portfolio. German companies spent some $4 billion on R&D activities in the US in 1995. This expenditure is correspondingly reflected in the number of patent applications that German companies developed in the US. Great Britain, France and Austria are other important R&D locations for German companies. Only a few German corporations have begun to conduct research in Japan (in chemical products). Approximately 3000 foreign industrial subsidiaries operate in Germany and more than half of them conduct research and development here. These

TABLE 7.4 – R&D expenditure of German firms abroad, by product fields, 1995

Product fields	%
Chemical products	31%
Pharmaceuticals	30%
Motor vehicles	20%
Electrical engineering	3%
Machinery	3%
Aircraft	3%
Telecommunications	3%
Measurement technologies	2%
Computer technologies	1%
Medical technologies	1%
Other	3%
Total	100%

Source: NIW/DIW/ISI/ZEW (1998).

subsidiaries spent DM9.5 billion on R&D in Germany in 1995, employing some 42500 persons (half of whom came from the US, which has approximately 25% of its foreign R&D capability stationed in Germany; the other half came from Europe). This places R&D expenditure by foreign companies in Germany approximately on a par with the R&D expenditure of German companies abroad. At 15%, the foreign subsidiaries' R&D expenditure here also parallels the 16% share that their combined workforce in Germany represents out of the total industrial workforce, a figure which has remained unchanged for the past several years.

As a consequence of faster growth in overseas markets, it can be expected that successful companies, in particular, will continue to increase their overseas involvement in production and concomitantly in research and development. Germany's R&D involvement abroad is expected to grow at a faster pace than foreign-owned R&D capability in Germany.

It is evident that the US is a significant recipient of foreign R&D expenditure. Figure 7.5 illustrates that the industrialised states have continuously extended their R&D activities there. At the same time, with the exception of the Netherlands, in all the countries examined here the growth rate for the R&D expenditures increased in the second half of the 1980s. In 1994, we find Germany, Great Britain and Switzerland in the lead, followed by Japan. Preliminary data of the US Department of Commerce for the R&D expenditures of foreign enterprises for the year 1995 indicate that German

FIGURE 7.5 – R&D expenditure of foreign enterprises in the US 1977-1994

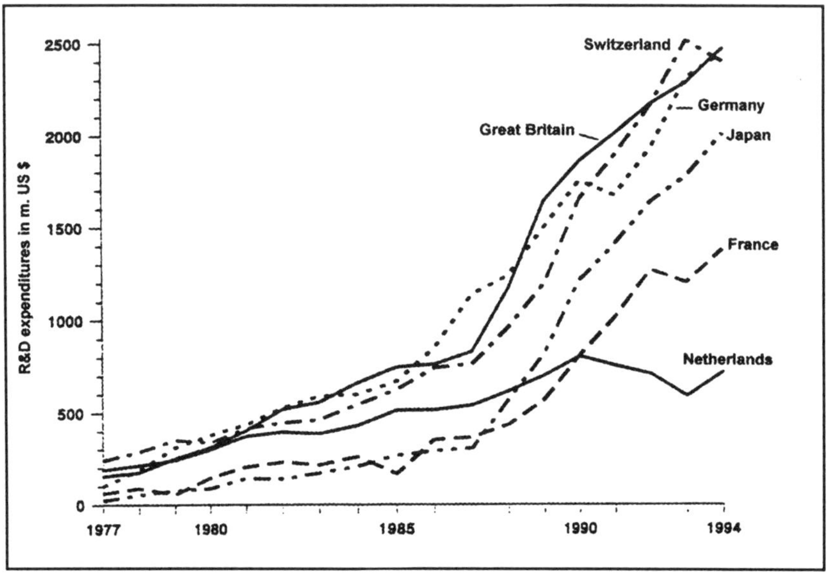

Source: Jungmittag, Meyer-Krahmer and Reger, 1999.

enterprises have yet again significantly increased their expenditure and now spend approximately \$3.9 billion on R&D (just \$2.5 billion in 1994). They thus have clearly the greatest R&D potential in the US, ahead of Swiss and British enterprises. The great increase in R&D expenditure by German enterprises in the US, from 1994 to 1995, is primarily explained by the boom in company takeovers, especially by important acquisitions in the pharmaceutical industry, in which the R&D capabilities also became part of the German enterprise (the following sections are based on Jungmittag, Meyer-Krahmer and Reger, 1999).

A similar picture is found when the number of independent research centres belonging to foreign enterprises in the US in the year 1994 is considered. Japan assumes the top position here, followed by Great Britain and Germany. If the technological foci of the foreign research centres are considered at the same time, it is seen that the enterprises conduct research in the US mainly in those fields in which they have their technological strengths in their home country. So the centres belonging to Japanese and Korean companies concentrate on the fields of computers and electronics. The European enterprises place their main emphasis in the sectors of chemicals, pharmaceuticals, biotechnology and new materials. Furthermore, German and Dutch enterprises perform research in the electronics sector, British companies in precision engineering, measurement and control technology, and Swedish companies in mechanical engineering.

The models of internationalisation of industrial R&D in three key technologies – pharmaceuticals, semiconductor technology and telecommunications – have been analysed in depth in a recent study carried out by ISI, DIW and ZEW (Reger, Beise and Belitz, 1999) for the BMBF. A number of results emerged from this study which advanced and complemented previous analyses and are summarised below.

The internationalisation of enterprises is more advanced in some branches than in others. Differences between sectors regarding the degree of liberalisation of international trade, the regulation of streams of direct investments, specific features of regional demand, economies of scale in production and the internationalisation of technological knowledge, result in different levels of internationalisation. The surveys in the three selected technology fields have shown that the internationalisation of R&D is mainly influenced by three factors, namely:

- early linkage of R&D activity to leading, innovative clients ('lead users') or to the 'lead market';
- early co-ordination of the enterprises' own R&D with scientific excellence and the research system;
- close links between production and R&D.

Our analysis showed that internationally active enterprises think in terms of value-added chains and process chains. Consequently, the criteria for selecting a location for R&D include not only factors of supply, such as a well-developed research infrastructure, but also demand factors, which are increasingly playing a more important part in the decisions of enterprises. Only by linking various value-added chains can (relatively) non-transferable "performance alliances" be created, establishing Germany internationally in selected fields as a location for competence centres which it would be difficult to transfer or duplicate elsewhere.

The importance of lead markets in anchoring existing industrial R&D activities and attracting new activities has increased. The market's function as a "lead market" is decisive for innovations which will become fully mature only when they come into close contact with demanding, innovative customers. In fields of technology that are strongly science-based, the results of scientific research constitute a driving force in the internationalisation of innovation processes. In both cases, regional proximity to external partners,

TABLE 7.5 – Determinants of the internationalisation of R&D in selected fields of technology

	Importance of R&D link to:	Lead market	Science/research system	Production
Pharmaceuticals	Pre-clinical research	Low	Very high	Low
	Clinical research	Very high	High	Low
Semiconductor technology	Process technology	Low	High	High
	Product development	Very high	Low	Low
Telecommunications technology	Hardware	Low	High	High
	Software	Very high	Low	Low

Source: Reger, Beise and Belitz (1999).

such as customers, competitors and scientific institutions, is an advantage. If there is a close interlinking of production and R&D activities, internationalisation of R&D follows internationalisation of production. The internationalisation of production is then the main driving force behind the internationalisation of R&D.

One central finding of this survey is that the determinants of internationalisation in the three fields of technology considered are different (Table 7.5). The dynamics of innovation concerning product development in semiconductor technology and in software telecommunications technology are largely driven by lead markets.

In process technology in semiconductor technology and in hardware for telecommunications, the linkage of production with R&D is also a significant factor. In the pharmaceutical industry a clear distinction has to be made between pre-clinical and clinical research: the innovation dynamics in pre-clinical research are driven by scientific excellence, whereas in clinical research it is the lead market that is the driving force. The link of R&D to production is very loose in this case.

It would be a fallacy to assume that national innovation policies will lose in significance as a result of globalisation. The international exploitation of innovations requires that national governments create conditions under which new technologies can be exploited within the respective countries. International technological co-operation is based on the national technological capabilities and possibilities available to the co-operation partners (Archibugi and Michie, 1995, p. 134). The international generation of innovations requires efficient national innovation policies, which meet the new challenges. This view is supported by new results obtained in innovation research, which show that, in the face of increasing international mobility on the part of enterprises and technologies, as well as the growing harmonisation of vital pre-conditions (infrastructure, human capital), efficient national innovation systems will have an increasingly significant role to play in the organisation and promotion of innovations (Porter, 1990, Lundvall, 1992 and Nelson, 1993).

7. CONCLUSIONS AND OUTLOOK

7.1 – Strengths and weaknesses of the German innovation system

Is the German innovation system still a successful model? The current picture seems still to be rather favourable, but fundamental structural and institutional changes are necessary to retain the system's long-term robustness. To cut a long story short, the German innovation system can be characterised in the following very simplified way:

- A strong role of private industry as a primary actor and a rather high proportion of industrially financed R&D with two clusters (in terms of Porter, 1990) of industrial R&D intensive sectors (mechanical engineering and the automotive industry, chemical and related industries); industry highlighting strengths in existing markets and weaknesses in new markets.
- A highly decentralised and a differentiated research system, but the institutional setting of the public research infrastructure does not fit modern modes of knowledge production; structural changes are on the top of the policy agenda instead of individual interventions on the level of research projects.
- Up to now Germany has a specific strength in "deepening" (see Ergas, 1987) further development and diffusion of key technologies compared to the US with its specific strength in "shifting" the research frontier is in many fields. The actors in the German innovation system are highly competent, but not flexible enough (no breakthroughs).
- The system is mostly manufacturing industry-based, but lacking in services. It is a growing economy but without additional jobs.
- Beside the Federal Government, the regional governments (because of their responsibility for universities and policies for small and medium-sized enterprises), local, European and public–private actors have become increasingly important.

The number of highly skilled employees is growing considerably faster than that of employees as a whole, workplaces with high qualification requirements are less subject to cutbacks than others. This creates severe scarcities of engineers and certain natural scientists on the German labour market, which is faced by a general unemployment rate of about 10%. New jobs and intensive utilisation of new technologies proceed in parallel. Innovative service companies in particular have much more favourable and, unlike industrial enterprises, positive employment expectations (see NIW/DIW/ISI/ZEW, 1998). The labour-saving effects of new technologies are unmistakable. Yet, many studies show that new technologies, as well as globalisation, primarily reduce the demand for less-qualified personnel. These persons are the real losers of technological change. Economic policy, not only S&T, has to develop new solutions, as in many OECD countries.

7.2 – Restructuring public research

An important consequence for science and technology policy is the necessary change of organisation, communication, interaction and motivation within the academic research world. The consequence for science-based technologies is the need for new ways of linking basic and applied research. Also trans- and interdisciplinarity afford better horizontal linkages between disciplines. Often in the literature the term "interdisciplinarity" is misleadingly used as a synonym for integration or for the mixing of disciplines. It is essential to understand that first-class trans- and interdisciplinary research is highly dependent on first-class disciplinary quality of the scientists involved in interdisciplinary research. Therefore, an efficient linkage between (and not the integration of) disciplines is crucial. Possible tools and mechanisms of such linkages are:

- Organisation of research: problem orientation in the case of well-definable social or industrial-technical problems. This requires, in contrast to the currently predominant internal objectives of science, new ways of project organisation and management.
- Better integration of the long-term application-oriented basic research in applied research would meet the future requirements better. This could be achieved, for example, by a better institutional network, co-operative research with specific subjects, new models of financing, improved communication and other assessment criteria.
- Team research must be strengthened besides the currently predominant orientation of academic research towards individualised research

setting. The promotion of research groups is a good beginning here, as well as the setting-up of institutes under co-operative management.

- Improved intra- and inter-sectoral mobility of researchers on an international level and also between science and industry.
- Increased flexibility of research structures: more rapid taking-up of new developments by: (i) flexibilisation of the present rigid public service rules and budget laws, (ii) deregulation of the academic administration, and (iii) networking of research institutions for a limited time, especially in an international framework ("virtual research institutes").

Applied research must be geared to a change in roles: it has not only to perform the (classic) transfer from basic research to industrial research, but also in the opposite direction, to transfer complex application problems to an attractive agenda of basic research.

A further consequence for science and technology policy relates to another traditional aspect of this area, university training. Science-based technologies as well as trans- and interdisciplinarity lead to an increasing demand for PhDs in engineering and natural sciences. This can lead to severe frictions and time-lags in the domestic educational system, which may give an incentive for firms to invest in those countries, in which such academic potential is available.

A final consequence is due to relations between industrial R&D and basic and academic R&D. In the past, government science and technology policy emphasised mainly basic and public R&D, assuming that the volume and structure of industrial and applied R&D is the responsibility of firms. The paradigms described lead to an increasing interdependence between public and industrial R&D, which may lead to the conclusion that industrial R&D cannot increase without an increase in public R&D. Up to now, there is not enough empirical evidence available to prove this. Nevertheless, it would have fundamental implications for future policy-making processes in science and technology.

7.3 – Political governance: changing roles and instruments

The changes of the context of S&T policy with growing environmental, economic and social problems, a growing systemic complexity and restricted budgets call for new ways for effective and efficient policies. The following changes in the policy-making context can be observed:

- innovation system: growing environmental, economic and social problems;
- knowledge production: trans- and interdisciplinarity, entwinement of basic and industrial research;
- policy: restricted budgets, need of effectiveness and efficiency.

Some indications exist of a change from a rational policy "decision model" (defining aims, making decisions, using well-known instruments, control of implementation), to "process organisation" (establishing visions, discussions and consensus platforms, assessing options, use of new sources of information such as technology assessment, evaluation, foresight).

Instead of "decision models" for policy-building and operation, a different model is needed which aims rather at organising societal processes to form a system of innovation to cope with the changed conditions. The German S/T system is characterised by a relatively high degree of institutional differentiation and autonomy of the major actors, one could call it a "multi-actor network" (Kuhlmann, 1998). Using evaluation procedures in such an environment is one means of creating more transparency in S&T policy. This implies an analysis of the frequently contradictory rationales of S&T policy players in order to facilitate a mutual critique and learning process. It also involves the assessment of indirect and unintended impacts of S&T policy initiatives in societal, economic and ecological spheres. This would require feedback of the knowledge gained through evaluation of the actor's networks and arenas.

In an excellent case study on interdisciplinary clinical research centres, which was deliberately designed as a policy moderation experiment, Kuhlmann (1998) shows that the underlying structure of the S&T policy arena and the present and near-future policy agenda call for advanced forms of mediated policy dialogue, and the conduct of policy evaluations producing actionable knowledge for policymakers instead of mechanistic judgements about their own performance. Ongoing analyses of the practical use of various evaluation concepts indicate that there is an important relationship between the nature of the related policy negotiation arena and the extent to which evaluation can be used as a moderation medium.

One eminent example of a new type of policy emerging from such considerations about changing policy contexts is the "BioRegio Competition". This promotional concept, announced in October 1995, also deals with a new, holistic approach to research and technology policy, which aims at integrating biotechnology capacities and scientific, economic and administrative activities. The aim is to push the commercialisation of biotechnology in

Germany, and thus create successful and also internationally acknowledged competence centres for biotechnology research and applications for the future in Germany. The object of support is the drawing-up of biotechnology development concepts, with which the regions should compete among themselves. The main focus here is the co-operation of all participants from science, industry and the public administration. Up to now this concept of pushing regional competence centres runs very successfully. It also demonstrates that the national level is no longer the dominant public player, regions as well as Europe are continuously increasing their influence. As in other European countries a new equilibrium in the division of labour between the European, national and regional levels, has to be placed on the future policy agenda.

7.4 – Internationalisation of R&D

In the context of regular reporting to the Federal Ministry of Economics on the structure of industry, a study by the HWWA – *Institut für Wirtschaftsforschung* (1995) comes to the conclusion that the globalisation of German industry (primarily with regard to production) implies a growing importance for industrial policy. With the increasing internationalisation of production, improving the quality of locations would mainly mean improving the qualification and flexibilisation of the workforce, promoting investment and accelerating public decision-making. According to this report, the financial support of domestic enterprises (i.e. enterprises with their headquarters in Germany), including the public promotion of technology, are increasingly missing the mark, since it is not certain whether these measures will generate income in national or regional locations.

An investigation by Gerybadze, Meyer-Krahmer and Reger (1997) comes to similar conclusions (for example on subsidising of R&D) with regard to this question of whether traditional technology policy is "on target" or not. However, it is precisely this circumstance which leads the authors to plead the case for a reformulated concept of technology promotion, namely: both to support national research institutions and enterprises on their path towards globalisation and, at the same time, to gain foreign research institutions and enterprises for the national innovation system and, in both cases, to attain synergy effects and spillover effects beneficial to the location. The fact that, on its own, technology policy will fall into an "inadequacy trap" under these altered circumstances needs to be emphasised again and again. Technology and innovation policy is a strategic, interdisciplinary task, and the effectiveness and success of this policy will depend, in large measure, upon whether it proves possible to establish internal networking in this field between policy areas, which have previously been fragmented.

It is decisive for the German innovation system that it should enable efficient transfer and rapid learning to take place through intelligent interlinkages, in order to pursue the strategies of the rapid second innovator, become seriously regarded as an international "player" and intelligently transfer structures, processes and framework conditions that foster innovations. This recognition brings with it a number of specific implications for technology policy, of which some examples are cited here. Some of these have already been elements of S&T policy for many years now, whereas others set new policy accents:

1. Supporting the international activities of national public R&D institutions and enterprises by: establishing international training/education and research programmes; fostering the international mobility of students and scientists as well as encouraging researchers and students from abroad to come to Germany; supporting the presence of domestic research institutions in other countries; supporting enterprises in their efforts towards a stronger global presence in R&D; building up technological competence and positioning as an international player to be taken seriously in areas which have not been among the country's classical strengths so far.
2. Supporting the location of foreign R&D establishments.
3. Two-way incentives such as promoting transnational projects; supporting the "brokerage function" of public research institutions, to support the international exchange of technology supply and demand; monitoring innovation-friendly structures in other countries and making use of this experience for national policy.

Our analysis of the innovation activity of transnational enterprises shows that they are increasingly thinking in terms of integrated process chains and lead markets, and are not primarily transferring their value-added to places which provide the best conditions for research alone (Gerybadze, Meyer-Krahmer and Reger, 1997). Market conditions obviously play an increasingly important role in R&D allocation decisions. The attractiveness of the national innovation system from this perspective is determined not so much by comparative, static competition factors such as costs and wages, as by its "dynamic efficiency".[1] This is largely dependent on the extent of social and

1. Economic theory differentiates between static efficiency – relating to one point in time – and dynamic efficiency – relating to a long-term development. It is quite possible for static and dynamic efficiency to conflict with one another.

organisational intelligence in the identification and acceptance of new structures and markets. Will complex system innovations (such as road pricing, product/service packages, closed-cycle economic concepts, new applications for information technology) be elaborated in Germany, for example, which will be used worldwide? Proactive learning through numerous field trials and pilot schemes for the identification of technical, economic, legislative and social solutions is important. Learning processes of this kind often take years. The innovation system that first succeeds in mastering these complex solutions gives participating enterprises competitive advantages, and appears more attractive to foreign investors. Candidates for such a policy approach in Germany are the fields of production, mobility, construction, health and environment.

REFERENCES

Amable, B. and R. Boyer, 1992, *The R&D Productivity Relationship in the Context of New Growth Theories. Some Recent Applied Research*, Paris: INRA/CEPREMAP.

Archibugi, D. and J. Michie, 1995, The globalisation of technology: a new taxonomy, in *Cambridge Journal of Economics*, 19, 121-140.

Arthur, W.B., 1988, Competing Technologies: An Overview, in G. Dosi et al., *Technical Change and Economic Theory*, London: Pinter.

BMBF (Bundesministerium für Bildung und Forschung) (Hrsg.), 1999, *Zur technologischen Leistungsfähigkeit Deutschlands*, Bonn.

Bräunling, G. and M. Maas, 1988, *Nutzung der Ergebnisse aus öffentlicher Forschung und Entwicklung, Forschungsbericht an die EG*, Luxemburg: Amt für amtliche Veröffentlichungen der Europäischen Gemeinschaften.

Cuhls, K., C. Uhlhorn and H. Grupp, 1996, Foresight in Science and Technology-Future Challenges of the German Science and Technology System, in W. Krull and F. Meyer-Krahmer (eds.), *Science and Technology in Germany*, London: Cartermill, pp. 63-81.

David, P. A., 1987, Some New Standards for the Economics of Standardization in the Information Age, in P. Dasgupta and P. Stoneman (eds), *Economic Policy and Technical Performance*, Cambridge: Cambridge University Press, pp. 206-239.

Ergas, H., 1987, Does technology policy matter?, in B.R. Guile and H. Brooks (eds), *Technology and Global Industry. Companies and Nations in the World Economy*, Washington, DC: National Academic Press, pp. 191-245.

Gehrke, B. and H. Grupp, 1994, *Strukturelle und technologische Position der Bundesrepublik Deutschland im internationalen Wettbewerb*, Band 8, Heidelberg: Physica-Verlag.

Gerybadze, A., F. Meyer-Krahmer and G. Reger, 1997, *Globales Management von Forschung und Innovation*, Stuttgart: Schaeffer-Pöschel-Verlag.

Grupp, H. (ed.), 1993, *Technologie am Beginn des 21. Jahrhunderts, Schriftenreihe des Fraunhofer-Institus für Systemtechnik und Innovationsforschung*, Heidelberg: Physica-Verlag.

Grupp, H. and U. Schmoch, 1992, Perceptions of scientification of innovation as measured by referencing between patents and papers: Dynamics in science-based fields of technology, in H. Grupp (ed.), *Dynamics of science-based innovation*, Heidelberg: Springer-Verlag, pp. 73-129.

Harhoff, D., 1991, *R & D incentives and spillovers in a two-industry model*, in Discussion paper series, Mannheim: Zentrum für Europäische Wirtschaftsforschung, pp. 30-32.

Harhoff, D. and H. König, 1993, Neuere Ansätze der Industrieökonomik - Konsequenzen für eine Industrie- und Technologiepolitik, in F. Meyer-Krahmer (ed.), *Innovationsökonomie und Technologiepolitik*, Heidelberg: Physica-Verlag.

Hippel, E. v, 1988, The Sources of Innovation, New York/Oxford: Oxford University Press.

HWWA-Institut für Wirtschaftsforschung, 1995, Grenzüberschreitende Produktion und Strukturwandel – Globalisierung der deutschen Wirtschaft, Forschungsauftrag des Bundesministeriums für Wirtschaft, Hamburg.

IFO Institut für Wirtschaftsforschung (ed.), 1989, *Innovation, Wachstum und Beschäftigung*, Berlin: de Gruyter.

Jungmittag, A., F. Meyer-Krahmer and G. Reger, 1999, Globalisation of R&D and Technology Markets – Trends, Motives, Consequences, in F. Meyer-Krahmer (ed.), *Globalisation of R&D and Technology Markets. Consequences for National Innovation Policies*, Heidelberg: Physica-Verlag, pp. 37-78.

Keck, O., 1993, The National System of Technical Innovation in Germany, in R. Nelson (ed.), *National Innovation Systems. A Comparative Analysis*, New York/Oxford: Oxford University Press, pp. 115-157.

Kline, S. and N. Rosenberg, 1986, An Overview of Innovation, in R. Landau and N. Rosenberg (eds), *The Positive Sum Strategy*, Washington: National Academy Press.

Koschatzky, K. (ed.), 1997, *Technology-Based Firms in the Innovation Process. Management, Financing and Regional Networks, Technology, Innovation and Policy*. Series of the Fraunhofer Institute for Systems and Innovation Research (ISI), Vol. 5, Heidelberg: Physica-Verlag.

Krull, W., 1991, Priority-Setting in R&D – Reflections on the Current Situation in Germany, in Danish Council for Research Policy (ed.), *Towards National R&D Specialization*, Conference Papers, Copenhagen, pp. 35-45.

Krull, W., 1994, Beyond Unification: German Science and Technology Policy in Transition, *Science and Technology Policy* Vol. 7, 3, 10-15.

Krull, W. and F. Meyer-Krahmer (eds), 1996, *Science and Technology in Germany*, London: Cartermill.

Kuhlmann, S., 1998, Moderation of Policy-Making? Science and Technology Policy Evaluation Beyond Impact Measurement-The Case of Germany, in *Evaluation*, 4, No. 2, pp. 130-148.

Kulicke, M., U. Bross and U. Gundrum, 1997, *Innovationsdarlehen als Instrument zur Förderung kleiner und mittlerer Unternehmen. Technik*, Wirtschaft und Politik Schriftenreihe des Fraunhofer-Instituts für Systemtechnik und Innovationsforschung (ISI), Band 27, Heidelberg: Physica-Verlag.

Lieberman, M. B., 1978, A Literature Citation Study of Science-Technology Coupling in Electronics, in *Proceedings of the IEEE* 66, pp. 5-13.

Lundvall, B.-Å. (ed.), 1992, *National Systems of Innovation: An Analytical Framework*, London: Pinter.

Malerba, F., 1994, Research Interfaces in Europe: The Role of Innovation Systems, Institutional Architectures and Technological Competences. CEPR/AAAS Conference, Stanford University.

Martin, B.R., 1995, Foresight in Science and Technology, *Technology Analysis & Strategic Management*, Vol. 7, 2, pp. 139-168.

Meyer-Krahmer, F., 1990, Science and Technology in the Federal Republic of Germany, in *Longman Guide to World Science and Technology*, Harlow: Longman.

Meyer-Krahmer, F., 1992, The Effects of New Technologies on Employment, *Economics of Innovation and New Technology* Vol. 2, pp. 131-149.

Meyer-Krahmer, F. and H. Wessels, 1989, Intersektorale Verflechtung von Technologiegebern und Technologienehmern: Eine empirische Analyse für die Bundesrepublik Deutschland, in *Jahrbücher für Nationalökonomie und Statistik*, Vol. 206/6, pp. 563-582.

Meyer-Krahmer, F. and K. Kuntze, 1992, Bestandsaufnahme der Forschungs- und Technologiepolitik, in Grimmer, Häusler, Kuhlmann, Simonis (eds): *Politische Techniksteuerung*, Opladen: Leske + Budrich.

Mowery, D., 1992, The US National Innovation System: Origins and Prospects for Change, *Research Policy*, 21, pp. 125-144.

NIW/DIW/ISI/ZEW, 1998, *Germany's Technological Performance*, Report to the Federal Ministry of Education, Science, Research and Technology, Hannover/Berlin/Karlsruhe/Essen/Mannheim.

Nelson, R. (ed.), 1993, *National Innovations System: A Comparative Study*, New York: Oxford University Press.

Pleschak, F. and H. Werner, 1998, *Technologieorientierte Unternehmensgründungen in den neuen Bundesländern*, Band 33, Heidelberg: Physica-Verlag.

Porter, M. E., 1990, *The Competitive Advantage of Nations*, London: Macmillan Press.

Reger, G., M. Beise and H. Belitz (Hrsg.), 1999, *Internationalisierung technologischer Kompetenzen. Trends und Wirkungen in den Schlüsseltechnologien Pharmazeutik,*

Halbleiter und Telekommunikation, Heidelberg: Physica-Verlag.
Schmoch, U. (ed.), 1996, *Technology Transfer in Germany*, Fraunhofer Institute for Systems and Innovation Research, Karlsruhe.
Schmoch, U. et al., 1996a, The Role of the Scientific Community in the Generation of Technology, in U. Schmoch (ed.), op. cit.
Schmoch, U. et al., 1996b, The Organisation of Interdisciplinarity-Research Structures in the Areas of Medical Lasers and Neural Networks, in U. Schmoch (ed.), op. cit.
Schumacher, D. et al., 1995, Technologische Wettbewerbsfähigkeit der Bundesrepublik Deutschland, in *Deutsches Institut für Wirtschaftsforschung: Beiträge zur Strukturforschung*, Heft 155, Berlin.

CHAPTER 8

THE UNITED KINGDOM NATIONAL SYSTEM OF RESEARCH, TECHNOLOGY AND INNOVATION

LUKE GEORGHIOU

1. OVERVIEW

1.1 – Introduction

In many ways, the central concern of United Kingdom policy for research, technology and innovation has remained constant throughout the 20th century, resting upon a preoccupation with the relationship between science and economic success. The origins of an explicit institutional structure for supporting science came during World War I, in reaction to a realisation of dependency upon the products of German industry. An advisory council, a Department of State, co-operative industrial research, and support for university research resulted from these actions.[1] Shortly afterward, in 1919, the distinguished economist Alfred Marshall was noting that the 'revolutionary discoveries' made by British scientists were being exploited by businesses in Germany and other countries 'where industry and science have been in close

1. For an account of the formation and subsequent operation of the Advisory Council on Scientific and Industrial Research, the Department of Scientific and Industrial Research, and the formation of the research associations see Gummett (1980).

touch with one another' (Marshall, 1919). At the end of the century the main thrust of science and technology policy remains that of finding better ways to exploit the nation's scientific capacity, this despite a series of classic studies[1] over the decades which have shown that, important as academic science is to the UK, it is not able, in a linear model mode, to transform a generally unsatisfactory performance in industrial innovation.[2] In a previous review of the British national innovation system Walker identified four persistent reasons for the erosion of the UK's industrial standing, in summary: the short-term outlook of capital markets, under-resourcing of education and training, weakness in co-ordination (including between science–industry, producer–user and banks–industry), and the loss of a strong technological culture in which engineers are accorded high status (Walker, 1993).

Even recognising the complexity of the science-industry relationship within the national system of the UK, the performance of the two sectors does tend to diverge. Hence, the UK is one of Europe's major scientific performers, both in terms of absolute expenditure and according to major output indicators. Recent bibliometric studies have shown a strong performance by England, second only to the US in many medical and life science fields, mathematics and physical sciences, but lagging in engineering (Adams et al. 1998). Earlier work by the Office of Science and Technology showed that the "productivity" of UK science was the highest in the world as measured by papers produced divided by millions of pounds spent on the science base in the three years previously (18.83 papers per million pounds). However, this statistic is symptomatic of a core problem as well as an achievement, the negative aspect being that a high-quality science base has been seriously underfunded for many years. Remedial action is only just beginning. Government investment in civil R&D in 1996 was about £600 million lower than in 1981,

---

1. As early as 1957, Carter and Williams in an investigation of hindrances to 'the speedy application of science' found the debate dominated by unsupported generalisations and that examination of examples 'makes the proposition that Britain is pre-eminent in fundamental science, but backward in applying it to industry, appear rather meaningless', p. 28. A breakthrough in understanding the deficiencies of linear mode thinking was provided by Langrish et al. (1972), a major empirical study which found that basic research only rarely was a driver of innovation, but was always providing industry with techniques of investigation, and people trained in techniques and scientific thought.
2. This poor performance in the period up to 1980 is thoroughly described in Pavitt (1980).

and as a percentage of GDP had fallen from 0.72% to 0.43%, in sixteenth position among OECD countries (Save British Science Society, 1998). The basic structure of research funding is shown in Figure 8.1.

Turning to the industrial sector, in 1997-1998 R&D performed by business enterprise accounted for 65% of the total, and government funded only 9.7% of this. Business spend is highly concentrated with three pharmaceutical companies accounting for 25% of the total and the top twenty companies accounting for 67%. There has been widespread concern about the innovative performance of UK industry outside the few leading sectors and firms. This deficiency forms part of a policy diagnosis addressing the wider competitiveness of industry. Direct measurement of innovation has proved elusive despite several efforts but proxy measures such as industrial R&D expenditure indicate the problem. Most recently, the 1998 R&D Scoreboard showed that while the world's top 300 spenders listed have increased R&D spending by an average of 10%, the 19 UK firms in this list have increased theirs by only 5% (DTI, 1998). This was from a baseline of 50% less aggregate R&D intensity than other G5 countries in the first place. The 1999 Scoreboard qualified these observations by noting that if the UK companies in the

FIGURE 8.1 – Research funding flows in the UK

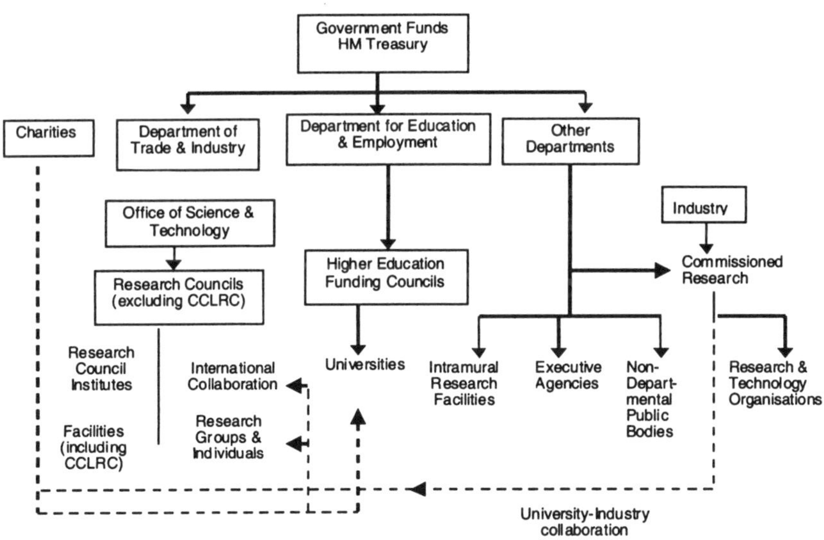

Source: Adapted from Cunningham and Hinder (1998).

international list were compared with others in the same sectors, their R&D intensity was only slightly below average. On the other hand, the exceptional performance of pharmaceutical companies raised the average considerably.

The challenges posed to the UK by these problems have resulted in the development of an innovative portfolio of science, technology and innovation policies. Reforms to the system have in one sense made the UK the largest 'laboratory' for applied science policy studies. The key developments will be described in the following sections after brief descriptions of the budgetary and policy-making structures.

1.2 – Funding patterns

UK Government expenditure on science, engineering and technology totalled £6372.4 million in 1997-1998. The breakdown of this expenditure by sector of government is shown in Table 8.1 and by Frascati type of research activity in Table 8.2.

Several points are worth noting or clarifying:

- Expenditure on defence remains high, though it has fallen as a proportion of the total in recent years and by 25.6% in real terms since 1986-1987 (section 7).
- Basic research is performed mostly in universities, which receive the bulk of their funding through the dual support system (section 3).
- Block funding, covering infrastructure and the proportion of salaries of academic staff allocated to research, is distributed to universities via the Higher Education Funding Councils (there are separate bodies for England, Scotland, Wales and Northern Ireland). This money is allocated by a formula, the largest element of which derives from a quality rating awarded in the Research Assessment Exercise, a panel-based rating of research performance carried out every four or five years.
- The larger side of the dual support system consists of funding for projects and centres, allocated competitively on the basis of peer-review by six Research Councils. One-third of this budget is for strategic research.
- Most of the Research Councils operate institutes in addition to their role in sponsoring university research. Together with universities these constitute the "science base" as a research-performing sector.
- UK-based charities account for an unusually high proportion of R&D funding for the science base, accounting for 12% in 1997-1998

TABLE 8.1 – Public research funding in the United Kingdom by sector of government (1997-1998)

Recipient	Expenditure 1997-1998	%
Civil government departments	1 361.8	21%
Science budget (mainly Research Councils)	1 331.1	21%
Defence	2 311.4	36%
Higher Education Funding Councils	1 032.7	16%
Contribution to EU R&D	335.4	5%

Unit: Million £.
Source: SET Statistics 1999.

TABLE 8.2 – Government R&D expenditure by Frascati type of research activity and sector of government % (1997-1998)

Type of research activity	Civil government departments	Science budget (mainly research councils)	Defence	Higher Education Funding Councils
Basic/pure	2.9	20.9	–	100.0
Basic/oriented	2.4	38.9	–	–
Applied strategic	41.2	31.2	6.7	–
Applied specific	46.4	8.2	21.6	–
Experimental development	7.1	0.8	71.6	–
Total	100.0	100.0	100.0	100.0

Source: SET Statistics, 1999.

(£392.7 million). The most prominent is the Wellcome Trust, a major shareholder in Glaxo-Wellcome, the UK's largest pharmaceutical company.

Government ministries (the civil departments shown in Table 8.1) have research budgets, which they use to pursue their missions. In the past much of this work was carried out in intramural research establishments. Most of these have passed through a process of becoming Executive Agencies dealing, in principle, at arm's length and on a competitive basis for the award of contracts. In recent years most of the largest and more applied laboratories have passed into the private sector, with a variety of ownership and management arrangements, but they remain largely dependent upon contracts from their former owners (see section 4).

The final element in the picture is the participation of the UK in the programmes of the European Union, notably the Framework Programmes for Research and Technological Development and more generally in internationalisation of R&D (section 6).

1.3 – *Institutions and directions in science policy*

The location of science policy in the UK government has changed several times during the past decade in response to the traditional tensions of science in government between linkages with education (particularly higher education) policy and linkages with industrial (particularly innovation) policy. Superimposed is the second tension, between cross-departmental co-ordination, carried out by a weak central agency, and associating science with a strong functional ministry at the expense of loss of credibility with other (rival) ministries.

Up until 1992, the science budget was a part of the responsibility of a Department of Education and Science (DES), which also provided block funding for research in Higher Education. Research Councils, which are autonomous non-departmental public bodies, had the bulk of their funds allocated by an Advisory Board for Research Councils (ABRC), the secretariat of which was in the DES. Co-ordination of science policy across ministries was the responsibility of the Government Chief Scientific Adviser, who operated with a small secretariat in the Cabinet Office (Prime Minister's Office). It remained government policy that co-ordination of matters relating to science across government departments did not require a dedicated science minister, but rather fell under the Prime Minister's general responsibility for trans-departmental issues. A consequence of this was that Margaret Thatcher chaired meetings of the Advisory Council on Science and Technology (ACOST) and answered questions in Parliament about broad S&T issues involving several departments (House of Lords Select Committee, 1982).

In 1992, the incoming Conservative Prime Minister, John Major, reversed the policies described above and announced the formation of the Office of Science and Technology (OST) in the Cabinet Office, together with the appointment of the first Cabinet level Minister for Science since the 1960s.[1]

1. In fact, the Minister, William Waldegrave, was responsible for the Office of Public Service and Science, the Department responsible for the Civil Service, in which OST was located. He held the title Chancellor of the Duchy of Lancaster, a position normally given to the Minister for the Civil Service within the Cabinet Office.

The Prime Minister retained overall responsibility for science. OST was created by combining the Cabinet Office S&T secretariat with the Science Branch of the Department of Education and Science. The latter brought with it responsibility for the Research Councils. OST was headed by the Chief Scientific Adviser, who also retained his cross-departmental function and position as Prime Minister's adviser.

In 1995, in a move which surprised even those most closely concerned, OST was moved to the Department of Trade and Industry (DTI). It was still to operate as a distinct entity within that ministry and the Chief Scientific Adviser retained his dual role. Science lost its dedicated Cabinet Minister, the new Science Minister being a junior minister within DTI, with science also being part of the portfolio of the Cabinet Minister in that department. Most accounts associate the change with a political deal that consolidated the ascendancy at the time of the incoming DTI Minister and Deputy Prime Minister Michael Heseltine (Flanagan and Keenan, 1998). The argument that this change would bring science closer to industry, though enthusiastically pursued subsequently, was seen at the time as an *ex post* rationalisation. As Flanagan and Keenan remark, the ring-fencing of OST within DTI significantly reduced the "better connection" rationale.

The Labour Government elected in 1997 retained this structure, preferring, it said, to concentrate on policies rather than structures. The main fault lines remain in place, with difficulties in co-ordinating both sides of the system for funding research in higher education and lack of full harmonisation between OST and the rest of DTI, which includes innovation policy in its remit. Co-ordination across government has been a high priority for the Labour Government across the full range of its policies. For science this has resulted in renewed efforts to engage ministers in trans-departmental scientific issues such as foresight. To some extent, though, Labour has introduced a new complication in that Her Majesty's Treasury has taken a much more active interest in policy for innovation, extending well beyond its traditional territory of the macro-economic and fiscal environments and including design of policies which DTI would see as being within its domain.

For the whole period described, the dominant policy trend has been towards stimulation of the relationship between the science base and industry. During the 1980s this was mainly promoted through academic–industrial collaborative research programmes, the largest of which was the £400 million Alvey Programme for Advanced Information Technology, in some ways the prototype of the large-scale European initiatives which, from a UK perspective, were later to replace it (Oakley and Owen, 1989).

The policy of improving connectivity within the system was set out in the Government's White Paper on Science, Engineering and Technology in 1993, the first such White Paper since the 1960s (OST, 1993). Research Councils were given mission statements, which made clear that their support for research and postgraduate training should be directed towards support for the national goals of increasing wealth creation and improving the quality of life. A major instrument announced in that White Paper was the Technology Foresight Programme. Though modelled upon the experiences of other countries, notably Japan, this Programme was designed to serve three objectives: informing priorities for public spending on science and technology, bringing together the science base and industry in new networks, and more generally, promoting a "foresight culture" in the UK. Implicitly, this recognised that the science base in the UK could no longer pretend to support the full range of opportunities, and hence that prioritisation would be necessary. The needs of user communities, whether expressed through Foresight or through direct engagement by the Research Councils, would be a major factor in setting priorities.

In parallel with the White Paper, DTI announced a reorientation of innovation policy. Key elements were an abandonment of support for national industry–industry collaboration and for single company support for all except SMEs, a focus upon helping firms to gain access to technology through a locally-based infrastructure, promotion of best practice in innovation, and support for forming partnerships. Academic–industrial collaboration would continue to be supported through the LINK initiative (see below) and support for EUREKA was maintained, though in both cases funding would not normally be available for large firms.

Under the Conservative Governments from 1979 to 1997, science and technology had not been exempt from the broader reforms affecting public sector activity. These included the requirement that all ministry programmes, including those of the Department of Trade and Industry, should be subject to a regime of testing the rationale for public intervention, setting verifiable objectives and subsequently being evaluated against these. A further manifestation of these reforms was the privatisation of government research establishments described below.

Despite the stated priority given to science throughout this period, and generally a protection of the level of Research Council budgets, other public sources of income for research declined. A series of reports from Parliamentary committees (House of Lords Select Committee on Science and Technology, 1986) and the ABRC (1987), among others, highlighted concerns about UK science in the face of financial cutbacks in universities and the declining volume of civil R&D. In 1986, a pressure group of several thousand concerned scientists, Save British Science, was launched, following an advertisement in *The Times*. These concerns continued through the 1990s, with declining infrastructure a particular concern. They were among many issues considered during 1996-1997 by a major inquiry into the future of higher education, chaired by Sir Ron Dearing and usually referred to as the Dearing Committee (see section 3.1).

The problem of infrastructure has only very recently begun to be alleviated. The incoming Labour Government had undertaken to adhere to the spending targets of its predecessor until it had carried out a Comprehensive Spending Review of all public expenditure. When the results of this were announced in the summer of 1998, science emerged as a major priority. Additional spending of £0.7 billion by the Government would be accompanied by a further £400 million from the Wellcome Trust (OST, 1998). Of this, £600 million is forming a fund for renewal of equipment and buildings. A combination of factors led to this outcome. In opposition, Labour had made much of the declining infrastructure resulting from Conservative underinvestment. An economic philosophy which distinguished between investment and consumption in public expenditure provided a suitable backdrop for a campaign which stressed the rewards available, and more pertinently the penalties of missing out on the wave of discoveries in genomics. A strong coalition on the outside, including pressure from industry and a public campaign, combined with a well-argued case from OST. This stressed that the Research Councils had put their own house in order, developing clear priorities through Foresight, cutting their own overhead costs to a very low level (under 4%), and performing an analysis of strengths, weaknesses, opportunities and threats for science.

At the end of 1998, the Labour Government published its Competitiveness White Paper, *Our Competitive Future: Building the Knowledge Driven Economy*. This is built around the thesis that sustainable growth and competitive advantage are dependent upon the successful exploitation of knowledge, skills and creativity. This document encapsulated the Government's thinking in much the same way as the 1993 Science White Paper. It consolidated some previously announced actions, together with several new initiatives. Measures are categorised as investing in capabilities, catalysing collaboration within regions and between companies, and providing a regulatory framework which fosters fair and open competition. To provide assurance that this was not simply a presentational exercise, the White Paper was followed in March 1999 by an implementation plan which summarised its commitments and provided a list of targets and milestones, with named officials attached to each commitment. Targeted at the full range of businesses in the UK, *Our Competitive Future* contains a wide range of measures but those concerned with technology and innovation play a central role.

2. NATIONAL PRIORITIES AND HOW THEY ARE FORMULATED

2.1 – Science policy advice

Advice on science policy in the UK is best described as pluralist, with a variety of committees and advisory groups acting in the system (Georghiou, 1998). This discussion will be confined to advice on policy for science as opposed to science in policy, though many of the same organisations and structures will be involved in the latter.[1] Only a few actors will be mentioned as almost any body engaged in scientific research or funding also has some role in expressing positions or influencing policy. Prior to the 1993 White Paper, a consultative exercise was unusually carried out on a set of key issues set out in advance by the Minister. Over 800 organisations and individuals provided evidence.

At the apex of the science policy system, the Chief Scientific Adviser (CSA), as mentioned above, provides advice directly to the Prime Minister and makes a direct input to Cabinet-level considerations of science issues. The other key figure in the Science Base is the Director-General Research Councils (DGRC) who is responsible for allocation of the Science Budget to the Research Councils. The CSA is also Deputy Chair of the Council for Science and Technology, a committee established in 1993 to provide advice on the balance of UK science, engineering and related issues to Cabinet Ministers. It is chaired by the Secretary of State for Trade and Industry and consists of members of the business and research communities. For much of its existence CST did not publish its advice, but the present Government, as well as changing the membership, has changed this practice.

---

1. The Chief Scientific Adviser has produced a note on procedures for the use and presentation of scientific advice in policy-making which recommends a presumption towards openness and debate. See May (1997).

A number of other official committees offer advice on specific scientific and medical matters. Individual ministries with a technical content to their remit generally have their own Chief Scientist and expert advisory committees. Research Councils may also act as a source of advice. All of the above constitute the official advisory system but the remaining bodies described below are also influential.

In Parliament, for many years the most active committee was in the Upper House, the House of Lords Select Committee on Science and Technology. Its 17 members include many eminent scientists and others with industrial or policy experience. The House of Commons Science and Technology Select Committee dates from 1992 but has raised its profile following the election of several scientists in the new Parliamentary intake. Neither committee has budgetary or legislative powers but their cross-examination of Ministers, senior officials and other actors provides the basis for some influential reports. Parliament is supported by an independent office, the Parliamentary Office of Science and Technology, which produces briefings, analyses and reports on issues of science policy or science in policy.

Among the professional and learned societies the Royal Society (and the Royal Society of Edinburgh) promote the natural and applied sciences and the Royal Academy of Engineering promotes engineering. Each of them frequently produces studies or statements intended to influence policy. Other bodies act as a platform for policy debate among their more general missions, examples being the British Association for the Advancement of Science and the Foundation for Science and Technology. As mentioned above, the Save British Science Society is an independent pressure group representing research scientists and engineers and engaging vigorously in policy debate. Policy research in the academic environment is carried out principally by PREST at the University of Manchester and SPRU at the University of Sussex. Some academic activities in the field are co-ordinated by the Science Policy Support Group, a non-profit company, which manages programmes and networks. In the consultancy sector there are some specialised firms that carry out science policy-related studies.

Clouding the distinction between policy for science and science in policy has been one of the biggest challenges to the scientific advisory system in the UK, an ongoing controversy over the safety of genetically modified foods and of trials for genetically modified crops. This has received prominent and often wildly exaggerated media attention but some real concerns have also been raised. Environmental activists have challenged the independence of scientific advisory groups, first objecting to industrial membership and more recently to academic scientists who have benefited from industrial funding. In a climate where scientists are being simultaneously exhorted to work with industry this is bound to create strains. It has also placed a barrier to some of the Government's plans to promote the development of gene-based industries. A further concern for the scientific community is that the negative associations for science could have long-term political repercussions at a time when there has been an unprecedented enthusiasm for science at high levels of government.

2.2 – *The Foresight Programme*

The UK Foresight Programme provides a rare example of foresight becoming deeply embedded and institutionalised within the national innovation system rather than operating only at a higher strategy-making level. The Programme began in 1993, having been announced as the centrepiece of a Government White Paper on Science, Engineering and Technology, which itself was seeking to greatly improve the connections between the national science base and its users, notably in industry. From the outset the Technology Foresight Programme (as it was initially known) combined this objective with that of informing priorities for public spending on science and, implicitly, promoting a 'foresight culture' more broadly.

The first cycle of the programme began in September 1993 with the appointment of a high-level Steering Committee. The pre-Foresight stage (to April 1994) involved identifying and briefing members of the 15 panels, which, together with the Steering Committee, would be the mainstay of the programme. Panels were drawn principally from industry, combined with academic and government membership. Indeed a notable feature of the Programme has been the high level of support and participation from industry throughout. It has been postulated that this resulted from the increasing dependence of firms upon external sources of technology to the point where formulation of strategy, previously an internal activity, must now at least in part be carried out in the public arena. By collaborating in their thoughts about the future, organisations may be better placed to anticipate the actions of their customers, suppliers and others such as regulators, who may influence the environment in which they operate (Georghiou, 1998b). Efforts were made at this stage to broaden the base of participation beyond that of regular advisers to government.

In the main stage of the Foresight activity, panels identified key issues and trends, developed scenarios for their sectors and consulted widely, by means

TABLE 8.3 – Foresight Criteria

- – Economic and social benefits;
- – Ability of the UK to capture economic and social benefits;
- – Likelihood of scientific or technological breakthroughs;
- – Strengths of the UK science and technology base;
- – Cost of investing in new science, engineering and technology; and
- – The time scale within which new technology is likely to become available.

TABLE 8.4 – Foresight Technology Priorities

I. Social Shaping and Impact of New Technology

1. Demographic Change
2. Risk Assessment and Management
3. Workplace and Home

II. Harnessing Future Communications and Computing

4. Communicating with Machines including interfacial software
5. Design and Systems Integration (including CAD/CAM)
6. Information management
7. Modelling, Simulation and Prediction of Complex Systems
8. Optical Technology
9. Software Engineering
10. Telepresence

III. From Genes to New Organisms, Processes and Products

11. Bio-informatics
12. Bio-materials
13. Genetic and Bimolecular Engineering
14. Health and Lifestyle

IV. New Materials, Synthesis and Processing

15. Catalysis
16. Chemical and Biological Synthesis
17. Materials
18. Materials Processing Technology

V. Getting it Right: Precision and Control in Management

19. Management and Business Process Engineering
20. Automation (including Robotics)
21. Process Engineering and Control
22. Sensors and Sensory Information Processing
23. Security and Privacy Technology

VI. A Cleaner World

24 Clean Processing Technology

25. Energy Technology
26. Environmentally Sustainable Technology
27. Product and Manufacturing Life Cycle Analysis

TABLE 8.5 – Foresight Infrastructural Priorities

I. The Skills Base	
1.	Training better SET teachers
2.	Communications skills
3.	IT competence
4.	Public understanding of science
5.	Business awareness
II. Research in the Science Base	
6.	Maintain support for truly excellent basic research
7.	New incentives for multidisciplinary research
8.	Incentives for universities and Research Councils to work with industry
III. Communications	
9.	Superhighway
10.	Science-watch function
IV. Finance	
11.	Development and encouragement of long-term finance for R&D and innovation
12.	Special incentives for SMEs
13.	Enterprise architectures
V. Policy and Regulation	
14.	Intellectual property rights
15.	Procurement by government
16.	Regulations on the environment, finance, communications, etc.
17.	A continuously updated scientific basis for standards
18.	Demonstrator projects (applications oriented)

of a Delphi survey (Loveridge et al., 1995), regional workshops and contact with other parties. The process culminated in the production of a report by each panel and a Synthesis Report by the Steering Committee (OST, 1995). The latter prioritised the recommendations of the sectoral panels in a matrix which plotted attractiveness of the potential economic and social benefits against the feasibility of the science and technology involved, with the ability of the UK to achieve the result a key consideration in both cases (see Table 8.3 for the criteria used). Priorities established were both technological (Table 8.4) and infrastructural (Table 8.5).

After the reports were published the Programme entered its second phase, in which the main objective was to disseminate and act upon the results through media, workshops, professional and trade associations among others. Some 600 events were held and 130000 reports disseminated. The panels were retained to drive this process, though with some adjustment in terms of reference and membership. The most direct follow-up was a new funding initiative, the Foresight Challenge fund (see section 5.1). Existing budgets for science were also influenced, though it is much harder to establish attribution here. An official estimate gave 54% of Research Council spending as being aligned with Foresight priorities of which £300 million was for new initiatives. Other government departments carried forward areas relating to their own remits, for example transport and environment (OST, 1996).

In 1997-1998 the third phase of the Programme gave top priority to the wider engagement of business, with the top priority being to increase business participation beyond R&D divisions to reach those involved in corporate strategy, marketing and finance. This was one reason why the term "technology" was dropped from the title. Other activities at this phase included efforts to improve co-ordination across government departments and to begin preparation of a new cycle of Foresight activity. Both the Steering Committee and the Panels were reconstituted as networks of implementing organisations. With the wide range of linkages across the whole of science and innovation policy, the Foresight Programme can be said to have been reinvented as a brand and binding agent for innovation policy in the UK.

In evaluating the successes and problems of the first Programme, on the positive side there was a clear success in attracting widespread support in industry and much of the science base. While the priorities which emerged were rather broad they have been generally accepted. There is a recognition that these are priorities for follow-up action, not an attempt to second-guess scientific creativity. The networking activities have universally been seen as beneficial and become more prominent as an objective over time.

Problems encountered included a rather rushed timescale for Phase 1, which inhibited the development of mature recommendations and the ability to take full advantage of the methodologies. While the panels mostly worked well, barriers of communication emerged between members and the rest of the community and between the panels themselves. This has led to the conclusion that the next cycle should have a more cross-sectoral approach and more permeable structures. In terms of participation there was a step change upwards in the numbers involved but these still were all in the expert community and did not include the wider public. Excessive focus on technology as a driver led to overemphasis on technical fix solutions relative to regulatory or social change.

A new cycle of foresight began officially on 1 April 1999 (Georghiou, 1999, see also http://www.foresight.gov.uk). Its goals remain similar though stated slightly differently, being to:

TABLE 8.6 – Foresight Panels

First round	Second round: Thematic	Second round: Sectoral	Underpinning themes
Agriculture, natural resources and the environment		Energy and natural environment Built environment	Sustainable development
Energy			
Construction			
Learning and leisure	Ageing population		Education, skills and training
	Crime prevention		
Defence and aerospace		Defence, aerospace and systems	
Manufacturing production and business processes	Manufacturing 2020		
Materials		Materials	
Chemicals		Chemicals	
IT and electronics		Information, communications and media	
Communications			
Transport			
Marine			
Food and drink		Food chain and crops for industry	
Retail and distribution		Retail and consumer services	
Financial services		Financial services	
Health and life sciences		Healthcare	

– develop visions of the future – looking at possible future needs, opportunities and threats and deciding what should be done now to make sure we are ready for these challenges;
– build bridges between business, science and government, bringing together the knowledge and expertise of many people across all areas and activities; in order to
– increase national wealth and quality of life.

The panel structure has been retained but with important changes. Consolidation of sectoral panels and a more supply-chain-based approach has reduced the number to ten (see Table 8.6). There is a tendency for these to be more application-oriented – for example the science-driven "Health and Life

Sciences" becomes "Healthcare". However, the sectoral panels represent the main element of continuity. An innovation is the introduction of the thematic panels. These address broad social and/or economic issues with cross-cutting implications for science and technology. The introduction of the thematic panels is symptomatic of a broader tendency to locate research in the context of socio-economic goals. Another example is the European Union's Fifth Framework Programme. The Thematic Panels will provide a more obvious interface to public policy – for example the Crime Prevention Panel is funded by the Home Office (the ministry responsible for policing) and will feed directly into its crime reduction strategy. Two devices exist to give additional dimensions to the structure: each panel is expected to consider the two underpinning themes of education, skills and training, and sustainable development; and underpinning technologies (IT and biotechnology) will be represented in the membership of appropriate panels.

In operational terms it is envisaged that panels will be strategic entities, which forward their agendas by establishing task forces to address specific issues. Panel chairs have a different profile from the first round where the predominant group was Research Directors of major companies. Following the trend of trying to get beyond R&D to the boardroom, the second round has enlisted several company Chairmen and Chief Executives. The task forces may include non-members of panels and may span the interests of more than one panel. Currently 56 such Task Forces exist, some of which are carrying forward the recommendations of the previous round (for example the Foresight Vehicle Programme and Clear Zones representing the Transport Panel's main recommendations).

In the present programme, Panels will not be asked to participate in common exercises such as the Delphi survey in the first cycle. They will, however, be supported by a common resource, the Knowledge Pool. This is a professionally managed library of strategic visions, views and information about the future available both on the Internet and in hard copy. Presently at a formative stage this includes results and other documents from past foresight activities in the UK and elsewhere and government reports relevant to science and technology policy from national, European and OECD sources. As the Programme proceeds its outputs will be added. All items contain details of authors and a record of comments made. The Knowledge Pool will also function as a discussion and consultation medium.

The ability of the Office of Science and Technology to support panels is limited by resources and hence a new mechanism has been devised – that of the Associate Programme. This effectively franchises external organisations to operate Foresight activities in partnership with OST and under agreed terms. Most of the operators are professional bodies or intermediary organisations.

Future evaluation of the Second Cycle will need to take account of both its intrinsic features and the changed environment in which it operates. In some senses the first cycle can be seen as a "heroic era" in which the successes were achieved through the driving force of key individuals championing particular activities. Already in the implementation phase there was a deliberate switch to a far more institutional perspective, reflected in a change in the nature of panel membership. The Steering Committee also evolved in this direction, with new members being selected to gain or secure the involvement of their organisations (for example the senior officers of the Confederation of British Industry and the Trades Union Congress). The growth of institutional linkages has continued in the second cycle. The aim has been to use the infrastructure rather than to cut across it. This strategy is not without risks. Institutions exist to promote a particular point of view and are often rather conservative in their outlook. The promotional dimension means that it will be difficult to produce a clear-cut list of priorities. Priorities within particular domains may emerge but the variety of origins of these does not facilitate any national level comparison. If such an exercise is attempted those involved will have to discount the effects of determined lobbies. There may well be a trade-off between prioritisation and more and better network-building.

It is of course too soon to draw any conclusions about the effectiveness of the new Programme. Some of the innovations, for example the thematic panels, promise a genuinely creative approach to the future. Changes in the structure offer the prospect of a broader base of participation. While the first Programme represented a steep change in the numbers consulted about priorities, it was still very largely confined to a technical elite. Coinciding with the Millennium and the associated upsurge in interest in the future, the new cycle has a real opportunity to capture the public imagination. Many interesting questions remain. For example, will the priorities established in the first cycle still be considered valid? What will happen if contradictory views emerge between past and present panels? Panels will start to produce findings from September 1999 to allow sufficient time for debate before they prepare their reports from April 2000 to November 2000. At that point the Steering Group will draw together the contributions into an overall report while the new implementation phase begins in parallel.

2.3 – Evaluation of research and innovation policy[1]

While not usually a tool for establishing scientific priorities, evaluation of research (and of related policies) has become a characteristic feature of the UK system and has certainly influenced the allocation of resources between actors. As in many countries evaluation in this domain is the outcome of a confluence between the internal quality control procedures of science, principally peer review, and the broader movement to make the public sector more accountable (Gibbons and Georghiou, 1987). The latter impetus came both from the Treasury, which promoted a Financial Management Initiative from 1982 intended to stimulate government departments to have clear objectives, together with a means of assessing their achievement (HM Treasury, 1983). In the domain of research, the early 1980s saw the foundation of a Science and Technology Assessment Unit in the Department of Trade and Industry (Hills and Dale, 1995) and the beginning of a large real-time evaluation of the Alvey Programme (Cameron and Georghiou, 1997). The latter exercise influenced the methodology of many subsequent evaluations.

A strong impetus to the spread of evaluation came when, in 1987, the Science and Technology Assessment Office was established within the Cabinet Office with the objective of ensuring that each department set up adequate evaluation approaches. It also published guidelines for evaluation (Cabinet Office, 1987). Continuing pressure from the centre of government has seen the general diffusion of an approach first used in the Department of Trade and Industry, known by the acronym ROAME or ROAMEF.[2] This required all new programmes (almost all research activity in government was moved onto a programme basis) prior to approval to produce a satisfactory statement including the reasons for state intervention, testable objectives and a plan for how these were to be subsequently evaluated. Accompanying this has been an increasing tendency towards the use of performance indicators. The latter has sometimes mistakenly been seen as a substitute for evaluation rather than as a means to motivate management. However, the evaluation culture appears to be deeply rooted and operates with a credible degree of independence, whether the evaluations are performed by in-house units or by external consultants with evaluation expertise.

---

1. Evaluation in this context is taken to mean *ex post*, interim or real-time evaluation of programmes or institutions. This discussion does not cover *ex ante* evaluation or project selection/appraisal.
2. Standing for Rationale, Objectives, Appraisal, Monitoring, Evaluation and Feedback.

Programme evaluations of this type have also become more common in the domain of science funding but the emphasis in Research Councils remains on peer review. Nevertheless, peer review itself has evolved considerably. At the beginning of the period the typical system involved a hierarchy of committees consisting mainly of academics allocating funds to applicants after consulting referees' reports. Pressure upon the Research Councils to become more responsive to user needs and to cut overhead costs has seen strong simplification of these structures and in some cases empowerment of programme officers. At the same time new approaches have been tried to broaden the base of referees. For example the Engineering and Physical Sciences Research Council assembles colleges of reviewers on the basis of nominations from past applicants, with templates to ensure coverage for expertise, users and younger scientists. Associate Programme Managers select three referees, two from the relevant college and one nominated by the applicant. An *ad hoc* group selected from the college then rank orders proposals (Nuttall, 1999). For Research Councils generally, feedback of referees' comments is given and an emerging trend for larger applications is to allow researchers to respond to referees' comments before the decision is made. Much greater attention than in the past is paid to follow-up at the end of the funding period, with final reports and outputs receiving a grade from postal reviewers, with implications for future applications.

It is slightly paradoxical that the evaluation exercise, which is most controversial in methodological terms, is also that which has the greatest effect, at least in terms of allocation of resources. Since 1986, four Research Assessment Exercises have been carried out by the bodies responsible for providing block funding for research in universities. In essence, this involves universities submitting information about their staff, research income, postgraduates and research context in a defined subject area (Unit of Assessment). Critically, under current rules, four research outputs (normally publications) published within the assessed period are required to be listed for each staff member. A specially convened panel then gives each submission a rating, on a seven-point scale in the most recent (1996) exercise. The rating is intended to reflect the proportion of international and national quality work in the submission. The large effort required to execute this exercise is justified by the use of the ratings to allocate resources to universities in the period until the next exercise. For a given Unit of Assessment, the available resource (known as QR funding) is divided between universities on the basis of a figure derived from the rating multiplied by the number of staff entered, with a further smaller credit for the number of research students and staff.

The bottom two ratings have a value of zero and the remainder increase on a rapidly ascending scale.[1] The effect is to concentrate resources heavily in highly-rated departments. Subjects are funded in bands related to research costs but proposals to apply a "policy factor" to reward areas which are internationally excellent or important have thus far foundered in the absence of an agreed method to achieve such a weighting.[2]

The detail of the RAE is constantly debated in the academic community and elsewhere. Commonly voiced criticisms include accusations that it rewards only conventional research, that its disciplinary structure is inherently biased against interdisciplinary work, that it fails to give due credit for applied or otherwise relevant research, that it has created a "transfer market" for distinguished academics at the expense of nurturing young talent, and that it drives academics to produce excessive numbers of journal articles as opposed to other forms of output or activity (including teaching). It is apparent that the criticisms listed above could be damaging even if they were untrue, in that perceptions would modify behaviour in undesirable directions. On the positive side, it is generally acknowledged that the RAE has substantially raised the profile of research in universities and improved management procedures.[3] Most of the academic community has come to accept that some form of selectivity has to be applied if standards of research are to be maintained.

Much less debate has taken place on the fundamentals of the exercise, perhaps because the academic community fears the prospect of an alternative, externally driven mechanism to create selectivity in funding. None the less, some fundamental questions can be raised from an evaluation perspective. The first is whether an exercise where an average of less than ten panellists meeting three or four times for one day (or occasionally two) to consider the work of a mean of 932 researchers in 48 submissions could be equated with the normal practice of peer review. Very few areas are sufficiently homogenous to make it likely that the expertise of the panellists could be sufficient to meet the refereeing requirements of a funding agency or journal. Furthermore, judgement of international excellence was made without

---

1. Grade 3b has a weighting of 1, 3a of 1.5, 4 of 2.25, 5 of 3.375 and 5* of 4.05.
2. The criteria thus far considered for weighting between units of assessment include national need and international standing as measured by citation analysis, patents and esteem indicators included in RAE submissions. HEFCE, *Research funding: introduction of a policy factor*, October 98/54 Consultation.
3. See for example a study on the 1992 exercise which consolidated research on its impacts (Higher Education Funding Council for England, 1997).

international panel membership (contrary to the practice of many countries in Europe). The system operated without any formal means of feedback other than the rating (with variable but generally very brief informal feedback available at the discretion of panels) and hence offered no formative learning benefits. It also offered no system of appeal against ratings. The next exercise, in 2001, will attempt to give a larger role to users (as a result of pressure from industry and government) but there is already a feeling that it will be followed by substantial change in the system. A review committee is examining options for the longer term.

3. ROLE OF THE UNIVERSITIES

3.1 – Changing demands

The higher education sector in the UK, consisting of universities and higher education institutes, has seen dramatic changes in the past 20 years with the number of students more than doubling. From an elitist system it is now the case that 1.6 million people are students in higher education, including about one-third of school leavers and a larger number of mature students. This has been achieved with a public funding increase of 45% in real terms (constant as a percentage of GDP) and a consequent 40% drop in the unit of funding per student. The consequences of what is termed the "massification" of higher education are felt upon research through pressure upon the time of staff and through a general financial weakening of universities, which has had severe consequences for their research infrastructure.

Universities are the main performers of basic research in the UK. While there are 111 universities in the UK, research is heavily concentrated. Hence the five largest by research income receive 25% of all research income and the top 15 receive about half of this income. The bottom 50% of institutions account for under 10% of research funding. This pattern is repeated for almost all research indicators and funding sources.

Higher education in the UK has recently been reviewed comprehensively by the National Committee of Inquiry into Higher Education (1997, the Dearing Report). While the most prominent outcome of the report was its recommendation for the introduction of student contributions towards tuition fees (now implemented), it covered the full spectrum of activities in higher education, including research.

Within the domain of research, the main recommendations of Dearing were:

- Concerns about the conduct of the Research Assessment Exercise, in particular the need for a study of the treatment of interdisciplinary research (see above), the need to have one or more international members on all panels, and that greater selectivity should be compensated by giving a limited sum for "scholarship" to those who opted out of the RAE.
- Establishment of a revolving loan fund of £4 to 500 million to address infrastructure and equipment deficiencies and financed jointly by private and public research sponsors.[1]
- That all projects and programmes funded by Research Councils should meet their full indirect costs and the costs of premises and central computing (the recommendation left open the question over whether this change should be funded by new money, by a transfer of resources from the block grant of the Funding Councils, or by the Research Councils issuing fewer grants).
- Establishment of an Industrial Partnership Development Fund with matching funds from industry to contribute to regional and economic development.

These recommendations have met mixed success. A study has been completed on the issues raised concerning the RAE but its results are inconclusive. The proposal for a scholarship fund has been generally rejected. The idea of a loan fund for equipment was judged infeasible but other elements of this suggestion have been incorporated in the new Joint Infrastructure Fund described above. The question of a "dual support transfer" has been avoided for the time being, which implicitly means that universities lose money when they accept research grants. Several actions on university–industry relations are now in process, though none so far match the Partnership Fund suggested. The present status of university–industry links is discussed below.

3.2 – University–industry links

There has been large-scale growth in recent years across the United Kingdom in the scale, number and variety of linkages between higher education and industry. This is detailed in a PREST survey on behalf of the Funding Councils for England, Scotland and Wales, which provides the basis for the data in this section (Howells et al., 1998). Examples of linkages include

1. The estimate was based upon a survey carried out by PREST/CASR: L. Georghiou, P. Halfpenny, M. Nedeva, S. Hinder and J. Evans, 1996.

research collaboration, provision of consultancy services, market transactions in the commercialisation of research and a growing involvement by industry as an interactive user of all types of teaching and training. Research funding by industry has grown by 30% over the past three years and by 11% in the past year. However, the level of income in 1996-1997 of £183 million is highly concentrated, with seven universities accounting for one-third of the total and half of universities for 8%. Income from intellectual property rights has also grown rapidly (by 58% in two years) but remains at a relatively low level (under 1% of all research income and under 10% of all research income from industry). Larger income streams are anticipated in the future (although about a quarter of universities spend more on intellectual property protection than they receive in fees) but the main message for universities appears to be that contracts from industry and collaborative research will remain for the foreseeable future the most attractive ways to enhance their income from government sources and their contribution to the economy.

There is some evidence that two decades of growing collaboration have resulted in significant learning benefits in terms of how to work with industry. Despite growing institutionalisation of such links in the form of technology transfer infrastructure, all concerned emphasise the central importance of good personal links and mutual trust. Problems arose where universities failed to provide a professional interface. The principal difficulty caused by industrial partners was the tendency to change the objectives or priority of the collaborative project as a result of change of corporate strategy.

Commercialisation is also manifested by the growth in the number of spin-out companies wholly or partly owned by higher education institutions (HEIs). The PREST survey indicated that half of the responding HEIs had established spin-out companies in which they held all or part of the equity (223 such firms were identified). These were of two types: umbrella organisations for holding intellectual property, and companies established to exploit the commercialisation of specific research results. Probably more important economically are spin-outs by staff in which the institution does not have a financial interest, this category being reported by 44% of institutions.

There is a developing infrastructure for the incubation and location of such companies. Probably too much emphasis has been given to the phenomenon of science parks. While these are generally doing well in the UK,[1] higher education provided only 15% of the finance going into science parks in 1997,

1. The UK Science Park Association had 48 members in 1997 with 1414 tenant companies employing 27 371 people and attracting £656 million of investment.

only 22% of tenant firms originated in universities and 3% are university-owned. A newer trend is for on-campus incubators and laboratories belonging to larger firms also located within the institution. These developments are based upon the belief that only the closest proximity is likely to produce the required cross-fertilisation. The principal barrier to commercialisation of university research is widely held to be a lack of seedcorn capital for the stage of development before the venture capital market is sufficiently confident to take over. New policies are addressing this deficiency (see section 5.2)

Industry–academy links also are growing in the context of teaching and training. Of most relevance for science and technology are schemes at graduate level. Two in particular have a high profile and penetration. The Teaching Company Scheme (TCS) was founded in 1975 and has been regarded as one of the greatest successes of UK HEI–industry links. The TCS was initiated by the DTI and it aims to develop active partnerships between universities and industry. The scheme sets up partnerships between firms and HEIs through the formation of teaching company programmes. Firms take on graduates, known as TCS Associates, to work full time on specific projects jointly supervised by the HEI and the company. Projects are intended to be closely linked to the interests of the firm and should be aimed at achieving a substantial and comprehensive change in the firm, for example in management and production techniques. A typical programme lasts for two years. The graduates have a science and engineering background and are recruited jointly by the partners. The Associates spend 90% of their time working in the company on specific projects and are paid at industrial rates. The remaining 10% of their time is spent within the HEI undergoing training. Well over 2000 TCS partnerships have been created since it was first established.

The Co-operative Awards in Science and Engineering (CASE) scheme is used to fund research students who are jointly supervised by academics and external sponsors who may come from industry or from public sector bodies. CASE awards are largely financed by the Research Councils, with some industrial finance for the student and academic department. They aim to encourage industrially relevant research projects by Ph.D. students at universities. Standard CASE awards are allocated to universities, typically by a quota allocation to a department. Industrial CASE awards are a variant in which the Research Councils allocate the awards to industrial companies to support projects at universities which they select.

An emergent phenomenon, echoing the situation in the US, has been that of courses tailored to the needs of single firms or groups of firms. A more extreme version of this occurs in the form of the "corporate universities" founded by some large firms. While some are little more than a grand term for basic vocational training, the more ambitious initiatives select modules and individuals from top universities to construct a curriculum reflecting their needs for higher-level managers and technical staff.

To summarise the situation, a complex mixture of collaborative and competitive relationships is emerging. Partnerships between universities and industry are evolving from one-off contracts to long-term strategic relationships. Universities also have a more competitive relationship with industry through the desire to raise revenue from the knowledge they generate. It will take some time before there is universal acceptance of a market for a previously free good. There is also increased competition among universities for industrial resources, driven partly by the matching funds requirements of many public initiatives but this too is counterbalanced by the emergence of several regional consortia of universities aiming to present a co-ordinated interface to industry.

4. NATIONAL LABORATORIES AND PRIVATISATION[1]

Civil government departments are major purchasers of research in support of their policy, regulatory, statutory or procurement functions. Results of this research may be consumed directly by the departments which commissioned it, as with forensic services, or as proxy customers for public goods (for example in procuring metrology services for industry or advice for farmers). The specialised, long-term and often sensitive nature of this work historically led to much of it being performed in government laboratories, typically a part of the department for which they worked.

From the late 1960s concerns began to be expressed in policy circles about the accountability of such arrangements. In the early 1970s, Lord Rothschild, head of the Central Policy Review Staff, proposed that the "customer-contractor principle" should govern the relationship for applied science. This marked the beginning of a process of differentiation between laboratories and research customers. Under the Thatcher Government, from 1979 onwards, reform on the lines of "New Public Management" began in earnest, with pressure for competitive tendering for funds, performance indicators and

1. This section draws heavily upon R. Boden, P. Gummett, D. Cox and K. Barker, 1998. See also K. Barker and A. Beasley, 1998.

a general decline in public funding. Between 1989 and 1996, 18 science and technology establishments were transformed into agencies with funding on an arms-length customer–contractor basis. This transition was insufficient to satisfy the ideological convictions of the Conservative Governments and plans were put in place to proceed to full privatisation. A series of reviews were initiated, the last of which concluded that laboratories belonging to the Research Councils (mostly the smallest or most fundamentally-oriented establishments) should not be privatised for practical financial reasons.

However, the larger and more applied laboratories were generally privatised in one form or another. Three main models were used:

- *Trade sales* – whereby the laboratory was transferred to a company via private contract. For example, the National Engineering Laboratory was transferred by DTI to a subsidiary of Siemens, though in this case the acquirer was paid £1.95 million to compensate it for an operating deficit. £30 million of work over five years was guaranteed by DTI, which remains the largest customer for the laboratory. The Laboratory of the Government Chemist was also sold but in this case to a management-led consortium supported by the Royal Society of Chemistry and venture capitalists. Again there were guarantees of work.
- *Companies limited by guarantee* – this a legal form for non-profit companies, generally operating in areas of mutual or public benefit. Examples here were the Transport Research Laboratory, now known as TRL Ltd. It is a wholly-owned subsidiary of a purpose-established foundation, including 86 transport sector interests and 25 employee representatives. The Building Research Establishment followed a similar pattern.
- *Government-owned, contractor-operated* – GOCO, as it is known, describes the situation where the laboratory remains government-owned but a private company is given a contract to operate it for a finite period. Employees and moveable assets are transferred to the contractor at a price. The main example here is the National Physical Laboratory.

The most important outstanding issue in terms of laboratory reform and privatisation concerns the nation's defence laboratories. Approximately one-third of the government's defence R&D budget is spent in these research establishments and the Defence Evaluation and Research Agency is the largest research organisation in Europe with more than 10 000 employees. Changes in DERA during the 1990s have already altered its relationship with defence companies as well as with the Ministry of Defence. The situation here is considerably complicated by DERA having a dual role as research performer and a funder of R&D. DERA subcontracts work to industry and academic institutions worth between 30% and 50% of the organisation's annual research funding and in 1996-1997 such extramural research cost DERA £162 million (House of Commons, 1998). This subcontracting is also seen as an important technology transfer mechanism. However, the move towards greater commercialisation of DERA's activities has generated some tension in this relationship. A 1995 joint report by the House of Commons Defence and Trade and Industry Select Committees noted that companies had three broad concerns about DERA: whether enough of its research was contracted out to industry; whether it duplicated industry's own research; and the availability of DERA's research to companies (House of Commons, 1995).

The Labour Government's policies for DERA address both the technology transfer issue and the processes of commercialisation and privatisation. The approach to technology transfer has been institutionalised through the establishment of the Defence Diversification Agency which aims to facilitate access to DERA's technologies and expertise, particularly for small firms, and at the same time to stimulate the inward transfer of civilian technologies (Ministry of Defence, 1998a). Science and technology parks adjacent to defence research establishments form one part of this strategy. In parallel with this initiative, falling defence research spending has put pressure on DERA, which operates as a government trading fund, to increase the figure of around 10% of its income that it currently receives from non-MoD sources. To give the organisation the necessary flexibility to do this, a Government consultation document proposes partial privatisation (Ministry of Defence, 1999). Some more sensitive activities (notably chemical and biological weapons defence) would be excluded but most of the Agency would be privatised as a single business within two years but with significant constraints and controls left in place. These would seek to maintain its independence and character, for example by preventing the defence manufacturing industry from holding shares. The House of Commons Defence Select Committee has already recorded its opposition to even a partial privatisation (House of Commons, 1998b). The principal arguments put against privatisation are that it would be a privileged competitor with industry for funding, the ownership of intellectual property rights arising form collaborative work, and whether collaboration with other countries (especially the US) would be jeopardised by non-governmental status.

In assessing laboratory reforms as a whole, Boden et al. (1998) distinguish between the transition to agency status, where the managerialist changes have been generally welcomed by the laboratories and their customers, and privatisation which has been a much more *ad hoc* process. They caution that when existing work guarantees expire or major capital items require renewal then the market may not be able to provide a solution in the public interest. The early evidence is that privatisation has led to efficiency gains, with laboratories reporting higher productivity and positive financial performance. On the other hand the profile of work in at least one case had shifted from research to dissemination of existing knowledge (Beesley, 1999). A further risk faced by all of the laboratories is that the more they move in a commercial direction, the more will they be exposed to competition from other providers. Other commentators have also raised the issue of whether privatised establishments, dependent for business upon government and upon private sector clients, will be able to give the appearance of sufficient impartiality if a major regulatory controversy of a scientific nature requires resolution. In a country still examining its scientific advisory structures in the light of the damage to their public reputation following the BSE disaster, this is not a minor issue.

5. INCENTIVES FOR INDUSTRIAL INNOVATION

A recent government consultative document (HM Treasury and Department of Trade and Industry, 1998) notes that:

- The amount of R&D undertaken by UK firms has declined relative to other G7 countries.
- Only in a few industries do UK companies spend comparable (or larger) amounts on R&D as a proportion of sales than other G7 countries. Falls in the R&D intensities of sectors such as metal products, machinery and equipment, and electrical apparatus have not been offset by increases in sectors such as pharmaceuticals;
- Expenditure on R&D by SMEs is generally low but those in high-technology sectors may have very high R&D-to-sales ratios;
- Fewer than half of all manufacturing SMEs employ qualified scientists and engineers and, of those that do, just under half employ any in an R&D capacity.

Concerns about the underperformance reflected in this analysis and matched by broader exercises in benchmarking competitiveness have caused the innovation performance of industry, and in particular of SMEs, to be a central policy concern of the present Government. In reviewing what is available to assist industry in meeting this problem, at first glance the situation in the UK is in some ways similar to that of other European countries in that existing policy mechanisms appear to continue while new ones are superimposed upon them.[1] However, this apparent persistence disguises some very major changes in the past and further changes in train at the time of writing.

5.1 – Priority programmes

At the beginning of the 1980s a system was in place for providing support for single companies in product and process innovation. Unpublished evaluation evidence suggests that in narrow terms this support was successful, in that companies were achieving their objectives with positive financial returns. However, near-market support was ideologically inconsistent with the philosophy of the Conservative Government. Instead it permitted the growth of "pre-competitive" collaborative programmes, beginning with Alvey, as mentioned above and a similar scheme for opto-electronics. At this point expenditure was on a large scale. Pressure on public spending led to a moratorium on support followed by a White Paper in 1988, which formally ended single company support (except on a small scale for SMEs) but allowed the continuation of collaborative research albeit on a reduced scale (DTI, 1988). At that point the main instrument for industry–industry collaboration was that of the Advanced Technology Programme, while academic–industrial collaboration was supported by the LINK mechanism. There was close, if not always easy co-operation between the Department of Trade and Industry, which supported industrial partners, and the Research Councils (mainly EPSRC), which supported academic partners (Cameron et al., 1996). Eventually, as described in the introductory section, support for large firms was ended in 1993. Budgetary pressures, partly driven by the relation to European funding described in section 6.2, were one factor but analysis in the DTI had also shown that certain notably profitable large firms had become regular users and were gaining the lion's share of this support.

1. This observation was made for Germany in Braunling's "shell model" which represented policies in the form of geological accretions over time.

The most prominent surviving mode for priority programmes is the LINK initiative. Also described here is the Foresight Challenge, set up as a deliberate follow-up of Foresight priorities but now integrated with the LINK structure. LINK's mission is to "offer a well established framework for collaboration between public and private sectors in support of science and technology (S&T) in areas of strategic importance to the national economy. LINK aims to enhance the competitiveness of UK industry, and quality of life, through support for managed programmes of pre-competitive S&T in market or technology sectors, and by encouraging industry to invest in further work leading to commercially successful products, processes, systems and services".

Currently 56 LINK programmes are sponsored by various government departments and Research Councils in a wide range of technology sectors. To observers from similarly-sized countries the most notable aspect is probably their small scale. Estimated government spend on these for 1997-1998 is £43 million, of which £16 million comes from DTI and £14 million from the Engineering and Physical Sciences Research Council. Each programme supports a number of collaborative research projects, which each last between two and three years. The Government funds up to 50% of eligible costs of a LINK project with the balance coming from industry. Because LINK's programmes focused on a particular technology or market area, the initiative became a good "vehicle" by which the Government could implement some of the recommendations coming out of the Foresight initiative. Since March 1995, the Government has announced 19 new LINK programmes, which are responsive to priorities identified under the Foresight initiative. These programmes will support projects costing up to £169 million over the next few years.

Among the follow-up measures to Foresight was a dedicated scheme, the Foresight Challenge competition, which was launched at the end of 1995 with the explicit aim of increasing interaction between industry and academia. Consortia of business and the science base were able to apply for matching funds for projects addressing Foresight priorities. In the first round, following a large number of applications, awards were made to 24 projects costing a total of £92 million, of which £62 million came from industry and £30 million from OST. The second round of the initiative, renamed Foresight LINK Awards, has £10 million of government funding available. SHEFC and HEFCW also provided funding for research projects reflecting Foresight priorities, awarding £7.5 million and £1 million respectively to support a total of 25 projects.

5.2 – Policies for start-ups, spin-offs and venture capital

The main emphasis in the Labour Government's policies has been upon the establishment of an entrepreneurial culture in which scientific capabilities are exploited through start-up and spin-off companies. Even the performance target for the Comprehensive Spending Review settlement for science has been expressed in these terms. An almost bewildering range of policies has emerged to address these issues. Among the most prominent, the Government announced in the March 1998 Budget the establishment of a £50 million seed capital fund to encourage the exploitation of scientific discoveries in universities. The Government has agreed to contribute £25 million to this 'University Challenge Fund' while additional funding will come from the Wellcome Trust, which is to contribute up to £18 million,[1] and the Gatsby Charitable Foundation (£2 million).

The aim is that universities will establish their own seed funds with the help of the University Challenge money and the seed funds will enable academics to scope out the commercial potential of research outcomes and take the first steps towards commercialising the research, thus addressing the "seedcorn-funding gap" mentioned earlier. The first round has 15 winners. While a variety of structures have emerged, typically universities will establish management boards with substantial representation from the business and venture capital sectors to manage their seed funds.

A second new initiative, with £25 million of funding, is the Science Enterprise Challenge. This is aimed directly at affecting cultural attitudes and will support up to eight new centres in universities to teach entrepreneurship and science commercialisation to scientists and engineers at all stages of their careers. A third initiative in the higher education sector is the Higher Education Reach Out Fund, sponsored by the Funding Councils and aiming to reward universities for interacting with business. Unlike the previous two initiatives, this will not focus on start-ups but is encouraging proposals across the full range of academic activities.

Other measures include a £20 million venture capital fund challenge fund within DTI's Enterprise Fund initiative, with six major financial institutions providing venture capital to small businesses with growth potential. Advisory services to start-ups will be up-rated. Changes to insolvency laws and intellectual property rights rules are also being considered.

---

1. See "University Challenge" Wellcome News 14, Q1, 2.

While the broad tax regime in the UK is generally considered favourable to industry, for many years the policy has been not to provide any specific fiscal incentives for activities such as R&D. After extensive inquiries, the 1999 Budget announced, on a consultative basis, new tax credits for small companies to perform R&D. Small firms (defined as having turnover less than £11.2 million, assets less than £5.6 million and not more than 250 employees) will benefit from about £150 million of such incentives. Eligible costs are direct staff costs plus the costs of consumables for R&D performed in the UK for the company's own benefit. 150% of R&D costs will be tax-deductible, giving a saving of 30% on R&D spending. Companies trading at a loss (a common situation for start-ups) will be allowed to reduce losses by 80% of the savings made under the credit if the company was in profit, through a subsidy. This R&D spend would not be tax deductible in future years. The total saving would be 24% of R&D costs. In addition, £20 of corporate tax relief will be allowed for every £100 invested in small high-tech firms. These investments may be offset against profits.

At a time of considerable policy innovation, perhaps the best course for an analyst is to recall Carter and Williams's dictum, that a wise policymaker "uses many instruments, and none to excess".[1] This holds true in the present phase but will need to be accompanied by careful evaluation so that successes can be grown and failures dropped.

5.3 – Regional policies

By international standards, the UK has been noted for its lack of regional policies, mainly because of the level of centralisation of government. The exception to this is in the technology transfer sector where policy delivery is focused at regional or local level, and where the smaller nations of the UK – Scotland, Wales and Northern Ireland – have been particularly active. Scottish and Welsh devolution are likely to produce more distinctive policies. For example, as noted above, there are already separate Funding Councils for higher education, which have varied their policies, on occasion, from those of their larger English counterpart. There is speculation that the Scottish Government might see higher education in a more explicitly economic role

1. Cf. C.F. Carter and B.R. Williams, 1964. Their explanation for this dictum is firstly that different circumstances require a different mix of remedies, and secondly that overuse could result in potentially harmful side-effects as when subsidies turn into protection.

than in the past, with implications for incentive structures. It has been agreed that Research Council funding will remain at United Kingdom level in order to maintain the critical mass needed for leading edge research.

In England, the newly established Regional Development Agencies have prepared economic strategies incorporating innovation strategies. These include linkages with higher education. A growing role in the delivery of innovation support is anticipated. A favoured concept is the promotion of the development of clusters. In some cases, notably biotechnology clusters, leadership is being taken at a national level. This will not be a policy for redistribution; the most likely location for a biotechnology cluster is around Cambridge. The science base there is very strong but overheating in the local economy and the desire to protect the remaining countryside from rapid urbanisation has provoked conflict with planning authorities.

6. INTERNATIONALISATION AND PARTICIPATION IN EU PROGRAMMES

6.1 – Internationalisation of UK research

The dimension of international co-operation in science outside the EU structures shows a sharp dichotomy between formal and informal arrangements. In the fields of "big science", particle physics, space and (to a lesser extent) astronomy the UK Government has often appeared as a reluctant investor. Cunningham, Nedeva and Reeve (1998) point out the lack of major international research facilities in the UK (with the main exception of the fusion energy research centre at Culham) as evidence of this tendency. Other formal links have also been on a smaller scale. For many years it was a policy to avoid, if possible, concluding framework scientific agreements with other countries.[1] This policy has now changed for two main reasons:

- the recognition that such agreements are a necessary precondition for co-operation with some countries, notably in East Asia; and
- with a trend towards collaboration in commercially exploitable areas of science, the benefits of a reference framework for intellectual property rights are now apparent (as evidenced in a recent agreement with the US).

1. Agreements between public agencies, particularly in the nuclear and defence research fields, have always been common.

Informal co-operation presents a very different picture. Partly for linguistic reasons and partly because of the quality of its science base, the UK is the second most popular destination for international movement of scientific fellows after the US. Using the measure of internationally co-authored scientific papers, in 1995 the UK shared with Germany the largest number of papers with the US and Japan respectively (Georghiou, 1998c). Traditional links with scientists in Australia, New Zealand and Anglophone Canada are also on a large scale, though these have declined in relative terms as these countries build up their links with other European countries and East Asia. The trend in the UK, as elsewhere, is for international collaboration in science increasingly to be used as a tool to pursue economic and foreign policy objectives, while not sacrificing a requirement for any sponsored work to be of high scientific quality.

The main body responsible for promotion of international linkages is the British Council, a registered charity which operates largely on the basis of grants-in-aid from the Foreign and Commonwealth Office and the Department for International Development. Its objective is to develop and promote educational, cultural, scientific and technological links between the UK and other countries. Science has been of increasing importance in the Council's activities. Research Councils, the Royal Society and other bodies also operate schemes to support fellowships and other forms of international co-operation.

6.2 – Participation in EU programmes

UK research organisations and industry are enthusiastic participants in the Framework Programme. In the Fourth Framework Programme, France was the most frequent partner country with a total of 4492 collaborative links, the highest pairing among any countries except for France–Germany with 4865 (CEC, 1997). In the year 1995-1996 UK participants received a total of £276m from the EU R&D budget,[1] a figure which broadly matches the country's contribution after allowance is made for funding to the Joint Research Centre and administrative overheads in the Commission. The most recent comprehensive evaluation of UK participation (Georghiou et al., 1993) concluded that EC-funded research was having an impact beyond that of its budgetary contribution for four reasons:

- the scarcity of research resources in the UK cause incremental funding to have a high marginal benefit;

1. Office of National Statistics and Higher Education Statistics Agency.

TABLE 8.7 – Percentages and cash attributions of European R&D

Departments	1995–1996		1996–1997		1997–1998	
	%	£m	%	£m	%	£m
Department of Trade and Industry (excluding OST)	57.7	193	53.6	183	52.3	191
OST	13.5	45	17.5	60	18.7	68
Department of Environment	8.0	27	8.3	28	8.5	31
Department of Agriculture, Fisheries and Food	3.9	13	3.4	12	3.7	14
Department of Transport	1.8	6	3.3	11	3.4	12
Department for International Development	–	–	2.7	9	3.0	11
Department of Health	2.7	9	2.6	9	2.8	10
Department for Education and Employment	1.5	5	1.5	5	1.6	6
Other departments	0.4	1	0.6	2	0.6	2
Overhead	7.0	23	5.2	18	4.1	15
Unallocated	3.5	12	1.4	5	1.2	4
Total	100.0	335	100.0	342	100.0	365

Source: Research Fortnight, 23 July 1997 based on Forward Look 1996 and SET Statistics 1997. Cash allocations derived from indicative contributions to EU R&D budget figures in SET Statistics 1997.

- EC funding is less tied to historical commitments and hence more focused on newer, fast-moving areas;
- through 50% funding and through its general influence on the contract-winning capability of organisations it mobilises more resources than are spent; and
- widespread participation and an expectation of growth focus attention in research-performing organisations.

Since that time the situation has changed to some degree. The expectation of growth was borne out but looks less likely in the future. Changes in national priorities and the influence of the Framework Programme's own historical legacies now mean that it is less valid to say that the EU Programme is more concentrated in fast-developing areas.

In one important respect EU participation has a greater effect upon the UK research system than in any other member state. Since 1984, the UK Government has operated a system of attribution, whereby the cost of financing growth in the EU budget is considered against the budget of the UK government department which would have made that expenditure had it been national. In effect, subject to Ministerial discretion, the attributed UK contribution (not receipts) is charged to the relevant national budget and may result in a deduction from national research. The logic of this system is that it is meant to make national officials consider EU spending as part of their own budget (manifested for example in UK opposition to excessive growth in the Framework Programme). Its weakness is that since other countries do not have this system, growth takes place anyway and national budgets are weakened in those areas. While this would be desirable if the EU programme was a superior substitute, there are other cases where an increase in national spending to take advantage of the European momentum would be the sensible course of action.

Historically, the Department of Trade and Industry has carried the main burden of attribution and this has been a contributory factor (along with ideological opposition and concerns about their effectiveness) to the large-scale decline in support for national industrial collaborative R&D programmes. Table 8.7 shows the allocations for the past three years. Growth in the basic research content of the Framework Programme has led to a rising share of the budget being attributed to the Office of Science and Technology (shown separately from the rest of DTI).

7. DEFENCE R&D[1]

7.1 – The UK defence industry

The UK has the largest defence industry in Europe and is the second-largest arms exporter after the United States. Since the 1980s, the defence industry has been privately owned. The two leading defence companies have been British Aerospace and GEC. British Aerospace is the leading defence and aerospace company in Europe. A manufacturer of military aircraft, missiles and munitions, it is the fourth-largest defence company in the world. In addition, the UK has a strong and diverse group of companies supplying subsystems and components. These include aero-engine manufacturer Rolls-Royce as well as a sizeable number of smaller suppliers and subcontractors (James et al., 1998). The defence industry continues to represent an important part of UK manufacturing activity. In 1996, the industry employed

1. This section draws heavily upon James (1998).

310 000 people. This represents roughly 7% of the UK manufacturing workforce and does not include indirect employment effects, estimated at about 300 000 jobs.

Historically, scientific and technological development for defence purposes has been organisationally (and often physically) separate from civil science and technology both in terms of its institutions (with dedicated defence research establishments and defence divisions of multi-business corporations) and a separate community of practitioners of scientists and engineers working on defence-related activities. The situation with regard to smaller suppliers is rather different and there is considerable evidence of dual-use technological activity amongst first- and second-tier suppliers.

The 1990s has seen considerable restructuring of the UK defence industry. This process began rather earlier than in most other European countries and has seen substantial job cuts, plant closures and a number of mergers and acquisitions. The outcome of the restructuring process has been a significant concentration in the UK defence industry. The largest and most recent element is the pending acquisition of Marconi (GEC's defence company) by British Aerospace, a move which consolidates the two largest players to achieve a world-scale company. However, the move has precipitated debate about future restructuring, with some seeing it as a barrier to future European mergers. Others argue the case for transatlantic companies to ensure continued access to the lucrative US market. In his analysis, James (1998) concluded that the institutions and relationships that contribute to the generation, application and commercialisation of technologies used in the UK defence industry are increasingly international in character and are likely to become more so in the decade ahead.

7.2 – Government expenditure on defence-related science and technology

The defence industry has always had a privileged position with regards to government science and technology funding, and a heavy commitment to defence technology and an unusually high proportion of funds spent on defence R&D have long been a characteristic of the UK's National Innovation System. From the late 1960s until well into the 1980s, half of government expenditure on research and development (R&D) went on defence, and the percentage was even higher in earlier years. Indeed, despite recent cuts, the UK continues to spend, by international standards, an exceptionally high proportion of public research and development funds on defence-related objectives. Amongst the leading industrial countries, only the US devotes a higher proportion of public R&D expenditure to defence, and government-funded defence R&D as a percentage of GDP is higher in the UK than in Germany, Japan or Canada – but lower than in France or the US and, in recent years, Italy.

Nevertheless, a number of recent official reports have noted that absolute UK expenditure is falling behind that of France, Germany and the US. Whilst government funding for defence R&D in the UK fell by 25% in real terms in the period 1985-1992, it increased in France (up 36%), Germany (up 24%) and the US (up 7%) (James and Gummett, 1998). At the same time, industry expenditure on R&D is declining not least as a result of the reform of the Ministry of Defence's procurement process, and in particular a shift towards more competitive Commercial Off-The-Shelf (COTS) procurement of defence equipment. Consequently, the 1995 Defence and Aerospace Foresight Report noted that declining government and industry R&D expenditure threatened UK competitiveness, and called on industry and government to reverse the decline in UK spending on research and technology demonstration.[1]

The Labour Government's policies have been set out in the form of a Strategic Defence Review. This sets out a strategy for modernising and reshaping the Armed Forces and their supporting infrastructure (Ministry of Defence, 1998b). Forming a part of the review is an essay on the impact of technology. This draws out key technological trends including the significance of advances in the civil market, the shorter life of some systems offset by a longer life for platforms, and the major change arising from the application of information technology to command and control. The essay then goes on to discuss the MoD's technology strategy and emphasises again the importance of interchange with civil technology. The policy towards DERA was set out in section 4. The importance of international collaboration is emphasised. The essay concludes that in the future technology issues will need to be considered from a politico-military as well as a technical perspective.

---

1. The fact that defence and aerospace had their own Panel is significant in itself, and says much about their importance in the UK National Innovation System. The inclusion of defence and aerospace as a specific area of attention made the UK Foresight exercise different to its German and Japanese counterparts (James and Gummett, 1998).

8. CONCLUSIONS

This chapter began by emphasising the long-term continuity in the concerns of the United Kingdom's science, technology and innovation policies. In the two decades covered by this particular review, the conundrum of harnessing scientific performance to economic success has been addressed in very different and sometimes conflicting ways. Some of the ideas behind the instruments appear to have been imported: collaborative programmes from Japan; foresight also from Japan and from Germany; and, most recently, the drive to create an entrepreneurial culture driven by technology start-ups has been fuelled by a succession of high-level missions to California. In each case, though, the transferred idea has crossed over only at an abstract level, with its implementation being uniquely British. This is not surprising, given the enormous differences in culture and infrastructure that exist. The nagging doubt is that good ideas promoted by government none the less have not penetrated to the core problems of the innovation system, in particular the attitudes towards risk-taking and investment in technology held at high levels in companies and in the investment community. To this may be added the problem of scale. The entire DTI innovation budget for 1999-2000 is £209 million, of which about one-third is for standards and infrastructure, leaving a relatively small sum for incentives. At this level of operation well-designed policies may none the less be sub-critical and disproportionate to the problems that they address.

In terms of content, the present portfolio of policies appears to be preoccupied with small firms, and within that group with technology start-ups. No one would deny the importance of these firms but it should also be remembered that after the 25% of R&D accounted for by the pharmaceutical majors, a further ten firms account for 33% of the R&D in the UK. This latter group is under-investing by comparison with its international competitors. In the small firm sector the challenge remains to raise the level of a very long tail of stragglers who have failed to grasp the opportunities presented by technological change. Both of these groups deserve urgent policy attention. Too much focus on start-ups risks repeating a past error, this being to present a science and technology policy as if it were an industrial policy and then subsequently to wonder why it did not fulfil expectations.

The very positive outcome for science from the Government's Comprehensive Spending Review has created a feeling of optimism in the scientific community unseen since the early 1970s, and a serious attempt to redress the problem of declining physical scientific infrastructure. However, this needs to be tempered by noting that it would take ten years of growth at this level to reach the OECD average for research as a percentage of GDP. Since the next spending review is already underway, well before most of the present investment has borne fruit, the arguments for science need to continue to be pressed.

Some other issues are still to be resolved. The university sector is under great financial pressure and unless Research Council funds meet their indirect costs by one means or another, taking on more research may weaken rather than strengthen its position. Industry believes the answer lies in greater selectivity and concentration, preferring a small number of world-class centres of excellence to what it describes as "large-scale mediocrity". However, funding is already so concentrated that there is little scope for significant transfers from the weaker end of the spectrum. Universities have successfully diversified their income by building up research contracts from industry but there is probably only limited scope for further large-scale growth. There also remains a major policy deficit in that the present financial system rewards traditional academic research through the RAE but offers no parallel incentives for academics to take their work into the economy or society.

The Comprehensive Spending Review result has also marked the culmination of a process which began with Foresight in terms of a shift towards biosciences. This is based upon several factors, including a strong relationship with the UK's most successful industry, the perceived opportunities at present, and the leverage exerted upon government by the Wellcome Trust. The current predominance of biosciences is so strong that the spending case for physical sciences is being made on the basis of the critical contribution they make to the advance of life sciences. Mindful of the UK's failure to capitalise upon early leads in computer technology, every effort is being made to secure a share of the future of this industry. The rhetoric is of establishing a "Genome Valley" in the UK, analogous to Silicon Valley. Realistically, UK expenditure continues to be a fraction of that in the US and at some point niches will have to be sought.

There is also some risk that the type of relationship which exists between science and industry in the biosciences could be assumed by policymakers to apply in all areas of science. Life science is an area where the route from laboratory to market can be very short conceptually if not temporally. While not exactly a manifestation of the linear model, this sector does not experience some of the difficulties of industries which are technologically-driven and where science has a more complex supporting role. Large, R&D intensive firms can use their academic links as a window through which they can from time to time reach and appropriate valuable intellectual property, but this option is not generally viable for less R&D intensive firms, including most SMEs.

A final point about the UK's innovation system is that from an ecological perspective, the net effect of past policies has been to reduce the level of diversity. Reforms and financial pressures have led actors in all of the main sectors to the market for contract research. Hence, former public laboratories compete not only for their own previous business but seek to diversify. Universities recognise that industrial funding has been the only real growth sector and compete for research contracts from government departments, including those which formerly went to the public laboratories. Research Council funds are now available for any non-profit research organisation to bid for. Even in industry, many former corporate laboratories now have an arm-length relationship with the rest of the company and compete for internal business while seeking outside contracts.

The overall phenomenon is one which I have termed "convergence", whereby all research actors are increasingly operating in an overcrowded "contract research space" (Georghiou, 1998d). The risk is that if they all respond to the same market signals, they will become more similar to each other, and to some extent consolidate, with an overall narrowing of capabilities in the system. The most likely loss would be of the activities which were originally most distinctive for each sector: creative basic research in universities, long-term strategic research directed towards new, innovative business in industry, and rigorous and unbiased long-term science to support policy and regulation in the former public laboratory sector. The policy challenge is to construct a market which rewards diversity and gives long-term signals.

REFERENCES

ABRC, 1987, *A Strategy for the Science Base*, London: HMSO, 1987.

Adams J., T. Bailey, L. Jackson, P. Scott, D. Pendlebury and H. Small, 1998, *Benchmarking of the International Standing of Research in England* – Report of a consultancy study on bibliometric analysis, Centre for Policy Studies in Education, University of Leeds, Institute for Scientific Information, Philadelphia.

Barker, K. and A. Beasley, 1998, La privatisation des laboratoires de recherche en Grande Bretagne, *Annales des Mines*, February.

Beesley, A., 1999, Strategic change in a government laboratory: the case of the building research establishment, Ph.D. thesis, University of Manchester.

Boden, R., P. Gummett, D. Cox and K. Barker, 1998, "Men in white coats ... men in grey suits", *Accounting, Auditing and Accountability Journal*, Vol. 11, No. 3, pp. 267-291.

Cabinet Office, 1988, *Policy Evaluation – A Guide for Managers*, London: HMSO.

Cameron, H. and L. Georghiou, 1997, Managerial Performance – the Process Evaluation in M. Callon, P. Larédo and P. Mustar, *The Strategic Management of Research and Technology*, Paris: Economica.

Cameron, H., L. Georghiou, T. Buisseret and T. Ray., 1996, *Evaluation of the Information Engineering Advanced Technology Programme: Summary Report*, Department of Trade and Industry and Engineering and Physical Sciences Research Council.

Carter, C.F. and B.R. Williams, 1957, *Industry and Technical Progress*, London: Oxford University Press.

Carter, C.F. and B.R. Williams, 1964, Government Scientific Policy and the Growth of the British Economy, *The Manchester School of Economic and Social Studies*, Vol. 32.

Commission of the European Communities, 1997, *Second European Report on S&T Indicators 1997*.

Cunningham, P. (ed.), 1998, *Science and Technology in the United Kingdom*, London: Cartermill International.

Cunningham, P., M. Nedeva and N. Reeve, 1998, International Dimensions, in P. Cunningham (ed.), op. cit.

Cunningham, P. and S. Hinder, 1998, *A Guide to the Organisation of Science and Technology in Britain*, 5th edition, London: British Council.

Department of Trade and Industry, 1988, White Paper *DTI – the Department for Enterprise*, Cmnd 278, London: HMSO.

Department of Trade and Industry, 1998, *The UK R&D Scoreboard 1998*, London: DTI.

DTI/OST, 1996, *Science, Engineering and Technology Statistics*, Cm3257-II, London: HMSO.

Flanagan, K. and M. Keenan, 1998, Trends in UK science policy in P. Cunningham (ed.), op. cit.

Georghiou, L., P. Halfpenny, M. Nedeva, S. Hinder and J. Evans, 1996, *Survey of Research Equipment*, PREST/CASR: Committee of Vice-Chancellors and Principals.

Georghiou, L. et al., 1993, *The Impact of European Community Policies for Research and Technological Development upon Science and Technology in the United Kingdom*, London: HMSO.

Georghiou, L., 1998, Science, Technology and Innovation Policy for the 21st Century, *Science and Public Policy*, April, pp. 135-137.

Georghiou, L., 1998a, *Science Policy Advice in the UK and the Work of PREST*, presentation to NISTEP Tenth Anniversary Conference, Tokyo, 16-17 October.

Georghiou L., 1998b, The UK Technology Foresight Programme, *Futures*, Vol. 28, No. 4, pp. 359-377.

Georghiou, L., 1998c, Global co-operation in research, *Research Policy*, 27, pp. 611-626.

Georghiou, L., 1999, Role and Effects of Foresight in the United Kingdom, paper for Conference *Forward Thinking: Keys to the Future in Education and Research*, Hamburg, 14-15 June.

Gibbons, M. and L. Georghiou, 1987, *Evaluation of Research – A Selection of Current Practices*, Paris: OECD.

Gummett, P., 1980, *Scientists in Whitehall*, Manchester: Manchester University Press.

HEFCE, 1998, *Research Funding: Introduction of a Policy Factor*, October 98/54 Consultation.

Higher Education Funding Council for England, 1997, *The Impact of the 1992 Research Assessment Exercise on Higher Education Institutions in England*, Ref: M 6/97.

Hills, P.V. and A.J. Dale, 1995, Research and Technology Evaluation in the United Kingdom, *Research Evaluation*, Vol. 5, No. 1.

HM Treasury and Department of Trade and Industry, 1998, *Innovating for the Future: investing in R&D* – a consultative document, March.

HM Treasury, 1983, *Financial Management in Government Departments*, Cmnd 2250, London: HMSO.

House of Commons, 1995, Aspects of Defence Procurement and Industrial Policy, First Report of the Defence Committee

and the Trade and Industry Committee, HMSO: London.
House of Commons, 1998, *The Defence Evaluation and Research Agency, Sixth Report of the Defence Committee*, HC 621, 6 July.
House of Lords Select Committee on Science and Technology, 1982, Science and Government – Government Observations on the First Report of the House of Lords Select Committee on Science and Technology (Session 1981-1982), London: HMSO July.
House of Lords Select Committee on Science and Technology, 1986, First Report Civil Research and Development *[HL 20-I, 20-II & 20-III]*, London: HMSO.
House of Lords Select Committee on Science and Technology, 1986, *Civil Research and Development*, First Report, Session 1986-1987, HL20.
Howells, J., M. Nedeva and L. Georghiou, 1998, *Industry-Academic Links in the UK*, Bristol: Higher Education Funding Council for England.
James, A., H. Cameron and P. Gummett, 1998, *Transfer and Civil Use of Defence-Related Technologies and Diversification of Defence-Related SMEs*, Final Report to the European Commission, Manchester: PREST.
James, A., 1998, The Place of the UK Defence Industry in its National Innovation System: Co-evolution of National, Sectoral and Technological Systems, *Workshop on The Place of the Defense Industry in National Systems of Innovation*, Cornell University, 16-18 October.
James, A. and P. Gummett, 1998, Defence in P. Cunningham (ed.), op. cit.
Langrish, J., M. Gibbons, W.G. Evans and F.R. Jevons, 1972, *Wealth from Knowledge: A Study of Innovation in Industry*, London: Macmillan.
Loveridge, D., L. Georghiou and M. Nedeva, 1995, *United Kingdom Technology Foresight Programme Delphi Survey*, Manchester: PREST on behalf of OST.
Marshall, A., 1919, *Industry and Trade*, London: Macmillan.
May, R., *The Use of Scientific Advice in Policy Making*, Office of Science and Technology, March 1997.
Ministry of Defence, 1998a, *Defence diversification: getting the most out of defence technology*, Proposals for a Defence Diversification Agency, Cm 4088.
Ministry of Defence, 1998b, White Paper, *Strategic Defence Review – Modern Forces for the Modern World*, London: HMSO.
Ministry of Defence, 1999, A Public Private Partnership (PPP) for the Defence Evaluation and Research Agency (DERA), United Kingdom Ministry of Defence Consultation Document, 5 May 1999.
National Committee of Inquiry into Higher Education, 1997, *Higher Education in the Learning Society*, ("Dearing Report"), London: HMSO.
Nuttall, W.J., 1999, Engineering and Physical Sciences Research Council (UK) Peer-review Procedures 1995-1996, *Research Evaluation*, Vol. 8, No. 2, pp. 132-140.
Oakley, B. and K. Owen, 1989, *Alvey – Britain's Strategic Computing Initiative*, Cambridge: The MIT Press.
Office of Science and Technology *Foresight – First Progress Report*, 1996.
Office of Science and Technology, 1993, *Realising Our Potential: a Strategy for Science, Engineering and Technology*, Cmnd 2250, London: HMSO.
Office of Science and Technology, 1995, *Progress Through Partnership – Report from the Steering Group of the Technology Foresight Programme*, London: HMSO.
Office of Science and Technology, 1998, *Allocation of the Science Budget 1999-2000, 2000-2001, 2001-2002*, Department of Trade and Industry, 27 October.
Pavitt, K. (ed.), 1980, *Technical Innovation and British Economic Performance*, London: Macmillan.
Save British Science Society, 1998, *A World Class Science and Engineering Base for the 21st Century*, April, London: Save British Science.
Walker, W., 1993, National innovation systems: Britain, in R. Nelson (ed.), *National Innovation Systems – A Comparative Analysis*, Oxford: Oxford University Press.

CHAPTER 9

THE NETHERLANDS: SCIENCE POLICY BY MEDIATION

BAREND VAN DER MEULEN AND ARIE RIP

1. INTRODUCTION

The Netherlands is a highly developed, medium-sized (in terms of economic activities and research activities) European country. It scores relatively high on R&D intensity, R&D expenditure and R&D output (measured as publications and citations). What is particularly striking is the number and strength of mediating institutions between the national Government and research-performing institutions, which form a densely populated "intermediary layer" (Van der Meulen and Rip, 1998). This contrasts with countries like the UK, with few mediating institutions and a much more direct exposure of research performance institutions to government measures. It also contrasts with countries like Germany and France, which had (however different they are in other ways) at least until recently, large parts of the research system functioning as semi-autonomous "blocks". Mediation allows mutual translation between non-scientific (public and private) goals and R&D agendas and R&D performance. Given the increasing importance of such translations, it is interesting to focus on the instruments and institutions involved, not just as a peculiarity of the Dutch research system, but as historically evolved modalities for linking science and society.

In all Western countries, in the decades after World War II, promotion of science without too many questions asked was the rule: a science policy regime which can be characterised by the phrase "Science, The Endless

Frontier", after the title of the 1945 report by Vannevar Bush to the President of the US. In the 1970s, relevance of science became a catchword, and a plethora of new policy instruments as well as new institutional forms appeared. In the 1990s, the (civilian) strategic importance of R&D is generally recognised, not just for economic productivity but also for environmental monitoring and improvement. "Strategic Science" could be the label for a newly emerging regime (Rip, 1997a).

But it is not just the science policy regime that has shifted. The modes, and the content, of knowledge production have been changing. Think of the path-breaking developments in the life sciences, of the advent of modelling and "paper" technology, of the changes enabled by information and communication technologies. And think of the loosening up of disciplinary modes of knowledge production, of the increasing public–private interactions, of consortia and research centres in which fundamental, strategic and applied research merge, and of "discovery in the context of application" becoming the rule rather than the exception (Gibbons et al., 1994). Think also of increasing public interest in, and critical scrutiny of, science and technology. In general, one sees increasing heterogeneity and a mixing up of earlier categories, in particular the simple cognitive and institutional distinctions of the Golden Age of science and science policy of the 1950s and 1960s.

Given these trends, national research systems need to develop institutional competencies to cope with heterogeneity. That is why the institutions and practices in the Dutch research system are of more than anecdotal interest. In an earlier study, we have distinguished two dimensions along which research systems can be characterised and which imply contrasting institutionalised capacities to handle heterogeneity (Rip and Van der Meulen, 1996 and 1997). One capacity is top-down steering, which homogenises the complexities in order to make research efforts fit the goals of a principal (for example, a national government). The other capacity is bottom-up aggregation, where local and specific activities and agendas are picked up and merged into a broader and forceful agenda. Mixed groups and hybrid fora can do such a job, and some national research systems are better in doing this than others. It may well be an accident of history (and political culture) that the Dutch research system has developed these institutional competencies. Whatever the causes and reasons, the actual accomplishment is interesting to other countries, even if they cannot just adopt Dutch approaches because they have their own historically-specific developmental trajectory.

In this chapter, we first provide a sketch of the Dutch research system and recent science policy, and then discuss mediating institutions. These often started their life as implementation of science policy measures, but became institutionalised and survived the policies (and policy purposes) that created them. We shall also indicate some recent shifts in the landscape of science and science policy, and conclude by offering an evaluation.

2. THE DUTCH RESEARCH SYSTEM, R&D EXPENDITURES AND PERFORMANCE[1]

National and international quantitative analyses of research and technology generally show a stable and positive result for the Netherlands. In the rankings, the Netherlands appears generally within a group of small, science-intensive countries like Sweden, Switzerland, Denmark and Belgium. R&D expenditure in the Netherlands fluctuates around 2% of GDP. The government and the private sector by and large equally share the burden of this expenditure. From an international perspective this implies that the private R&D sector is relatively small. This is probably due to the existence of a strong service sector, with hardly any visible R&D activities, of a strong agricultural sector that relies to a large extent on an excellent public R&D infrastructure, and to a lack of larger industries filling the gap between SMEs and multinational companies.[2] In this section we provide a rough overview of the flows of public and private research funds (see Table 9.1) and of research performance. We note that there is little boundary crossing: most private

1. This section draws on the data from the Central Bureau of Statistics (Van Deventer et al., 1999), and for the scientific performance data on the 1998 report of the Netherlands Observatory of Science and Technology (Tijssen et al., 1998). Tables 9.1 and 9.2 as well as Figure 9.1 are based on data available in CBS Statline (http://www.cbs.nl/ nl/statline/index.htm). A useful source of statistics on the science system is a facts and figures document published by the Ministry of Education, Culture and Sciences (OCenW, 1998). Figure 9.2 is based on data from this document.
2. In the second half of the 1990s, several studies considered whether the relatively low R&D intensity was explained by the sector structure, with diverging conclusions. Verspagen and Hollanders (1998) argued on the basis of statistical analysis that sector structure explains only part of the low R&D intensity so that a substantial part of the gap with other countries must be 'intrinsic'. Snijders (1998), however, after tracing the specific situation in different sectors, concluded sector structure explained the gap fully. He claimed there is no problem: the Netherlands has the appropriate R&D level given its sector structure. Verspagen, in contrast, maintains that a sector structure that leads to low R&D intensity, is by itself a problem for economic growth. One should consider, however, that R&D intensity is only one indicator; the existence of relevant "bridging institutions" is also important (cf. Martin and Scott, 2000).

TABLE 9.1 – Research financing in the Netherlands (1997)

Research performance organisations	Firms	Non profit organisations	Government institutes	Universities	Total funding
Total R&D Expenditures	4695	257	1516	1876	8 352
Source of funding (in %)					
Firms	7.5%	1.4%	13.6%	4.3%	7.7%
Non profit organisations	0.1%	17.4%	1.5%	6.4%	2.3%
Government institutes	1.3%	2.3%	9.4%	6.9%	4.0%
Universities	0.0%	0.4%	0.4%	0.3%	0.2%
Foreign organisations	14.1%	1.6%	4.1%	0.1%	8.8%
Total contract funding	23.0%	23.1%	29.0%	18.0%	23.0%
Government funding	3.0%	12.7%	63.7%	78.7%	31.3%
EU funding	1.0%	0.5%	2.6%	3.3%	1.6%
Total external funding	26.7%	36.3%	95.3%	100.0%	55.9%
Own funding	73.3%	63.7%	4.7%	0.0%	44.1%
	100%	100%	100%	100%	100%

Unit: MECU.

TABLE 9.2 – R&D expenditures by firms with own research personnel (1997)

Personnel	15 – 100 employees	50 – 200 employees	200 and more employees	Total
Industry	100	292	2 420	2 812
Services	138	148	402	688
Others	12	32	170	214
Total	250	472	2 992	3 714

Unit: MECU.

expenditure is on R&D performed in private R&D institutions, and similarly for public expenditure.

Most of the shifts in total expenditure derive from changes in R&D expenditures by industry. Until 1990, industrial R&D expenditure was dominated by a few multinational companies in chemistry and electronics (the 'Big Five') and sensitive to their economic performance. In the early 1990s, total industrial R&D expenditure decreased, mainly due to the reductions at Philips. In the same period, the share of R&D investments of the five large multinationals decreased from 58% in 1990 to about 41% in 1997. (In 1970 this percentage was still above 70%). Despite the decline of the role of the five large

FIGURE 9.1 – R&D investments by firms in technology fields

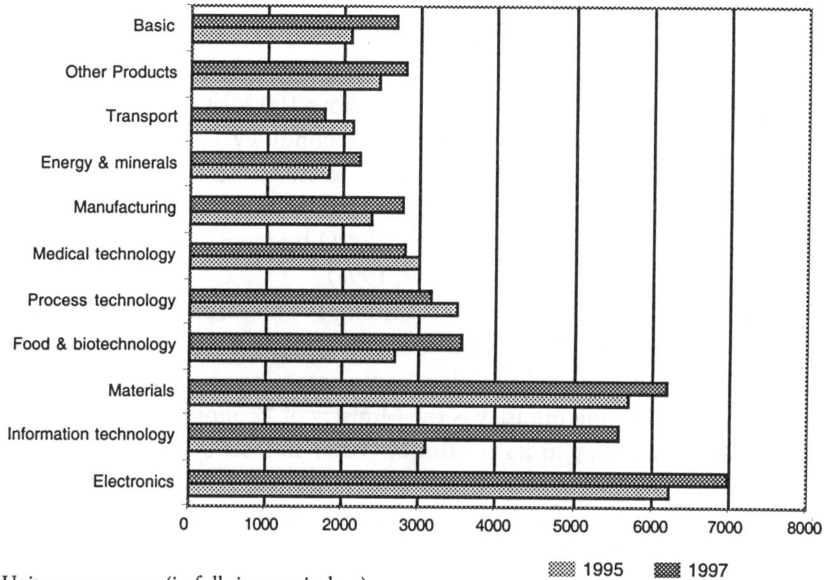

Unit: person-years (in full-time equivalent).

multinationals, they still count for the largest share in industrial R&D expenditures. What has changed is that larger Dutch firms are becoming more important in the overall picture (see also Table 9.2). In terms of sectors, the role of the chemical (especially end-products) and electronic sectors remain strong; they continue to be the most R&D-intensive sectors. Their R&D intensity, in terms of their R&D expenditure as a percentage of realised added value, is above 20%, while for the other sectors this does not exceed 6%. If we look at technological fields (R&D investment in 1995 and in 1997), the dominance of the electronics and chemical sector is also visible (Figure 9.1). Electronics and materials are the main technology fields. But also visible is the growth of R&D investments in information and communication technology, in food and biotechnology, and the decline of R&D investments in process technology.

Figures about investments in technology fields show that total expenditures have increased again, probably due to the combination of better economic performances and better tax facilities for industrial R&D. In 1994, Parliament accepted a new Law on Industrial R&D Support, which emphasised general incentives to industrial R&D by decreasing its costs. Under the Law, firms can submit R&D projects for eligibility. The project descriptions are assessed by SENTER, the technology agency of the Ministry of Economic Affairs, on the kind of work implied and the innovativeness of the work. Firms can get a reduction on the employer's part of wage taxes and social insurances of R&D personnel involved in the projects. Indicative for the success of the law is the budget growth from about 100 MECU in 1994, to 230 MECU (about half of the budget for technology stimulation in the Netherlands) in 1997. In an early evaluation study, R&D-intensive firms' responses indicated that the tax facilities enabled them to keep their R&D budget at a high level; firms with small R&D budgets turned out to have increased their budget (Bureau Bartels, 1996).

The lion's share of government funding for research comes from the Ministry of Education, Culture and Sciences, which is responsible for governmental higher education policy and for governmental science policy. The other main research-funding ministry is the Ministry of Economic Affairs, which has technology policy in its portfolio. Other ministries with substantial research budgets are those for Agriculture, Nature and Fisheries, for Defence, for Transport and Public Works, for Health, Welfare and Sport and for Housing, Spatial Planning and the Environment (see Table 9.3). Compared to other countries, there is little defence-related R&D, and government funding of privately-performed R&D is only 10% of total government R&D expenditure (excluding effects of fiscal measures).

Despite ongoing budget cuts in (direct) government funding of universities and public research institutes, government R&D expenditure as a percentage of GDP tends to be stable. Apparently, budget cuts have been

TABLE 9.3 – R&D expenses of Dutch Ministries (1998)

R&D expenses per Ministry	%
Education, Culture and Sciences	61%
Economic Affairs	19%
Agriculture, Nature Management and Fisheries	6%
Transport and Public Works	3%
Defence	3%
Foreign Affairs and Development Co-operation	3%
Health, Welfare and Sport	2%
Housing, Spatial Planning and Environment	2%
Other	1%
Total	100%

Source: Science Budget Progress Report, 1998.

implemented together with a formally unrelated increase in indirect funding, like the tax facilities for industry mentioned above, the Technological Top Institutes that will be described in the next section, and the selective use of so-called windfall profits from natural gas. The latter (and peculiarly Dutch situation) is an example of how in the 1990s investments in science became linked to economic policy. In the early 1990s the government separated part of the profits from its shares in natural gas exploitation to invest in the economic infrastructure, the so-called ICES scheme. Initially, the focus was on new industrial areas and improvement of transport structures. The Ministry for Education, Culture and Sciences and the Ministry of Economic Affairs successfully argued that investments in the knowledge infrastructure were at least as important for the quality of the economic infrastructure. Investments under this scheme now include programmes for developing biotechnological methods for soil cleaning, for improvement of high-performance computing infrastructure, support for a research centre on nanotechnology, for a new Technology Park near Amsterdam and a joint R&D organisation of universities and technological institutes in the field of hydraulics and civil engineering.

The public R&D sector comprises the university sector, and other academic research institutes for basic research (overseen by the Royal Netherlands Academy of Arts and Sciences, KNAW, or by the national funding agency, NWO), and applied and/or policy-oriented research institutes, mainly of two kinds: institutes related to a government ministry, and a small number of big public technological research institutes and TNO, the general organisation for applied research. As Table 9.1 shows, the universities get about 1500 MECU in government funding, and the government institutes about 1000 MECU. About two-thirds of this 1000 MECU goes to institutes related to government ministries. TNO and the big technological institutes obtain about as much funding from private sources as they get from public sources.[1]

There are two modes of funding of universities and public research institutes: direct and indirect funding. Direct funding for the universities comes in the form of a block grant, the size of which is related to student numbers and to the number of researchers in accepted programmes (the definition

1. We give only orders of magnitude because it is notoriously difficult to obtain reliable numbers. The data in different publications are not always consistent, even if they draw on the same sources (for example, OCenW, 1998 and Van Deventer et al., 1999 on the public research institutes).

of what is accepted tends to follow what exists already). The well-developed assessment process of university research has no immediate effect on funding; it affects the status of research groups and influences actor strategies (including attempts at improvement) (Westerheijden, 1997). Direct governmental funding of public research institutes depends on their mission: it varies from over 80% for basic research institutes and governmental research institutes to about 20% for public technological institutes like TNO, the marine research institute MARIN and the air and space laboratory NLR. While the direct funding used to come as a block grant (and comprised a larger percentage), in the late 1980s and early 1990s, the government implemented a policy (also for the governmental research institutes) by which institutes and government strategically allocated the direct funding to specific areas.

In addition to the block grant to universities, they acquire research funding from the so-called second stream, competitive grants for projects and programmes awarded by the national funding agency NWO. In terms of full-time equivalents R&D personnel, the size of second-stream funding is about 30% of the first stream (Van Deventer et al., 1999, p. 79). Money-wise, however, the proportion is not much more than 10%, which has to do with the limited overhead made available to the universities by NWO.[1] On both counts, the proportions are small compared with countries like the US where first-stream funding is less extensive. Privately-funded R&D in public research performance institutions occurs primarily in the public research institutes. Universities do obtain a sizeable part of their research budget (up to 30%) from other sources than direct and indirect (second-stream) governmental funding, but only 20% of this "third-stream" funding comes from the private sector.[2] The remainder is research commissioned by public bodies including government ministries, by medical charities and by international bodies, in particular the European Union. Universities and government institutes now often collaborate and join forces in research programmes, centres of excellence and in those graduate schools which are based in more than one university.

The 2% of GDP for R&D expenditures corresponds to less than 1% of worldwide R&D expenditure. Research productivity is high: the Netherlands

---

1. Data from Hackmann and Klemperer 2000. They report a stable ratio of first-, second- and third-stream research funding in Dutch universities in the 1990s, of 72:8:20.
2. This amounts to about 5% of the universities' research budget, which is the average percentage in OECD countries (OECD, 1998, p. 38).

FIGURE 9.2 – Share in international publications (1995-1996)

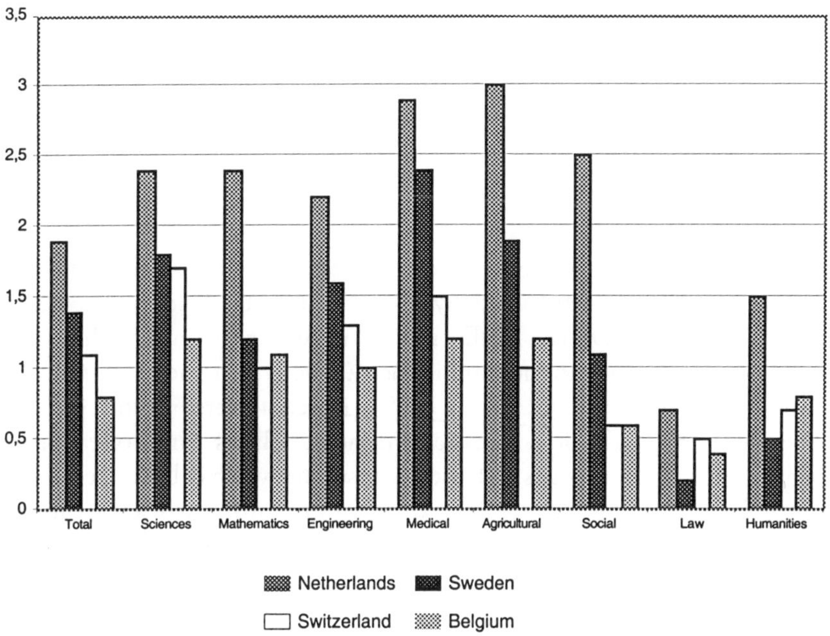

produces more than 1% of the total of international publications, and the average citation impact of these publications exceeds the average citation impact. Relative to the size of the country, the performance of Dutch R&D in terms of scientific publications and scientific impact has been consistently high, including a relatively large number of Nobel Prize winners (Willink, 1991). If we compare scientific output with European countries that have a similar pattern in R&D investment and a similar socio-economic structure, Dutch scientists appear to be very active in the international scientific publication world. In all areas, the Netherlands leads other medium-sized countries (see Figure 9.2). Relatively strong areas are agriculture and medical sciences while law and humanities are weak in terms of international publications. Also in the sciences, engineering sciences, mathematics and social sciences, the contribution of Dutch researchers is substantial. An important caveat, of course, is that such measures, based on ISI and similar publication indexes, are biased against non-English language publications which are important in the social sciences and especially the humanities.

In terms of impact of the publications, the Netherlands are at the top for the sciences, with a citation impact considerably higher than the world mean. Again agriculture happens to be very strong, but also the veterinary sciences, physics, chemical technology and the earth sciences. The social sciences have an average impact, even with their present active publication strategy.

As in the case of publications, the Netherlands' share in total patents in the database of the European Patent Office is somewhat above 2% and is comparable to Italy. If the number of patents is 'normalised' to the number of R&D personnel, the Netherlands do better and are the second West European country in patent activity, just behind Switzerland. Most patents come from the Dutch R&D intensive multinationals within the electronics industry and the food industry. The Dutch chemical industry appears to have a rather low patent profile. Dutch universities and research institutes traditionally do not have active patent policies. Firms often maintain the patents. Universities and research institutes which collaborated in the research on which the patent is base are usually rewarded with a share in the royalties.

3. MEDIATION IN THE DUTCH R&D SYSTEM

To understand the present Dutch research system and its functioning, it is necessary to pay attention to its institutional development. This is not just a matter of writing its history (Blume, 1986). Indeed, new organisations were formed over the years, and existing organisations adopted new responsibilities and practices. There were struggles between various actors: government Ministries among themselves, central government with the national funding agency NWO and with the universities (jointly and several). This will continue, but it is possible to discern certain patterns.

When the Netherlands entered the 1990s, its science policy at the top level was fragmented (struggles between different Offices within the Ministries of Education and Science, struggles between this Ministry and the Ministry of Economic Affairs, and incidental attempts to keep up the co-ordination role, all this in the face of decreasing budgets and no one Minister with an overarching responsibility). Science policy for the science basis was, in a sense, predicated on this situation. For the universities, the emphasis was, and continues to be, on excellence, and the new organisational form of Research Schools was intended to selectively stimulate excellence

(Van Kasteren, 1993; Rip and Van der Meulen, 1995).[1] But this was only possible because the orientation towards relevance was, by then, firmly entrenched at the performance and intermediary levels. Assessment of quality of research is done through disciplinary panels, organised by joint universities (Verkleij, 1999). Budget cuts (of limited extent, but more appear to be coming) derive from overall retrenchment of government expenditures, not from specific measures to influence the science base.

A key feature of the Dutch research system is its densely populated layer of intermediary organisations. In addition to the national funding agency NWO, and the Royal Netherlands Academy of Arts and Sciences KNAW, the relation between the science base and the government (as representative of society) is mediated by advisory (sectoral) research councils, the Association of Dutch Universities VSNU, disciplinary subcommittees of the Royal Academy, and formal negotiation fora aiming at consensual strategy development in specific areas. The members of these organisations interact frequently and are, as it were, a community or world by itself, which is slowly changing through new entrants. Members of programme committees which guide R&D programmes and their implementation are often drawn from this world.

These phenomena are particularly striking in the domain of chemistry, with organisations and committees with overlapping membership: ACC in the Royal Academy, the VSNU disciplinary council for chemistry and the assessment panel, the panel in the foresight exercise (described in the next section) which was an initiative of the chemists, and the various boards and panels linked to chemistry within the funding agency NWO. Initiatives, consultations and reports emanate from this world, and there is regular interaction with societal actors, in particular, the Netherlands Association of Chemical Industries VNCI. The attempt to have an "own" advisory council for chemical research (see the next section for a discussion of such councils) failed, but they were successful in getting an area council specifically for chemistry in the reorganisation of the funding agency NWO – a reorganisation which aimed to reduce the number of divisions in the organisation.

---

1. The convoluted history of attempts of the Minister (of Education, Culture and Sciences) to select for excellence, and the resistance of the university system, has led to the recognition of a few Research Schools which are more excellent than others, as well as the recent decision to spend the further funds originally set aside for stimulating excellent Research Schools on innovative work, in particular through "young blood" schemes.

Responsibilities for important policy processes like programming, evaluation, foresight, strategy development and networking are delegated by government ministries to intermediary organisations. Because of the horizontal linkages and dependencies, individual intermediary organisations cannot just maximize their own interests, however. Successful policy development depends on the ability to translate other organisations and, thus, relates their own strategic goals to those of others.

An example is the implementation of the Technological Top Institutes. The idea for new technological institutes, to be co-financed by industry and government, was initiated by the Minister of Economic Affairs as a way to orient public research toward industry needs, and in such a way that the laborious and inconclusive processes of regular science policy could be circumvented. Initially the aim was to raise new institutes, in research fields suggested by industry, and with funding sufficient to launch them into world top rankings. Opposition from academe (including NWO and KNAW) and, less explicitly, the Minister for Science Policy (who argued successfully that research excellence cannot be made but has to be nurtured through time and that one should take advantage of the existing strengths of public sector research) as well as the necessity to have some selection and implementation procedure, forced the Minister of Economic Affairs to be more co-operative and link up with usual procedures. Also the available budget was too small for three new initiatives. Consequently the Minister of Economic Affairs in co-operation with the Ministers of Agriculture, and Science Policy solicited proposals from consortia of industry, academe and public research institutes. After a double evaluation process, with evaluations by a consulting firm and by a 'traditional' peer committee, the Minister decided to finance institutes for food technology, metallurgy, polymer research and for telematics. The Ministry of Economic Affairs could thus only succeed in its objective to create new university–industry centres of excellence by linking it to the ongoing aims of the Ministry of Education and Sciences to create centres of excellence within the universities and to the strategy of the Royal Academy to protect existing scientific quality against too radical interventions.[1]

1. The Food Technology Institute in Wageningen appears to manage the combination of relevance and excellence, and its director links its success with the trend towards outsourcing of industrial R&D: Unilever Research Vlaardingen in 1997 spending 12% of its annual research budget in universities compared with only 2% in 1987 (Nieuwenhuis, 1999). The Polymer Technology Institute is not so successful, however.

Initiatives from the top level aimed at improving the strategic orientation of the science base, turn out to focus on three entrance points: the mission of the funding agency NWO, the possibility of identifying overall priority themes (see the next section on foresight exercises), and stimulation of new research centres, in a focused approach as with the Technological Top Institutes, or by supporting initiatives from below, as with the knowledge centre BioMade (BioOrganic Materials and Devices) in Groningen getting funds from the infrastructure scheme ICES (set up with natural gas profits, see above).

While NWO's charter was changed in 1987 to allow funding of strategic research, it took some years before it embraced research programmes as a proactive policy instrument. This shift was stimulated by the transfer of the Special Programmes from the government Office of Science Policy to NWO, in the early 1990s.

By now, NWO also engages in consortium-funded programmes with government Ministries (one interesting example being the programme on Ethics and Policy) and with other actors, for example the energy R&D agency NOVEM in a programme on interdisciplinary (natural and social science) energy research. In being proactive in this way, NWO also holds off attempts at strategic direction by the Minister of Education, Culture and Sciences which is its principal.

As part of the overall shift in the position of NWO, the Minister has aimed at an increase of the NWO budget at the expense of the universities. This policy (as well as earlier initiatives in the same direction, since the early 1980s) faltered, and when a new Minister was appointed in the Cabinet formed after the 1998 election, the idea was scrapped – for the time being. The details of the story offer an interesting view on arguments and interactions in the Dutch research system. It is a clear example of institutional inertia, and we have argued that such inertia is particularly strong in a system with high mediation (Van der Meulen and Rip, 1998).

Thus, there are good reasons to analyse the operation of the intermediary layer in the Netherlands in some detail. We will do this by looking at instruments and institutions as institutional competencies in aggregation or in steering (Rip and Van der Meulen, 1996, 1997). Agenda-building and concomitant interest aggregation are the main features of aggregation processes. These happen all along, but are getting more attention "at the top" because of the need for strategic direction of the research system. In the Netherlands, steering is limited to structural (and infra-structural) measures and strategic R&D programmes.

4. AGGREGATION

While priority setting and the recent foresight exercises can be found in almost all countries, the Advisory Councils on Research (ACRs), focusing on specific domains ("sectors") are peculiarly Dutch. They were set up in the 1970s to increase the relevance of the science base, with the existing Council for Agricultural Research NRLO and its tripartite composition (researchers, users, government) reflecting the structure and context of agricultural research generally, as the exemplar.[1] Their main task is to "programme" research for the mid-term and long term (we use quotes because the ACRs can only advise, and depend for their eventual influence on being convincing and on creating appropriate network linkages). They are also involved in processes which monitor and stimulate research programming by others, like foresight and evaluation.

The ACRs are interesting because they embody an institutionalised competence in aggregation, and one which reflects both the needs of certain sectors, and Dutch political culture. The fact of having these four sectors and not others (agriculture and health research are traditionally public responsibilities, while environment and development research are new concerns for which government sees itself responsible) is also interesting. Originally, the idea was to cover all sectors (and to overcome compartmentalisation between government ministries). Of the five councils created in the 1970s, the Council for Energy Research was not able to find a viable niche, and was stopped around 1980.[2] The possibility of a Council for Industrially-Relevant Research was explored, but nothing came of it. Recently, the Ministry of Education, Culture and Sciences has raised the possibility of setting up new ACRs again, for example for education, for work-related research, for legal affairs and for traffic and transport, but only for the latter two have discussions been started

1. They were given a legal basis under the *Wet Sectorraden* of 1984. There are ACRs with legal status: NRLO for agricultural research, RMNO for environment and nature research, RAWOO for development-related research, and RGO for health research. Another advisory council, on physical planning research, operates, and participates in joint activities.
2. Occasionally, an Advisory Council for Energy Research is proposed again. In the most recent attempt, the working party set up to explore the pros and cons concluded that for energy research it was useless to have a co-ordinating body without a research budget (thus going against the rationale used for the existing ACRs). Also, NOVEM, the implementation and programme management agency for energy research, with a considerable budget, is already fulfilling a co-ordination function. Clearly, there are other ways to handle aggregation, and their viability is different in different sectors.

with the relevant ministries.[1] The situation by the end of the 1990s is quite different from the 1970s and 1980s, however, in particular because other organisations with an aggregation and co-ordination role have emerged, often with a budget of their own (as in the example of energy research). The ecology of the intermediary layer is different now.

The basic set-up of the ACRs reflects a belief in aggregation processes for their own sake: different kinds of actors are members of the council (and further actors join in working parties and advisory groups): research users and representatives of groups interested in the effects of research, researchers and representatives of research organisations, and representatives of government agencies. The ACRs have no research funds to allocate and thus depend on quality of their advice and their relationships with funders (that is Ministries with a research budget for the respective sector). They have to prepare quadrennial documents on overall policy and research directions, but their main impact is through a variety of reports published in the meantime, and the interactions and networking that are part of the preparation and dissemination of the reports.

NRLO, the ACR on Agricultural Research, continues to be successful. It has a range of study groups and programming groups, which monitor developments, discuss among themselves, feed insights back into their respective constituencies, and occasionally produce programming documents (which are often forceful, because they reflect the evolving consensus in the area). These activities, and their success, are predicated on the existence and continued functioning of the agricultural network. The conservative element of this set-up was visible in the early 1980s, when the sudden advent of modern biotechnology had the network drawing back on itself, refusing to co-operate with a new R&D stimulation programme on biotechnology (Rip and Nederhof, 1986). It recovered rapidly (as did the sector), however, and by now, biotechnology is an integral part of its activities. Recent NRLO activities focus on the restructuring of the agricultural research sector, and on its role in strategic science policy-making. The investment in foresight exercises is one example, and its interest in partnering with similar organisations in other countries is an interesting further attempt.

In fact, NRLO shows how an ACR can be spokesperson for a sector, and influential in such a role depending on its position in the evolving and

1. See the Evaluation Report on the Consultative Committee of the ACRs (Deetman et al., 1998) and the reaction by the Minister for Education, Culture and Sciences to the Consultative Committee, in his letter of 30 August 1999.

interacting networks, the style of key individuals (like the chair and secretary of an ACR), and the overall ecology of the intermediary level. At present, for example, the interest, and actual achievements, in foresight exercises of the ACRs creates tensions with the science policy advisory council AWT which has now been charged with the responsibility for science and technology foresight (see further below).

The other two visibly successful ACRs, RMNO on environment and nature research and RGO on health research, reinforce this general picture. RGO has been very close to the medical research world, and now also the health research world, and has been instrumental in efforts at quality assessment and scenarios for healthcare. It has played a key role in the establishment of ZON, an agency for healthcare research, taking over from the government of Public Health the responsibility for commissioning healthcare research, as well as in ZON's close collaboration with the Medical Science Area Council within the national funding agency NWO.

RMNO is positioned differently. It has built on the increasing interest in environmental research, linked to the rise of environmental issues on the societal and political agendas, and the position of its chairpersons in relevant networks, and as spokespersons for the relevance of environmental and nature research has been important. Its success also derives from the way it has organised its activities: a series of (active) programming and studying groups, with a lot of freedom to act (and thus to reflect what is happening, rather than implement some terms of reference); a proactive secretariat, and the capacity to fund special studies (after agreement from Council members); and a series of study reports and research memorandums which cover topics of interest, rather than specifically research programming tasks.

One conclusion is that the success of ACRs very much depends on the sector and how they can position themselves. There is no general recipe. They have played an important role in developing methods of agenda-building, foresight and assessment of societal quality, and putting the importance of such methods, and the issues addressed by them, on the agenda of other bodies in the research system. The annual workshops of the Consultative Committee of the ACRs devoted to such issues have been fora for discussion at which representatives of all actors in the research system tended to be present.

While the Ministry of Education and Sciences can draw on the work of the ACRs in its occasional attempts to have a framework of overall science policy priorities, in the early 1990s the then Minister instituted a new scheme of foresight exercises and in 1992 appointed the OCV, the Consultative Committee on Foresight, consisting of researchers in key positions in the research system, and some industrialists, to do a first round and develop advice, both on priorities and on a definitive scheme for science and technology foresight exercises. As it turned out, the OCV developed an approach which drew on the institutional competencies for aggregation in the Dutch research system.

The focus of this committee was on separate foresight studies of specific fields of research or societal domains, rather than a comprehensive Delphi-type study. To link research efforts and opportunities to societal needs and developments, the Committee originally emphasized scenarios as an important approach, but in practice, this approach was not always followed by the panels.[1]

The Committee organised, supervised or joined in 17 foresight panels and about 20 background reports, some of which would have been done by others anyhow, and could be reframed as an OCV exercise, while others were taken up by the field in question after some action from the Committee. It is striking how the Committee has built on ongoing and earlier foresight or foresight-related activities in the several areas that it has selected, and thus helped to articulate and systematise bottom-up foresight and strategic priority setting activities which were happening already.

In 1996, the Committee published its final report, *A Vital Knowledge System*, and defined 14 priorities for science policy (Table 9.4). Ten of these priorities are defined as strategic areas considered by the committees to be 'vital' for meeting the challenges of 'the information and communication society', 'a sustainable economy', 'internationalisation and regionalisation' and 'quality of life'. The other four priorities are vital to maintain a healthy science base. As the committee emphasised, the priorities are not meant to be implemented top down, but should be adopted in the strategies by research organisations and funders within a framework of autonomy and accountability.

The follow-up confirmed this advice. The government accepted the priorities as authoritative in its next annual Science Budget, but did not introduce special funding (as in the UK) or other measures conducive to implementation. The main driver towards implementation is the additional legitimation of proposals when they refer to the priorities. One can see this

1. The exercise should start with societal scenarios (for example the ones drawn up by the Central Planning Office at the time), and articulate how science could contribute. This avoids the customary science-push bias in foresight exercises, even if it did not always work out that way. But the possibility of using scenarios is now generally recognised.

TABLE 9.4 – Priority themes from *A Vital Knowledge System*

- – Long-term research on Electronic Highways
- – Learning and Human Capital in a Knowledge-based society
- – Quality of Agriculture and Food
- – Research for the Service Sector
- – Ecological modernisation
- – Integral utilisation of space
- – Business activity and innovation
- – Processes of Internationalisation and regionalisation
- – Social Cohesion
- – Healthresearch on quality of life, and efficacy of healthservice

- – Develop co-ordinated strategies for the natural and engineering sciences
- – Organise multidisciplinary programmes in the social sciences
- – Improve co-ordination of humanities
- – Improve co-ordination of research in medical technology

happen in a few government initiatives, in some NWO priority programmes, and in the adoption of themes/priorities in universities and other research organisations. The themes of "social cohesion" and "ecological modernisation" have been taken up more systematically – but not necessarily because of their appearance in the list of priorities.

The OCV list of priorities based on foresight exercises and the deliberations of the Committee is actual part of larger processes of heterogeneous aggregation. The Committee's advice to organise a multidisciplinary research effort on the theme of social cohesion, for example, drew on a Report on a strategy for the social sciences, commissioned by the Committee. But the choice for the theme came in a meeting with the Secretaries-General of all government Ministries, organised by the Committee a few weeks before it issued its final report. The Secretaries-General made it clear that they expected the Committee to include social cohesion in its list of priorities, because they considered it one of the key political issues. The subsequent implementation was similarly heterogeneous. In the 1997 Science Budget, the Minister for Education and Sciences announced that there would be a Stimulation Programme on the topic, under the responsibility of NWO. Over the next two years, NWO organised such a Programme with joint funding by the Ministries of Education, Culture and Sciences, of Justice, of Health, Welfare and Sport, of Social Affairs and Employment, and by NWO itself. NWO also redefined two researcher-initiated and competitively selected programmes (on Multicultural and Pluriform Society, on Societal Participation, Binding and Commitment) as now being part of the Stimulation Programme.

As an overall approach, foresight exercises in the style emphasized by the OCV are now generally accepted, and are being done by various actors for their own purposes. The national-level foresight exercise and the articulation of priorities for science and technology policy is now the responsibility of the Advisory Council for Science and Technology Policy AWT, but it has had difficulties in positioning itself. Part of the problem is that it is not clear how it should add to the activities going on already.

5. STEERING

In Dutch political culture, top-down steering is not prominent, even if Ministries, acting on their responsibility as a principal, do attempt to "steer", directly and indirectly. In the two examples we mentioned, the Ministry of Economic Affairs setting up Technological Top Institutes and the Ministry of Education, Culture and Sciences deciding to shift some of the university funding to NWO, their action was refracted by the existing institutions. Like light by a polyhedron, the intervention was scattered by the involvement of many actors and instead of making a real change, the ultimate result merely reflected existing institutional patterns. There have been various solutions (or attempts at solutions) to the game between principals and agents in science policy (Van der Meulen, 1998). In the Netherlands, since the 1970s, strategic research programmes have been the instrument of choice, and a definite institutional competence has evolved. The other main approach in science policies of the last decades, dedicated research centres, has been relatively neglected in the Netherlands, but is now acquiring prominence.

A range of programmes emerged after the National Research Programmes starting in 1974, which marked the transition from the problem-solving mission-oriented programmes of the 1960s and 1970s to strategic research programmes.[1] From 1979 onwards, a scheme of Innovation-Oriented Research

1. The Programmes were dedicated to complex societal problems: energy, environment, labour and demography, but had to mobilise the potential of the research system for these issues, rather than emphasise problem-solving. In the 1974 Memorandum on Science Policy, the scheme of National Programmes then already underway, was introduced as an explicit policy instrument, linked to policy goals: "Mobilisation of university research potential for fundamental investigations as part of programmes aiming at the solution of societal problems is an important objective of science policy". By the late 1970s, the Ministry of Economic Affairs took over the responsibility for some of the specific National Research Programmes.

TABLE 9.5 – Examples of Strategic Research Programmes

- Catalysis Research (IOP)
- Advanced Materials (IOP)
- Separation Technology (IOP)
- IC-technology (IOP)
- Industrial Proteins (IOP)
- Biotechnology (IOP)
- Language and Speech technology (PRI)
- Non-Linear Systems (PRI)
- High Performance Computing (PRI)
- Information Technology and Law (STI)
- Dutch Culture in European Context (PRI)
- Ethnic Minorities (PRI)
- Environmental Economics (STI)
- Antarctica (STI)
- Tropical Forest (STI)
- Chronic Diseases (STI)
- Gerontology (PRI/STI)
- Human Genome Analysis (PRI)
- Toxicology (STI)

IOP: Innovation-Oriented Programmes (Ministry of Economic Affairs)
PRI: Priority Programmes (National Funding Agency ZWO, now NWO)
STI: Stimulation Programmes (Office of Science Policy, Ministry of Education and Sciences)

Programmes (IOP), announced in the 1979 Policy Memorandum on Innovation (which was prepared by the Office of Science Policy of the Ministry of Education and Sciences, in close collaboration with the Ministry of Economic Affairs), was established under the Ministry of Economic Affairs, where it continues, and successfully, until today.

In the early 1980s, the Ministry of Education and Sciences initiated so-called Stimulation Programmes, to be able to intervene strategically (one example at the time was toxicological research). The National Funding Agency ZWO (now NWO) set up various Priority Programmes, partly in response to pressures that it should be more active in linking basic science to socio-economic objectives. Recently, the Ministry of Education and Sciences has moved its Stimulation Programmes to the reformed National Funding Agency NWO. Table 9.5 lists examples of strategic research programmes, as these have been established under the several schemes.

By the late 1980s, there was general recognition of the nature of these programmes, their goals, and how to go about them. Three characteristic objectives can be set out:

- development of new knowledge and methods relevant for a certain socio-economic field or problem;
- improvement of knowledge infrastructure which will be maintained and continue after the end of the programme – called *verankering* (anchoring);
- transfer and dissemination of knowledge leading to concrete innovations or other practical uptake.

Evaluations of the strategic research programmes started to trace the achievements of these three objectives (Rip and Van der Meulen, 1995). Overall, they have been successful on the first objective, and done reasonably well on the third. *Verankering* is difficult to ensure. Some IOPs were quite successful in this respect, but worked on themes that were already being taken up by the relevant communities without the incentive of an IOP; biotechnology is a clear example (Rip and Nederhof, 1986). Stimulation Programmes were initiated and orchestrated by the Office of Science Policy of the Ministry of Education and Sciences and did introduce a strategic shift – for the period the programmes ran. It turned out to be difficult to create permanent effects. The objective of *verankering* continues to be important when setting up and legitimating strategic research programmes, and as long as themes and approaches draw on bottom-up aggregation, there will be some implementation.

While attempts at steering through strategic programmes are refracted by existing institutions, they do mobilise researchers and research institutions. The net effect is "induced" aggregation, as one can also see happening when research performance indicators are introduced (Rip, 1997b), and a (sometimes appreciable) shift in positions. There are various feedbacks and "feedforwards" (i.e. strategic anticipation) involved. The Dutch approach to articulation and selection of programme themes and implementation through formal and informal consultation offers ample opportunities to have these. And steering actors use such routes to get their way.

An interesting example is provided by the situation around catalysis research. An IOP Catalysis ran from 1989 on, with a last funding round in 1997. The theme was a runner-up for a Technological Top Institute; it did not make it partly because firms like DSM and Shell gave priority to a Top Institute on Polymers. Since then, however, the existing Research School Catalysis was selected to be one of the Top Research Schools, with special funding of Mfl. 43 over seven years.[1] As a Top Research School, it focuses on scientific breakthroughs, not on relevance for industry (even if there are traditionally good links with industry). In a Top Institute Catalysis, industry could have steered the research programme. However, with the conclusion of the IOP Catalysis, the current concern is that industry has no regular way to collaborate programmatically with universities. The Ministry of Economic Affairs had already included catalysis among its selection of core technologies for the Netherlands (in its *Technologie Radar*), and announced its intention to start discussions between industry and the knowledge infrastructure to see what bottlenecks are apparent and possibly fund new initiatives to overcome the short-term focus in industry. In October 1999, and building on an earlier "roadmapping" workshop, the Ministry sent a letter to key figures in catalysis inviting them to contribute to a joint strategic (research) policy agenda of firms and knowledge institutions. The chemistry magazine *Chemisch Weekblad* regularly reports on the progress of this example of "induced" aggregation.

Clearly, there are institutional competencies in steering with the help of strategic programmes, in the Dutch way. Present initiatives build on these competencies, and further variety is introduced, for example the increase in consortium-funded programmes (involving a number of government Ministries and often led by NWO), and a stronger participation of stakeholders of various kinds.

6. INSTITUTIONAL COMPETENCIES AND NEW DEVELOPMENTS

The Dutch research system is strong on "aggregation" and relatively weak on "steering" (Rip and Van der Meulen, 1996). This has allowed the sectoral advisory councils (ACR) to be effective, and gave rise to a specifically Dutch version of science and technology foresight exercises. The importance of strong aggregation is visible in the major role of bottom-up initiatives and in the wide consultation when steering actors prepare a strategic research programme. While we concentrated on science rather than technology policy,

---

1. In his drive for excellence, the previous Minister for Education, Culture and Sciences set aside money from the universities' block grant for a few Top Research Schools. The combination of protest from universities and lack of political support made his successor stop the scheme after the first round, and shift the funds to a young-blood scheme, the *Vernieuwingsimpuls* (Impulse for Innovation).

the same typically Dutch features are visible there. In an overview of cluster-based (technology) policy-making in OECD countries, Roelandt identifies (within an overall value-chain approach) the following policy initiatives and principles (Roelandt, 1999; Roelandt and Den Hertog, 1999):

- dialogue in specific platforms;
- brokerage and network policy;
- public consultancy;
- providing strategic information (for instance through technology foresight studies);
- renewal in procurement policy;
- deregulation and competition policy.

While these features of the Dutch research system are related to Dutch history and culture, including its peculiar "consociational democracy"[1] (Lijphardt, 1968), this is not the whole story. In the case of science policy, principals face the additional problem of the open-ended character of research and the indirect linkages with societal uptake. Most science policy goals are therefore open-ended, directed towards groups and developments that cannot be controlled by the policymaker, and about which he has insufficient expertise.

From this perspective, the creation of intermediary bodies like research councils, and subsequently the further filling up of the intermediate layer by organisations and committees for programming, funding or evaluating research, is the creation of an incentive structure for research performers. It may not always happen intentionally, but the emergence of an intermediary level has in principle offered governments a way of respecting open-endedness and autonomy without giving up on steering: they do not force scientists, but they address (and create, and recreate) intermediary bodies, programmes and committees. In the UK, the Government is prepared to "force" the intermediary bodies to change. In the Netherlands, such attempts turn into a long drawn-out process, but may have effects, as with the reorganisations of TNO and NWO.

---

1. In political science the concept of consociational democracy is used to characterise political systems in which certain rights to ethnic groups are guaranteed, under the assumption that ethnically divided states have to live with ethnic cleavages rather than wish them away. In the Dutch political culture the cleavages were not between ethnic groups, but between societal "pillars" with religious or political identities. Although these pillars have dissolved by and large, the idea that minority groups should not simply be overruled by a majority (or coalition of other minority groups) is still alive in Dutch governance.

There is another aspect to the situation. Intermediary organisations are not just conduits for governmental policy goals. They also aggregate and represent what is happening in the worlds of science. Within scientific communities, but also in research institutions and the research parts of industrial firms, there is a cultural repertoire in which articulation occurs of what is important and why, how mature some areas are, how far progress is made along promising research avenues. This is an ongoing *de facto* science and technology watch, and one that can be tapped for other purposes – as when strategic research programmes are prepared, and when foresight exercises are done.

The ACRs in the Netherlands have become experienced players in these arenas, and their relative success builds on how they can tap the repertoire of the *de facto* science and technology watch and connect it with societal issues. Government Ministries are more distant, and depend on the extent to which their staff is linked up with the terrain. A policy, as in the Dutch Ministry of Economic Affairs, to regularly shift staff destroys such linkages, even if it avoids staff being captured by specific interests. This introduces an element of chance: programmes and interventions succeed or fail depending on the combination of Ministry staff and the situation out there, in the field. In the Netherlands, linkages and competence in consultation are particularly important because of its neo-corporatist political culture in general, and because of the crowded intermediate level of its research system, including shared and therefore diffuse responsibilities.

In the Netherlands, mediation and the reliance on aggregation (including induced aggregation) rather than steering, has created a research system which is productive in many ways, and is characterised by a densely occupied intermediary layer with many lateral interactions. This layer is a buffer zone, refracting steering attempts from above. There is change, but it is gradual because of general contextual changes and the demography of the research system. With the increasing importance of strategic science, the scientific establishment resisting calls for relevance by the Young Turks in the 1970s has had to accommodate them. These Young Turks have been active since in programmes, council activities and working parties. Their voices are now heard and listened to. In fact, they are the new elite in the regime of Strategic Science.

From these considerations a general (and first-round) conclusion can be drawn: historically evolved institutional patterns enable and constrain research performance as well as policy action. Such patterns can be seen as an institutional competence to do certain things well – in the Netherlands consultation and mutual accommodation – but also as institutional inertia – as when the Dutch Ministry for Education and Sciences tries to shift part of the block grant for the universities to the National Funding Agency NWO, but to no avail. Thus, research systems which are strong on aggregation are cost-effective, because priorities are articulated at the intermediary level, and with an eye to implementation. At the same time, such systems are conservative: receptive only to those goals which help actors in their strategic games (among themselves and/or in relation to external audiences).

In the longer term, changes do occur. And there is an additional phenomenon, which will lead us to a second general conclusion. When we consider recent developments in the research system in the Netherlands, we see a proliferation of local initiatives, increasingly picked up by public actors (a number of government Ministries) and private actors. The buffer layer also created freedom to move, at the research performance level, with respect to new challenges.[1]

Many new ventures occur in universities, initiated by the scientists but increasingly also by the administration. The traditional Agricultural University merged with the research institutes of the Ministry of Agriculture, and is now the Wageningen Agricultural University and Research Centre. The University of Twente labelled itself as an "entrepreneurial" university at an earlier stage, and has consolidated this drive in a European consortium of innovative universities (Clark, 1998) and in investing in linkages with the Information and Communication Technology sector, as well as participating (with the other Technical Universities) in the new public–private R&D company Avantium, building on scientific excellence in process technologies, catalysis research and nanotechnology. The Groningen expertise centre BIOMADE, with its national (ICES) and regional support, is now international through participation of the UK venture capital firm 3i.

Such initiatives flourish, not because of a market-orientation as in the US (even if the US is often referred to as a role model), but because of a mixed market and state approach. Spin-off companies in ICT can profit from the Twinning Programme (with public and private funding), and the Ministry of Economic Affairs has just announced a Life Sciences Action Programme to support incubators and start-up companies in biotechnology.

The changes in the research landscape are broader, and include investment in 3M sciences (measuring, monitoring, modelling) relevant for earth,

1. This argument is the macro-version of the observation about strategic research characterised by gaps in the innovation chain bridged by promises (Rip, 1997a).

environment and climate issues, and in the interest in knowledge or expertise centres which should serve government Ministries and/or sectoral interests. At the same time, concerns are emerging about too close links between such a knowledge centre and their clients, leading to biases in the research.

These recent developments mix categories (of fundamental, strategic and applied research) and actors (public and private). Institutionally, the existing buffer layer between the Government and the research performance levels provided the necessary freedom of movement, in spite of its inherent conservative bias. Establishment actors like NWO, KNAW and VSNU now respond (with some delay) to the changes which occurred in spite of them, and try to create workable arrangements by mixing first-, second- and third-stream funding of university research. This cannot really contain the heterogeneity.

What is also visible now is the tension between the international (global) character of many of the new developments, and the necessarily national orientation of institutions in the intermediary level, between the national Government and the research performance level. Our analysis of the Dutch research system, and the advantages and limitations of its historically evolved institutional competencies, shows its capacity to handle and contain heterogeneity and make it productive at the national level. Mediation works – but will it also work in the diffuse and uncertain contexts of the global world?

ABBREVIATIONS

ACC	Academy Committee for Chemistry, a standing committee in KNAW
ACR	(sectoral) Advisory Council on Research (*sectorraad*)
AWT	Advisory Council for Science and Technology Policy (the official advisory council on science and (parts of) technology policy to the government)
ICES	Interministerial Committee on the Improvement of the Economic Structure (Committee responsible for advising the government on allocation of the "ICES funds")
IOP	Innovation-Oriented Research Programmes
KNAW	Royal Netherlands Academy of Arts and Sciences

MARIN	Maritime Research Institute Netherlands
NOVEM	Netherlands Organisation for Energy and Environment (a public agency, managing government programmes, but also taking up other broker and manager responsibilities)
NLR	National Aerospace Laboratory
NRLO	National Council for Agricultural Research (an ACR)
NWO	Netherlands Organisation for Scientific Research (the national funding agency for basic and strategic research)
OCenW	Ministry of Education, Culture and Sciences
OCV	*Overleg Commissie Verkenningen*, Consultative Committee on Foresight Exercises (temporary committee established by the then Ministry of Education and Sciences)
RGO	Council for Health Research (an ACR)
RMNO	Council for Environmental and Nature Research (an ACR)
SENTER	Semi-independent implementation agency for the Ministry of Economic Affairs. Handles subsidies, credits and tax regulation and R&D programmes
STW	Foundation for Technical Sciences (part of NWO)
TNO	Netherlands Organisation for Applied Scientific Research
VNCI	Netherlands Association of Chemical Industries
VSNU	Association of Dutch Universities
ZON	*ZorgOnderzoek Nederland*, Healthcare Research Netherlands (semi-independent agency for the Ministry of Public Health to commission healthcare research)

REFERENCES

Blume, S., 1986, *The Development of Dutch Science Policy in International Perspective, 1965-1985*, RAWB background report 14, Zoetermeer: Ministerie van Onderwijs en Wetenschappen.

Bureau Bartels, 1996, *Evaluatie Wet Bevordering Speur-en Ontwikkelingswerk (WBSO), (Evaluation of the Law on Industrial R&D Support)*, Report to the Ministry of Economic Affairs and Ministry of Finance.

Clark Burton, R., 1998, *Creating Entrepreneurial Universities. Organisational Pathways of Transformation*, IAU Press/ Pergamon Press.

Deetman, W.J., M. Epema-Brugman and A. Rip, 1998, *Rapport van de externe evaluatiecommissie mbt de Commissie Overleg Sectorraden*, The Hague: Office of the Mayor.

Dits, H., 1988, *Turn to Coal. Mission Orientation of Academic Research*, Ph.D. Thesis, University of Amsterdam.

Gibbons, M., C. Limoges, H. Nowotny, S. Schwartzman, P. Scott and M. Trow, 1994, *The New Production of Knowledge.*

The Dynamics of Science and Research in Contemporary Societies, London: Sage.

Hackmann, H. and A. Klemperer, 2000, *University Research Funding: An international comparison*, Enschede: University of Twente. A report commissioned by NWO.

Lijphart, A., 1968, *The politics of Accommodation: Pluralism, and Democracy in the Netherlands*, Berkeley: University of California Press.

Martin, S. and J.T. Scott, 2000, The Nature of Innovation Market Failure and the Design of Public Support forPrivate Innovation, *Research Policy*, 29, pp. 437-447.

Nieuwenhuis, H., 1999, Voedingsindustrie heeft langetermijnonderzoek nodig, *Chemisch2 Weekblad* (9 October), pp. 12-13.

OCenW, 1998, *Education, Culture and Science in the Netherlands. Facts and Figures*, The Hague: SDU Service Center.

OECD, 1998, *University Research in Transition*, Paris: OECD.

Rip, A., 1997a, A Cognitive Approach to Relevance, *Social Science Information*, 36, pp. 615-640.

Rip, A., 1997b, Qualitative Conditions of Scientometrics, the New Challenges, *Scientometrics* 38, pp. 7-26.

Rip, A. and A.J. Nederhof, 1986, Between Dirigism and Laisser Faire: Effects of Implementing the Science Policy Priority for Biotechnology in the Netherlands, *Research Policy*, 15, pp. 253-268.

Rip, A. and B.J.R. Van der Meulen, 1995, The Patchwork of the Dutch Evaluation System, *Research Evaluation*, 5, pp. 45-53.

Rip, A. and B.J.R. Van der Meulen, 1996, The Post-Modern Research System, *Science & Public Policy*, 23 (5), pp. 343-352.

Rip, A. and B.J.R. Van der Meulen, 1997, The Post-Modern Research System, in R. Barré, M. Gibbons, Sir John Maddox, B. Martin and P. Papon (eds), *Science in Tomorrow's Europe*, Paris: Economica International, pp. 51-67.

Snijders, H., 1998, Nederlandse R&D, het valt best mee, *Economisch Statistische Berichten*, 83, pp. 112-113.

Tijssen, R.J.W., T.N. Van Leeuwen, B. Verspagen and H. Hollanders, 1998, *Wetenschaps- en Technologie-Indicatoren 1998*, Leiden/Maastricht: Centrum voor Wetenschaps- en Technologie-Studies/ MERIT.

Van der Meulen, B.J.R., 1998, Science Policies as Principal-Agent Games: Institutionalization and Path-Dependency in the Relation between Government and Science, *Research Policy*, 27, pp. 397-414.

Van der Meulen, B.J.R. and A. Rip, 1998, Mediation in the Dutch Science System, *Research Policy*, 27, pp. 757-769.

Van Deventer, J.M.M., L. Klomp, J.J.M. Pronk and M.J. Roessingh, 1999, *Kennis en economie 1999, Onderzoek en innovatie in Nederland*, The Hague: Elsevier bedrijfsinformatie/Centraal Bureau voor de Statistiek.

Van Kasteren, J., 1993, Research Schools in the Netherlands: the Midfield Player's Dilemma, *Science Policy*, 15 (4) pp. 16-20.

Verkleij, A.C.L., 1999, Self-Evaluation and External Review. A Contribution to the Workshop Qualitätsförderung durch Evaluation, Wissenschaftszentrum Berlin für Sozialforschung, 20-21 May.

Verspagen, B. and H. Hollanders, 1998, De Nederlandse Innovatie achterstand, *Economisch Statistische Berichten*, 83, pp. 290-291.

Westerheijden, D., 1997, A Solid Base for Decisions, *Higher Education*, 33, pp. 597.

Willink, B., 1991, Origins of the Second Golden Age of Dutch Science after 1860: Intended and Unintended Consequences of Educational Reform, *Social Studies of Science*, 21, pp. 503-526.

CHAPTER 10

SCIENCE, TECHNOLOGY AND INNOVATION POLICY IN FINLAND

ERKKI ORMALA

1. INTRODUCTION

Traditionally science and technology policies have been concerned mainly with knowledge generation, and with how to strengthen the national or global knowledge base. In the 1980s increasing attention was paid to the issue of technology diffusion. Gradually, new concepts such as the interactive model of innovation, the absorptive capacity of firms and national systems of innovation were developed. Better understanding of the nature of knowledge, particularly the distinction between codified and tacit knowledge, clarified further the conditions of innovation. This again drew attention to the ability of organisations and individuals to acquire and upgrade knowledge and to the characteristics of "learning organisations".

The development of science, technology and innovation policies in Finland followed the same path, at the same time new policy concepts were also adopted. The national system of innovation was introduced in 1990 to provide an overall framework for the policy formulation process. New institutions were established and old ones renewed to respond to the new challenges. These initiatives were strongly supported by growing evaluation activity, which addressed both the contents of the policies, as well as the institutional setting in which the policies were implemented. Gradually science, technology and innovation policies expanded beyond their traditional boundaries.

The conditions for innovation are closely linked to many different policy areas. Therefore, science, technology and innovation policies have become interrelated with other relevant policy areas such as: industrial policy, economic policy, and regional policy. At present, Finland has been able to create a set of policies which are mutually consistent and which provide a solid foundation for a continuous development of national innovative performance.

This chapter describes briefly the development in Finland, a story of one country's response to the challenges of technological revolution and globalisation. In the final section, discussion centres on some of the future needs of this development. The main lesson from Finnish experience can perhaps be summarised as follows: the effectiveness of any innovation system is only as good as its weakest link.

2. THE DEVELOPMENT OF SCIENCE, TECHNOLOGY AND INNOVATION POLICY IN FINLAND

2.1 – National system of innovation

Over the last 20 years, the development of science, technology and innovation policy has been dramatic, with Finland becoming a leader in the 1990s, having previously been a follower. The beginning of the 1980s was in many respects a decisive period in this development. Many discussions within industry at this time concluded that industrial development could no longer be based on resource-based industries and on the expansion of the Eastern markets and that new markets should be exploited. Under these conditions competitive advantage should no longer be based on cost-efficiency but rather on knowledge-intensity and technological superiority. These considerations led to a very rapid increase in business sector R&D expenditure in the 1980s. These developments were accompanied by initiatives in the public sector. Finland's Technology Committee completed its work in 1981, laying the foundation of the national technology programmes. The implementation of these programmes together with the management of the public contribution to industrial R&D were delegated to a new organisation: the Technology Development Agency, TEKES, which was established in 1983.

The next main step was taken in 1987 when the S&T Policy Council was established. Through this decision science policy and technology policy were combined. The result of this merger was, however, a much larger policy area than that previously covered by science policy and technology policy as separate entities. Finland was the first of the OECD countries to adopt the concept of the national system of innovation as the framework for its science, technology and innovation policy formulation. This concept was introduced to the debate in 1990 in the triannual review produced by the S&T Policy Council. The concept has been refined and developed further in all the Council's reviews published in the 1990s. The national system of innovation has become an established framework for S&T policy planning in Finland.

The use of the national system of innovation has been very pragmatic. The system involves all those elements which contribute to the generation, diffusion and application of new knowledge (Figure 10.1). Along with the development of knowledge and changes in the conditions, the composition of these factors change. The concept is adapted accordingly. This flexible approach has turned out to be very practical for policy purposes although it has raised some concerns among researchers.

FIGURE 10.1 – Components of an innovation system

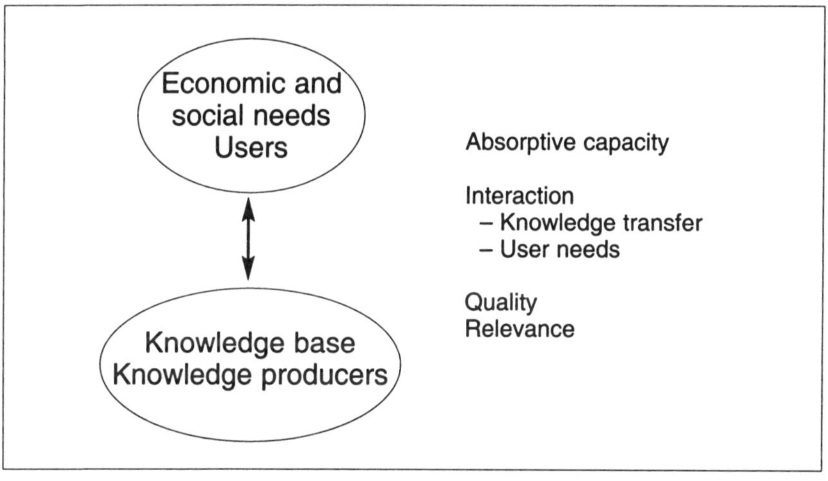

In the middle of the 1990s the national innovation system approach was extended beyond solely national elements. Finland's membership in the EU in 1995 brought the problems of European science, technology and innovation policy into the picture. The systemic approach has been instrumental in the process of defining Finland's views on European developments. At the European level, discussions on the First Action Plan for Innovation in Europe and on the preparation of the Fifth Framework Programme have made contributions to the development of the system approach.

In another direction, the innovation system approach also became relevant in the context of regional development. The network of the Finnish universities, technology centres and the Centre of Expertise programme had, among other factors, developed the innovation infrastructures to the extent that it became legitimate to talk about regional innovation systems. In 1995, the EU's Structural Funds brought additional resources and an international perspective to the development of regional innovation systems. The need to co-ordinate national and regional actions led to a review of and recommendations for the development of the synergy between actions at these two levels.

Throughout the 1990s education policy, industrial policy, employment policy were among the policy areas developed in co-operation with science, technology and innovation policy in order to develop a balanced mix of policies which reinforce each other and contribute both to innovative performance and to the solution of sector-specific problems. The same endeavour to improve synergy has encouraged the development of public–private partnerships in all areas of the innovation system. The S&T Policy Council has facilitated all these developments over the years.

2.2 – Additional public R&D funding

R&D funding has developed steadily since the early 1980s (Figure 10.2). The business sector led the development, but public investment was also growing. The national R&D effort grew at a rate of about 10% in real terms throughout the 1980s. Although it ceased to grow during the severe recession in the early 1990s, it did not decrease and has continued to grow at an even faster rate since 1994. The national R&D effort reached the level of 2.9% of GDP in 1998.

In the middle of the 1990s the development of the national system of innovation entered into a new phase. In autumn 1996 the Government decided to increase its research funding with a view to raising the national research input to 2.9% of the GDP by 1999. Government R&D funding would be increased by 25% from 1997 to 1999, meaning that the level of public R&D expenditure would be increased from 6 billion in 1997 to 7.5 billion in 1999, i.e. by 1.5 billion FIM. This, together with an increase in private sector funding would raise the national R&D effort to the target level by 1999 (Figure 10.2). The government invited the S&T Policy Council to prepare plans for the allocation of the additional funding. In 1996, the Prime Minister characterised this decision as "the most important industrial policy decision his government has made".

FIGURE 10.2 – Share of R&D in GDP in some OECD countries

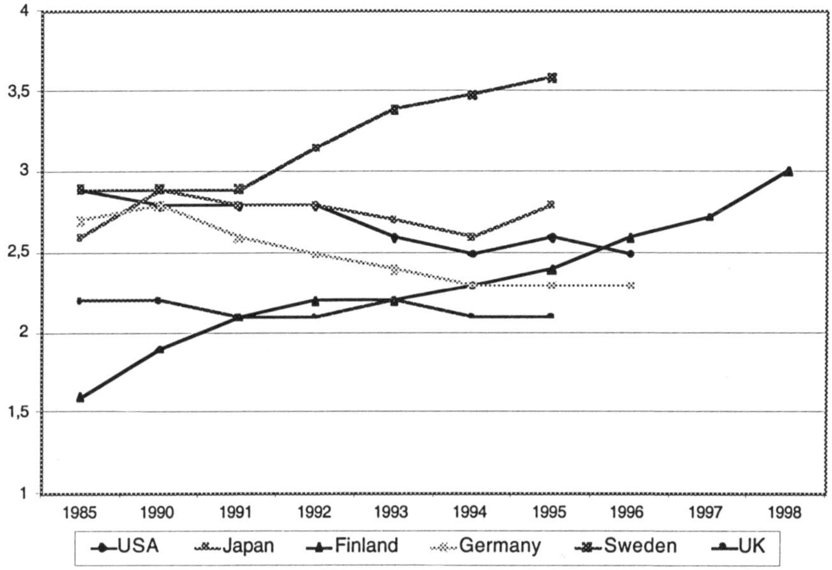

It is not possible in this context to give a full account of the various arguments and discussions preceding this decision, however, the following observations are impressions which characterise the debate and which highlight the nature of the argumentation.

INDUSTRIAL STRUCTURE. Finland has been able to initiate a strong renewal process in its industrial structure, with a move towards specialisation in knowledge-intensive, high-growth sectors and products. This process, which started in the beginning of the 1980s, has accelerated during the last few years (Figure 10.3). Empirical evidence available describes the characteristics and prerequisites of this development.

A more diversified industrial structure is vital for Finland's economic stability under the conditions of a single currency. The structure of the Finnish economy has traditionally been different from the structures of the large European economies. This implies that the decisions of the European Central Bank do not necessarily reflect the needs of the Finnish economy. The growth in knowledge intensity is seen as the most important process for economic diversification. On the other hand, the decision to increase public R&D funding was released with a clear statement that the necessary funds had been

FIGURE 10.3 – Finnish exports by industry 1960-1997

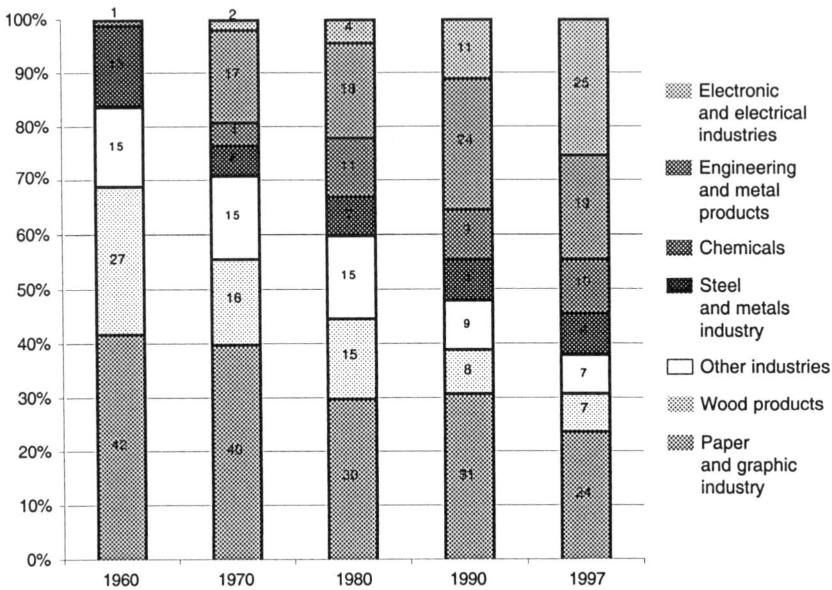

raised from the sales of state owned companies and would not effect Finland's endeavour to meet the convergence criteria.

JOB CREATION. The high unemployment rate (about 16% of the labour force, about 400 000 people in 1995) made job creation a vital policy issue in Finland. Knowledge intensive firms were able to increase employment throughout the recession at the beginning of the 1990s. This process has accelerated since 1994. For example, the electrical and electronics industries have created 30 000 new well-paid jobs in Finland during the last five years. The S&T Policy Council has estimated that the direct net effect of the growth in knowledge intensity on employment will be 30 000-50 000 new jobs by the end of 1999. This estimate does not yet include jobs to be created within subcontracting and related business services. Some empirical evidence suggests that these indirect effects should generate 0.5 to 1.5 jobs for each job generated directly.

INCOME. From the point of view of the national economy, knowledge-intensive jobs are of vital importance. The bulk of knowledge-intensive output is exported and the relevant jobs are well paid. Thus this growth adds significantly to the revenue from abroad and taxes for the economy, and creates considerable demand for domestic services.

In the debate other arguments were also presented. For example, knowledge-intensive growth is seen to contribute to balanced regional development and to the prerequisites for environmentally sustainable development.

Innovation was seen as the driving force of the knowledge-intensive growth. Innovative firms have an ability to adjust to changes in the global marketplace, to increase their productivity and to create new jobs. These firms are, however, in many ways dependent on a favourable environment, facilitating the emergence and growth of innovation and new business.

The national system of innovation is the main policy instrument to enhance innovation. The examples above reflect the context in which innovation policy is discussed in Finland. Effective innovation policy is seen as highly relevant for economic stability and for many other major policy concerns. Innovation policy is a part of the Government's overall strategy to respond to future economic and social challenges.

The decision to invest new resources in R&D rests on the understanding of the generic role R&D is playing in the innovation system. R&D is not only a way to generate new knowledge, it is an essential element of a learning organisation, integrating the different types of knowledge needed for innovation. It opens up channels for external sources of knowledge and facilitates participation in strategic alliances and knowledge networks. Finally, R&D seems to strengthen an organisation's absorptive capacity and ability to apply new technology. Therefore, the Government's decision to increase R&D activity affects the core of the innovation system and is expected to strengthen the innovative performance on an economy-wide basis.

The particular aim of increased research funding by the Government was to develop the national system of innovation to the benefit of the economy, business and employment. This, together with the systemic view, provided a clear focus for the allocation. The first principle was to allocate financing to end-users on the basis of competition. The new resources were used to increase non-committed competitive funding. Another principle was to allocate the bulk of the money to technology, targeted basic research and education. Therefore, the main recipients of the new money were TEKES, the Academy of Finland and universities. Some funds were also allocated to sectoral Ministries for the so-called cluster programmes and to the Ministry of Labour for the development of new models for high performance work places. The allocation plan is presented in detail in the S&T Policy Council's review 1996 (Table 10.1).

TABLE 10.1 – The allocation of the additional public R&D funding

	1997	1998	1999
Commitments	*520*	*775*	*1 140*
Academy of Finland	170	190	270
Technology Development Centre (1996: 170)	350	585	870
Appropriations	*240*	*315*	*360*
Ministry of Education, Science and Culture			
- Universities	175	210	250
- Academy of Finland	5	5	5
Ministry of Trade and Industry			
- The Ministry	10	10	10
- Technology Development Centre	10	20	25
- Technical Research Centre		20	20
Ministry of Agriculture and Forestry	5	10	10
Ministry of Transport and Communications	10	10	10
Ministry of Social Affairs and Health	10	10	10
Ministry of Labour	10	10	10
Ministry of the Environment	5	10	10

Unit: Million FIM.

The new resources changed the balance between committed and non-committed R&D resources within the public sector. While the share of basic funding of the research at the universities and research institutions of the total public R&D funding was 75% in 1997, the increase in non-committed funding means that this share has dropped to 60% in 1999. This change of the structure of public research funding was intended to bring about systemic changes in the institutional structure of the innovation system.

Despite a rapid growth in GDP, the aim of reaching a 2.9% share of the R&D expenditure has been achieved. Development both in traditional and new areas is proceeding well. New activities, networks and research areas have been opened up. The whole system seems to be developing dynamically. On the other hand, there is some turmoil within the system in particular in those parts, which were not fully involved before.

In the following sections a brief description of each of the main actors of the innovation system is presented. In the same context the role and intended use of the additional R&D funding is described.

3. THE MAIN ACTORS OF THE INNOVATION SYSTEM

3.1 – The institutional setting

The institutional structure of the Finnish R&D system is presented in Figure 10.4. The bulk of research and development in Finland is carried out by business enterprises (about 70% of the national R&D effort) with their own funds. In the public sector, the two important Ministries are the Ministry of Education Science and Culture (ME) and the Ministry of Trade and Industry (MTI). Under the ME are all the universities and the Academy of Finland, which is composed of the national research councils. The Academy is the central financing and planning body in basic research. The Technology Development Agency (TEKES), which is located in the administrative domain of the MTI, has a corresponding position in the planning and financing of technical research and development. Under the MTI is also the Technical Research Centre of Finland (VTT), which is the largest and most important research institution in the country.

Other Ministries bear responsibility for research, which serves the development of their fields. Research institutes under the Ministries carry out most of this so-called sectoral research. Ministries may also finance research

FIGURE 10.4 – The Finnish R&D system

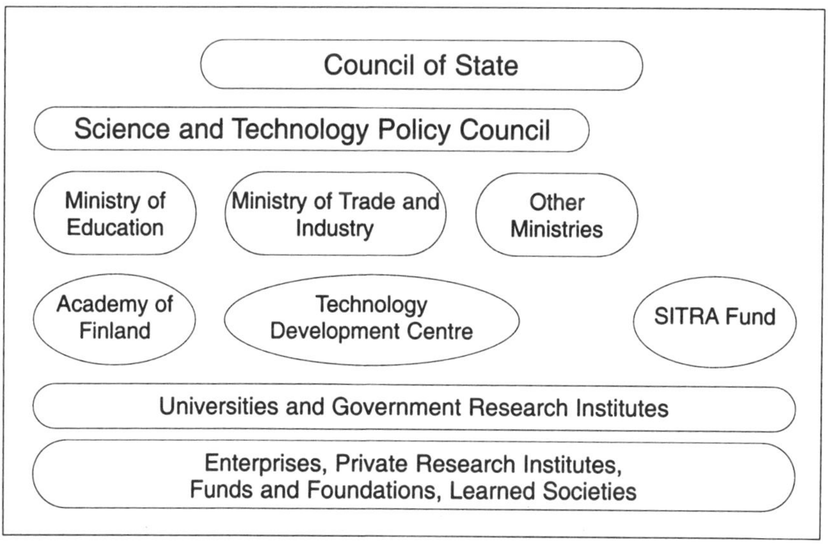

TABLE 10.2 – Budget appropriations to the main actors in 1998-1999

	1998	1999	Growth
Research organisations	*3 039*	*3 152*	*3.7%*
Universities	1 838	1 879	2.2%
Government Research Institutes	1 201	1 273	6.0%
VTT	295	321	9.0%
Appropriations	*3 051*	*3 370*	*10.4%*
Academy of Finland	847	925	9.1%
Tekes	2 204	2 245	11.0%

undertaken in universities, business enterprises and research institutes in other administrative domains. The budget appropriations in 1998 and 1999 to the main actors are presented in Table 10.2.

The structure of Finnish R&D is basically similar to R&D systems in many other countries. There are, however, two additional organisations, which are particular to Finland: the Science and Technology Policy Council and the Finnish National Fund for Research and Development, SITRA. These two organisations have brought to the system a certain flexibility and co-ordination which has been unique in the development of the Finnish situation.

3.2 – S&T Policy Council

The S&T Policy Council is an advisory body to the Government and administration in issues relating to science, technology and innovation policy. The Council was established in 1987 to continue the functions the Science Policy Council established originally in 1963. The membership of the Council consists of: the Prime Minister, who chairs the Council; the Minister of Education and Science; the Minister of Trade and Industry; the Minister of Finance; and two other ministers, as well as ten other members well versed in science and technology. Industry and the labour market are also represented in the Council. The Government appoints the members for a three-year period.

The Council has two subcommittees, one for science policy and the other for technology policy, chaired by the respective ministers. The Council's secretariat consists of two full-time chief planning officers.

The Council has a very broad mandate. It issues statements and recommendations covering all the issues, which it finds relevant for science, technology and innovation policy. During the last few years, more and more operational duties have been assigned to the Council by the Government, such as: the preparation of the horizontal R&D budget, which it does together with the Ministry of Finance; steering the development of sectoral research; and the preparation of the allocation plan for additional R&D financing. In addition to separate statements, every third year the Council issues a policy review, which considers past developments and current challenges and defines the guidelines for the development of policy for the next few years. The Government approves this review, which then becomes a part of the Government's official strategy.

The Council meets about four times a year. One of these meetings is usually a half-day workshop to discuss some important specific issue. In the past these issues have included technology and employment, European research and innovation policy, the development of the university system and the development of the Finnish information society.

The development of Finland in general and in science, technology and innovation policy, in particular, has increased the role of the Council during the last few years. Today the Council has an important role in developing the national system of innovation in a co-ordinated way, across sectoral boundaries, and in improving the synergy between science, technology and innovation and the other relevant policy sectors including economic, industrial, education, regional, environmental and energy among others. The Council has had a leading role in the efforts to develop science, technology and innovation policies to pave Finland's way towards a knowledge-based society.

3.3 – The Finnish National Fund for Research and Development, SITRA

SITRA is a specific fund subordinate to the Parliament. Essentially, SITRA has an experimental role in the national innovation system. It explores future challenges, experiments with new policy instruments, and then if they turn out to be successful, it transfers them to either business or the public administration. In the past SITRA has played a significant role in promoting activities, which are currently managed by other organisations. For example in the early 1970s, SITRA developed environmental policies, which were later moved to the new Ministry of Environment, and energy policy, which is currently in the responsibility of the MTI. Later in the 1970s SITRA concentrated on the promotion of industrial R&D, responsibility for which was taken over by TEKES in the mid 1980s.

In the 1980s, SITRA started to develop Finland's venture capital business. It created a network of venture capital funds and launched a systematic programme to gain experience and develop competence, which did not exist in Finland at that time. In 1997 the whole venture capital business was reorganised and a process of privatisation launched. The public promotion activities of the venture capital business were delegated to SITRA. SITRA's corporate financing is organised into a separate holding company, to which all public money involved in the venture business is transferred. In the long term, public funding is limited to financing at the very early stage. The matching service and the seed-financing scheme remains a part of SITRA's main activities and are funded from SITRA's own funds. All the rest of the venture capital businesses are structured into four funds each of which will be managed by a separate management company. These are the Biotech Fund, the Fenno Fund for financing interesting ventures in the transition countries, the Technology Fund and a Collection of Regional Funds, an arrangement bringing together all the existing regional funds. Currently, all the funds are partnership funds, involving both public and private capital. In the longer term, (within five years) public investments should be sold and the whole system will then become private (with the two exceptions mentioned above).

AT present, SITRA is moving into further new areas. Globalisation, the commercialisation of R&D, innovation system studies and various future studies are now on the agenda. In its experimental role SITRA has made significant contributions to the development of the Finnish innovation system and to society as a whole.

3.4 – Technology Development Agency, TEKES

The primary objective of TEKES is to promote the competitiveness of Finnish industries (in both manufacturing and services) through technological means. Activities should diversify production structures, increase production and exports and create a foundation for employment and social wellbeing. TEKES' main instruments are applied technical research financing, industrial R&D loans, industrial R&D grants and capital loans for R&D.

In 1998, TEKES financed 2454 projects with a total value of 2144 million FIM. Industrial R&D (in the form of grants and loans) was given 1312 million FIM, while universities and research organisations received 832 million FIM. 75% of the industrial financing was allocated to companies with less than 500 employees. TEKES' technology experts consult companies, universities and research institutes, evaluate their development projects and help find suitable partners to carry out the projects.

National technology programmes have turned out to be an effective mechanism to promote networking and co-operation for companies and the research community. They are used to promote the development of a specific area of technology or industry, and to transfer research results to industry. About 50 comprehensive national programmes are currently being implemented and the annual level of finance of these programmes is about 600 million FIM. There is a strong orientation of these programmes towards the rapidly developing new technologies: IT, biotechnology, integrated manufacturing systems, new materials and their application, on the one hand, and towards developing technologies which upgrade the products and processes in traditional industries, on the other.

The distribution of TEKES funding is presented in Table 10.3. The traditional instruments of industrial R&D (grants and loans) have recently been complemented by new instruments, which are specifically aimed at developing the commercialisation of research such as the Equity Capital Loan scheme and the Technology Clinics scheme. These are instruments, which are intended to bridge the gap between research and business.

TABLE 10.3 – The allocation of TEKES funding in 1999

	Million FIM
Capital loans	200
Industrial R&D loans	270
Industrial R&D grants	970
Technology programmes and projets	880
Total	2 320

The Capital Loan scheme is an R&D loan, which is given to a start-up company instead of a project. It is designed in such a way that it can be treated as equity in the firm's financial statement. The condition to pay back the loan is dependent on the success of the firm. This instrument strengthens a firm's position in the financial markets and increases its credibility. While granting a loan, TEKES also rates the technological potential of the recipient firm, which is also available to other financiers.

The Technology Clinics scheme is a consultant programme provided by TEKES for SME's. It is intended to lower the barriers to the use of external

R&D, to enhance technology absorption within firms and to help R&D service providers work with technology start-ups.

TEKES' contribution is very strongly based on its expertise in technology, which is very highly appreciated by the industry. Projects to be funded are planned in close collaboration with TEKES' experts. Through this process, the quality of projects is significantly improved. TEKES' central administration and the main part of its intellectual capital are located in Helsinki. In order to provide a regional contact, TEKES has small technology units in all government Employment and Economic Development Centres in different parts of the country. TEKES also has a network of technological overseas offices mainly outside Europe.

Additional resources, based on the 1996 decision, allocated through TEKES (900 million FIM of the total of 1.5 billion) were intended to diversify TEKES' activities both upstream and downstream. New targeted basic technology programmes, conducted jointly with the Academy of Finland, are intended to improve the relevance of the national knowledge base in terms of industrial development. In the other direction, the rapid expansion of the capital loan instrument should facilitate the commercialisation of research results. Additional resources were also allocated to improve the leverage of TEKES operations by facilitating the participation of new partners in knowledge networks, increasing the interaction between large and small firms, and providing incentives for regional actors to generate sufficiently qualified development projects to become eligible for TEKES funding. TEKES activities were also expanded towards cluster programmes and technology diffusion and innovation in the service sector. TEKES has an extremely strong position in the Finnish innovation system. About one-third of all the government R&D appropriations are channelled through this one organisation.

TEKES is an expert organisation. Most decisions to finance specific project are made by TEKES' own personnel. Only decisions regarding the largest projects, national programmes and strategic directions are taken by the board.

3.5 – The Academy of Finland

The Academy of Finland is a collection of public research councils. It funds basic research, mainly in universities, but also in government research institutes. It has research programmes, some of which are collaborations with TEKES, sectoral ministries or industry. The Academy has four research councils for culture and society; natural sciences and engineering; health and environment; and natural resources. The research councils fund all the various activities of the Academy in their specific fields. The Academy has a very distinct role in the system, and represents the autonomy of science. Scientists are taking decisions in the research councils about projects and programmes and are evaluating their peers. This specific role of the Academy is highly appreciated in Finland. The Academy has been able to attain credibility to represent the view of the scientific community and to make decisions, pertaining to the self-governance of science.

During the recent years, the Academy has been very active in developing the research system and research environments. The Academy has created a system of the centres of excellence, strengthened the graduate schools in the universities, developed evaluation and created a postdoctoral education system in Finland. The aims of these initiatives are to improve the quality of research and increase professionalism in research careers and specialisation in the research system. Where possible, the thresholds between different scientific disciplines, universities, research institutes and private sector research units are lowered.

The first round of the centre of excellence programme expired at the end of 1999. The programme included 17 centres. A new six-year programme started in 2000, with 26 centres receiving their first special funds (altogether 43 million FIM – with some of the centres funded by TEKES). The centres represent all main scientific disciplines. Their role is to provide a benchmark for the whole research community to promote scientific quality and reward specialisation.

Additional R&D financing allocated to the Academy (270 million of the total of 1.5 billion FIM) was needed to launch strategic research programmes, strengthen the centres of excellence, establish a system for postdoctoral education and otherwise promote young Ph.D.s' research careers and to expand international science co-operation. In 1998, the Academy published its report on "the State and Quality of Scientific Research in Finland", which was a major "science watch" analysis. According to the report significant progress has been made during the last decade. In general, the quality of research has reached the international level and has attained the top level in certain selected fields (many of which are industrially relevant). For example scientific performance (measured in terms of bibliometric indicators) in electronics and information technology is at the highest international level. There is a clear development within the knowledge base towards improved quality and relevance.

3.6 – Universities and polytechnics

There are 20 universities in Finland, located in different parts of the country. Ten of these universities are multi-faculty institutions and the rest are specialised institutions, of which three are universities of technology, three are schools of economics and business administration and four are art academies. All universities are engaged both in research and education and they all have the right to award doctoral degrees.

The university system underwent a period of expansion between the 1960s to the 1980s, and there was rapid growth in resources and an increase in independence from the mid-1980s and the 1990s. In the 1990s the functional independence of universities has continued to grow, while the attention has focused on improving performance and quality. The Government has announced an action programme for structural development aimed at rationalising higher education.

Universities are the main performers of basic research in Finland. They also have an important role to play in applied research and development. The share of university research is about 17% of the national research effort. Increasing attention is being paid to the relevance of university research from the point of view of the business sector and society. The emphasis in university research and in its funding is on the high quality of the research and

TABLE 10.4 – Funding of university research activities in 1995

Funding sources	% of total funds
Budget financing	60%
External financing	40%
Total	100%
Ventilation of external financing	**% of total funds**
Own funds	3%
Academy of Finland	37%
Tekes	16%
Other public financing	20%
Charities	3%
Industry	14%
International funding	7%
Total	100%

on the development of internationally competitive centres of excellence and researcher training.

On average, the universities fund 60% of their research with their own funds and 40% with external financing. The distribution of external funding by source is presented in Table 10.4. In some of the technical universities the share of external research funding is well over 60%.

Engineering is very strongly represented both in university education and research. In this regard, the Finnish University system differs from the European average. The share of technology in university research is about 25% and the share of the number of graduates in engineering of the total number of university graduates is also 25%. The strong position of engineering is being strengthened further following the implementation of a government decision in 1998 to increase the education of the IT workers at the higher education level. About 13% of 18- to 20-year-olds will receive IT-related higher education. This figure is the highest share in the OECD area and reflects Finland's very strong commitment to the development of a knowledge-based society.

Universities are strongly involved in adult education. Adult education at the university level consists of seven complementary areas, which together form a complete entity: (a) professional continuing education; (b) employment training; (c) open university; (d) regional and organisational development projects; (e) development of teaching materials; (f) research and publication; (g) careers services. The main activity has been in the first three areas, but the trend from the universities' point of view can be described as a qualitative change from being arrangers of courses towards being experts in the planning, development and application of education. The underlying principles in all areas of activity are lifelong learning, learning organisation and open distance learning.

Universities have strongly developed their interaction with the users of their research, leading to an increasing number of joint projects in which knowledge generation and utilisation proceed as an integrated process strengthening each other. The road the universities have chosen in managing this collaboration has been to establish intermediary organisations to facilitate the interaction. A recent study listed about 200 of these kind of intermediaries affiliated to the universities. Now, the big question is whether this approach is viable in the long term or whether there should be a completely new approach, integrating the utilisation of research into the core of the university activity as a third basic mission, in addition to research and education. The latter option would require substantial changes in university management and the development of new management tools. This issue will be discussed profoundly during the coming years when the new role of the universities will be explored in depth.

Another issue relating to the new role of the universities is their contribution to regional development. Recent studies in Finland have indicated very clearly that new jobs and economic growth are concentrating in those regions where there is a university. The Centre of Expertise Scheme (discussed in section 8) has enhanced this development further.

Additional R&D funding for the universities (250 million of the total of 1.5 billion FIM) was allocated to strengthen the best graduate schools and to establish new ones in selected fields of science and technology, renew equipment and other research prerequisites, develop networking and transfer mechanisms in order to intensify the diffusion of research results and technologies to industry and other users and increase education of experts in the fields of science and technology in response to the needs of working life.

The non-university sector of the higher education has gradually been developed during the 1990s. Around 25 new institutions, "polytechnics", are being created by upgrading the specialised institutions which previously offered vocational higher education and merging them to form multidisciplinary institutions. The overall aim of the development of the higher education system is to offer openings in higher education to 60-65% of the age group.

3.7 – Research institutes

The largest research institute in Finland is the Technical Research Centre of Finland (VTT). VTT is a multidisciplinary expert organisation that produces technical and techno-economic research services in order to improve the competitiveness of companies, to diversify the structure of Finnish industry and to develop the infrastructure of society. VTT's research covers all the main areas of technical research with the exception of paper and pulp technology, which is covered by a separate private research institute. VTT has nine research units representing different technologies. VTT employs about 2700 people and its annual budget is about one billion FIM; 70% of the funding comes from external sources and 30% from the Government.

As a result of an evaluation in 1993, VTT was completely reorganised in 1994. The aim of the reorganisation was to increase the transparency and flexibility of the organisation. The decision-making system was also amended.

A board consisting mainly of customers and other interested parties now governs VTT. The old system, in which the division directors formed the board, turned out to be inappropriate in an environment where radical reorientation of activities and reallocation of resources are necessary in order to maintain and increase the relevance of the research.

Sectoral research, i.e. research which serves the development of different policy sectors, is mainly conducted in government research institutes. Over 20 research institutes are working in nine administrative fields. Their combined research volume is about 14% of the national R&D effort. The institutes neither form a cohesive group nor have been developed as an entity like the university system. The S&T Policy Council has contributed to the steering of relevant developments for a number of years. The point of departure in the development of research institutes has been to develop them both as a part of the research system and as a part of the activities of the relevant sector.

The S&T Policy Council's work has concentrated on three critical issues relating to the development of sectoral research. Firstly, the competence of the Ministries to manage research and exploit its results need to be developed. The Ministries will have to learn to use research as a strategic tool for the development of the sector. Secondly, sectoral research needs to be reorganised according to the general model for sectoral research developed by the Council (Figures 10.5a and 10.5b). In this model, research institutes, other research organisations, the Ministries that steer the research institutes and the research institutes' clients operate in an open network. Research organisations both compete for research assignments and financing, and form partnerships. This trend means that the strong demarcation lines between different administrative fields and between Ministries' "own" research institutes and other research organisations will disappear.

Thirdly, the action programme for structural development is implemented. The first step in this process is the evaluation of all government research institutes. This comprehensive effort is described in more detail in section 5.3. The next step in the development is to look for opportunities to reorganise the research institutions according to the functional task entities assigned to them. The Government commissioned the S&T Policy Council to conduct this assessment. The reorganisation should take place both at the national and regional levels.

Additional R&D financing brought a new element to this development in 1997. The sectoral ministries received some additional funding for cluster programmes. The programmes cover telecommunications, foodstuffs, transport,

FIGURE 10.5 – The general model of sectoral research from a Ministry's point of view (5a) and from a research institute's point of view (5b)

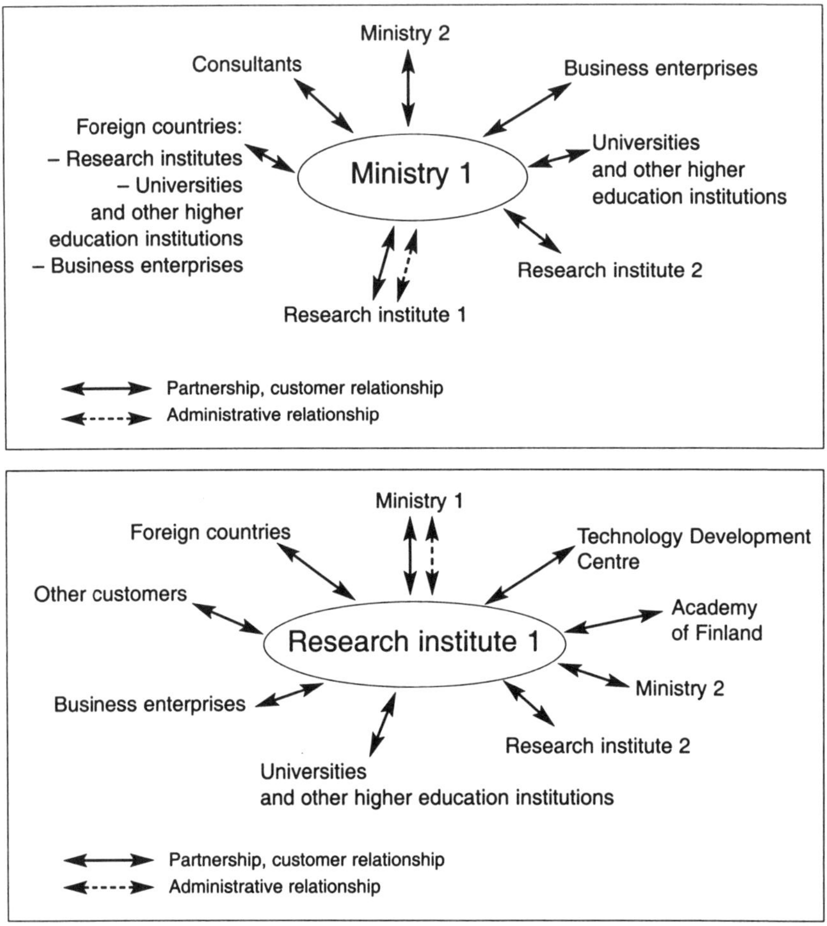

environment, forest and welfare clusters. These programmes, which are to be conducted and financed jointly by the relevant Ministry, TEKES, the Academy and industry, are intended to create new co-operation between scientific and technological organisations, industry and government organisations. This co-operation should strengthen the synergy between innovation policy and other relevant policy sectors. This will provide new opportunities, such as the development of regulatory frameworks that are more conducive to innovation, and generate demand for new innovations in the public sector. On the other hand, these programmes should also improve the contribution of science, technology and innovation to the solutions of sector-specific problems.

Some funding was also allocated to the Ministry of Labour for the development of high-performance workplaces, which can be used as "benchmarks" for an economy-wide organisational reform. One of the aims of this exercise is to study organisations which facilitate learning and strengthen the absorptive capacity of firms.

4. INDUSTRIAL R&D

Industrial R&D has grown very rapidly during the last 20 years, particularly since the middle of the 1990s: R&D expenditure increased by 14% annually between 1995 and 1998. Despite the growth in the public R&D expenditure, the share of industrial effort has been increasing steadily. In 1998 industry's share was 67% of the national R&D effort.

Electronic and electrical industries account for about 50% of overall industrial R&D. The growth in that sector has been very rapid: approximately 25% annually between 1995 and 1997. Large corporations dominate industrial R&D. The ten largest R&D performers carry out about half of the industrial R&D. In 1993 this share was 33%.

This rapid growth in industrial R&D expenditure is partly due to changes in the Finnish production structure. Knowledge intensive sectors both in manufacturing and services have grown very rapidly during the 1990s. The exports of high-tech products have grown annually by 35% over the 1990s. Finland is one of the few OECD countries in which the trade balance of these products is clearly positive.

Knowledge flows in industry have also been studied extensively. Enterprises acquire a significant proportion of their technologies outside their own fields, but this varies greatly between fields. For example, the forestry industry procures over 70% of the technologies it applies mainly from the engineering, electronic and electrical industries. The other extreme is the telecommunications industry, which produces over 80% of the technology it needs. These observations have significantly contributed to the overall understanding of the relevance of the systemic approach in S&T policy. They have also helped to lay the foundations for the cluster programmes funded from the additional R&D resources. Knowledge clusters and their promotion through public intervention is one of the principal challenges of future innovation policy in all countries.

5. EVALUATION

Comprehensive high-quality evaluation has become an invaluable instrument in the development of the national system of innovation. Evaluation provides feedback on the quality and effectiveness of S&T policy instruments and institutions.

Responsibility for evaluation follows the overall structure of the R&D system. The S&T Policy Council provides guidelines and monitors evaluation activities in various parts of the system. Ministries organise evaluations of the agencies in the their domains, while agencies organise evaluations of programmes and research fields.

5.1 – Evaluation of basic research

The Academy of Finland occupies a central position in the evaluation of basic research. Since 1983 the Academy has evaluated about 25 scientific disciplines, some of them twice. These evaluations have followed the traditional peer review procedure, with selection of examiners, collection of material, site visits and reporting. The basic feature of these evaluations is that the examiners are leading experts in the field and are usually from abroad. The evaluations are directly utilised by the respective research council for its planning and financing activity.

As mentioned above, in 1996-1997 the Academy carried out a comprehensive review of the state of the art of science in Finland. The research councils were responsible for evaluation in their respective fields, with each council selecting its own approach. Some councils (such as the Research Council for Health) relied heavily on bibliometric approaches, while others (such as the Research Council for Natural Sciences and Engineering) exploited earlier disciplinary evaluations and surveys among researchers and their clients. This evaluation also made strong reference to the international level by assessing the development of scientific performance in Finland in comparison with other countries. Despite some inconsistencies due to the lack of data or differences in approach, this review drew a very comprehensive and clear picture of the strengths and weaknesses of science in Finland.

The process (both the review and its application) revealed a number of problems in the system. Firstly, there is no systematic database available that includes main indicators of scientific performance. However, such data is available, covering the educational performance of the universities. The Academy has launched a project to create a similar database for research.

Another problem relates to the implementation of findings in decision-making. While positive observations could easily be utilised in new programmes, such as the programme to promote centres of excellence, the graduate school programme or postdoctoral education programme, negative observations did not lead to any significant action.

In 1993, the Ministry of Education commissioned foreign experts to conduct an evaluation of the Academy of Finland itself. This evaluation resulted in a reorganisation of the research council structure of the Academy, with a reduction in the number of councils from seven to four.

Basic research evaluation is supported by the overall evaluations of universities. While the emphasis in these evaluations was on teaching, they also produced valuable information about the development of research environments and the attitudes of the research community. University evaluations are tailored to each situation. They draw heavily on self-evaluation, science and education indicators and the use of external panels.

5.2 – Evaluation of technical R&D

Technical R&D supported by TEKES is evaluated systematically. National technology programmes are usually evaluated by foreign experts, either mid-term or at the end of the programme. These evaluations differ from the traditional peer review in that the evaluators are not necessarily scientists in the field. An ideal evaluation panel is composed of independent scientists, users of the research results and some experts in evaluation or technology policy. These evaluations are supported by a monitoring system, including a number of key indicators describing the effects of the programmes. The monitoring system also covers the industrial R&D projects supported.

A comprehensive evaluation of industrial projects supported by TEKES was conducted as a part of the overall evaluations of TEKES. The Ministry of Trade and Industry commissioned this exercise. Two distinguished foreign experts were invited to assess the performance of TEKES in 1995. Their work was supported by a comprehensive survey of the recipients of funding. More than 1000 projects were examined. The evaluation, while giving a positive overall assessment, suggested that the TEKES should be developed from a traditional R&D funding agency towards an innovation service organisation with a more diversified mix of policy instruments. With the additional public R&D financing these changes were implemented in 1996-1997.

Systematic evaluation in TEKES is the responsibility of an independent unit directly under the Director-General. There is a clear division between internal assessment and external evaluation. The costs of evaluation do not usually exceed 1% of the costs of the programme.

A continuous process of technology foresight supports evaluation of technical research. This is an iterative process involving industrial organisations, research institutes, universities and technology administration. Rather than creating some sort of a master plan for technical development, the aim is to establish a systematic dialogue between the various partners in order to facilitate each participant's own priority setting. As far as the public sector actors are concerned, evaluations are in place to explore how well they succeed in this effort.

5.3 – Evaluation of sectoral research

All government research institutions have been evaluated during the 1990s. In accordance with a recommendation by the S&T Policy Council, the Government decided to launch the process as part of the renewal of the central administration. The evaluation was to be conducted under the supervision of the Council. In addition to a code of conduct of a good evaluation practice, the guidelines issued by the Council involved a number of practical suggestions to facilitate both evaluation and the use of its results. As an example of good practice it was suggested that the Ministries should always nominate a high-level working party consisting of the representatives of the institute, relevant administration and user community to assess the conclusions and to convert them to political or administrative decisions. The Council supervises both the evaluation phase and the utilisation phase.

The basic approach adopted was to use international panels. Usually one or two members of the panel were, however, from Finland to guarantee that the national context could be understood correctly. Sometimes the work of a panel was supported by empirical surveys performed by professional evaluators.

The evaluation of the government research institutes has turned out to be a success. In some cases radical changes have been implemented either to renew the administrative structure or to reorient the research agenda (see the discussion of the evaluation of VTT above). The process has also significantly strengthened Ministries' competence to both manage and utilise research as a strategic tool for the development of the specific sector. It has also improved horizontal co-operation between administrative fields in research. A new round of evaluation of sectoral research is being planned. While the first round assessed institutions as independent entities, the second round will look at government research establishments as a whole to explore possibilities for improved synergy, decreased overlap and merger.

5.4 – Evaluation of international co-operation

International R&D co-operation (for example the EU RTD Framework Programme and EUREKA) is evaluated as a part of the national evaluation effort, and similar methods and approaches are applied. A recent evaluation of the Framework Programme completed in spring 1998 used five different sources of data: (1) the basic data of the Finnish participants; (2) a survey and interviews with the participants; (3) interviews with the Technology Directors of the major Finnish companies; (4) a survey of those who did not participate; and (5) a survey among the Finnish delegates in the programme committees. Some parts of the evaluation were carried out in co-operation with Norway and Sweden and some comparisons were made with the results of similar studies in Germany and the UK.

The evaluation revealed that the Finnish participants, in both the public and private sectors, had been able to benefit from co-operation. New channels and partners were found. On the other hand, the review concluded that there are significant costs and delays involved. The importance of participants' own efforts to implement the results was also confirmed. Participation did not imply major changes among the Finnish partners, mainly due to the fact that many of them had previous experience of the Framework Programme.

5.5 – Evaluation of the additional research financing

As a response to the Government's request to evaluate the effectiveness of additional R&D financing, the Council suggested the launch of a three and a half year continuous evaluation process. This is based on the work of a high-level independent evaluation panel involving representatives from science, economic policy, industry and administration and top level foreign experts in evaluation and innovation policy. The panel has at its disposal a permanent secretariat and a sufficient amount of money to finance all the necessary studies required to give a credible response. The panel will have access to and be able to exploit all research or evaluation data from different sources.

Ultimately the panel should be able to assess the impact of R&D on economic development, structural renewal and job creation. At the same time it should be able to explore what kind of systemic effects the rapid increase of funding is generating. It is a challenging task, which will require significant development of the evaluation methodologies and practices. A separate research programme of technology studies supports the work. Within this programme new methodologies and approaches will be developed and tested. The idea is to generate a set of qualitative and quantitative data, which provide partial indications of the impacts at different levels. The highly developed statistical data in Finland will allow, for example, the examination of the relationships between R&D investment, innovation and economic performance or employment at the establishment level. Studies should also look at how both firms and research organisations are reacting, in terms of management and strategy, to the changing conditions. The panel is then supposed to use this proxy data and its own judgement to arrive at a well-justified opinion about the impact of the financing.

This exercise is, as a whole, an experiment representing not only a new way of carrying out evaluation, but also a new way of policy-making. Policies are perceived as adaptive entities, which, after having been launched and implemented for some time, are evaluated and adapted as a response to the findings. This approach is becoming more and more common in public policy-making in Finland.

5.6 – The Finnish evaluation system

In Finland, R&D evaluation has become an established support tool for both administrative and political decision-making, and is at the vanguard of evaluation practice. At present when evaluation is being introduced to other policy sectors, experiences gained in R&D evaluation are widely exploited. The practices applied meet, to a large extent, the OECD "best-practice" evaluation principles as described in a recent OECD report.

R&D evaluation with its clearly defined guidelines, responsibilities and practices constitutes an evaluation system. It is important that this system allows continuous development of evaluation methods and practices and identifies weaknesses or gaps whenever they occur. It should also provide a means to react to these deficiencies.

Conditions for this development have been created over the years through an interactive process in which research on evaluation methods and practices goes hand in hand with extensive evaluation activity. International experience and co-operation have played a significant role in this development: first, between the Nordic countries, and later more widely both at the EU level and globally. Sufficient competence in methodologies is necessary but not enough. Perhaps, even more important is the development of organisational conditions, which allow continuous learning and efficient exploitation of evaluation results in the decision-making processes. Evaluation also requires a positive attitude. It is necessary that all the parties involved find evaluation a legitimate and reliable instrument that improves the quality of decision-making at all levels.

6. INTERNATIONALISATION

Internationalisation of R&D co-operation has been expanding rapidly. In the middle of the 1990s about 30% of industrial R&D was conducted abroad. On the other hand, internationalisation of industrial R&D has so far been driven by the natural need to expand research near the market. No major R&D activities have been transferred abroad. On the contrary, some multinationals have strengthened their R&D activities in Finland.

The integration of the Finnish research into international R&D co-operation structures has also proceeded rapidly. During the ten-year period from 1986 to 1995, Finland became a full member of all the major European co-operation arrangements: EUREKA, CERN, ESA, EMBL, ESF and the EU RTD Framework Programme. Finland has traditionally been involved in the COST system, in the Nordic R&D co-operation and in bilateral arrangements, in particular, with China, Japan, Russia and the US. Finland has very clear policy guidelines for the development of all these different areas of R&D co-operation (see for example the S&T Policy Council's Review 1996). It is possible in this context to present only very general remarks on these guidelines.

International science and technology co-operation is a central development object of the innovation system. Most of the knowledge needed in Finland is developed abroad. One of the foremost goals in international co-operation is to acquire knowledge needed in Finland. Effective co-operation is based on mutually beneficial interaction, which requires that Finland must have a strong national knowledge base.

International S&T co-operation serves the same aims as domestic research. They are not mutually exclusive, but instead complement each other. In decisions on the steering of research and the selection of projects, weighting must be given to that research which is best realised through national efforts and that which should be undertaken through international co-operation. Decision making on these issues is decentralised to a wide range of actors in the national innovation system. There are no separate organisational arrangements for international S&T co-operation. Each area of co-operation is managed by the organisation responsible for that research area in Finland.

The architecture of European R&D co-operation is complete (see the list above). No new major arrangements are needed. The schemes available are independent and form a complementary entity. They all have their distinct role in the development of the European knowledge base. The emphasis on the development of the international scientific co-operation is on free-form interaction. There are good reasons to strengthen the co-operation in science in Europe, but it does not require any official agreements. The primary means is to develop co-operation between national science organisations and their own international organisations, such as the ESF.

Finland has succeeded well in participating in EU research. On the whole the experiences gained are positive. Finland has supported the endeavour to strengthen RTD activities within EU. On the other hand, there are some concerns regarding current developments. While European industries are specialising in medium-tech and low-growth sectors, the Framework Programme should be able to strengthen European competitiveness in the high-tech, high-growth sectors. In addition, European effort is still reflecting quite strongly the linear view of innovation and efforts to build a European innovation system have not been too successful. One of the major problems seems to be poor co-ordination between the different policies delegated to the Union level. Horizontal innovation policy has not yet been able to gain sufficient support from the Union structures. Another concern relates to the quality of the EU research in the longer term. If top-level research and top-level partners can not be found in European research consortia, the interest of the most advanced firms and research groups may be lost.

As for EUREKA, Finland has emphasised its role as an independent near-the-market scheme, the flexibility of which should be developed even further. In ESA, Finland is actively taking part in scientific and technological work and concentrates its efforts on the so-called ground segment of the activities.

In the area of bilateral co-operation, recent debate has concentrated on bilateral co-operation with the US and Japan.

Overall, Finland's response to the challenges of technological progress and globalisation has been very consistent. Instead of trying to concentrate on protective measures Finland has actively opened its markets, increased competition, strengthened the knowledge base and tried to benefit from the new opportunities offered by these developments. This approach has so far been very successful in many respects. The big question is, however, how far this kind of an approach can carry a small and remote but highly developed country.

7. INTERACTION WITH OTHER POLICIES

Throughout the 1990s there has been a strong trend of integrating innovation policy with other relevant policy areas. This trend has been driven by the finding that innovation performance is dependent on contributions from many different policy sectors. At the same time, solutions to the problems within other policy sectors become more and more dependent on new knowledge and innovation. It is not possible in this context to go into details of this interaction. It is sufficient to say that today innovation policy closely interacts with economic, fiscal, industrial, education, regional, environment, energy, employment, health and welfare policy, among others. The S&T Policy Council has played a significant role in promoting this interaction.

There are a number of joint instruments in each of these areas, but there is still a great deal of work to be done in order to be able to improve interaction to the extent required. These instruments include the cluster programmes, the Centre of Expertise Programme, the education programme of the IT workers, and the new arrangements for the venture capital business. The following examples describe developments in some key sectors:

Economic development and financial stability are becoming more and more dependent on the innovative performance of the economy. A new balance between fiscal consolidation and the resource requirements of technology policy need to be found. The structures of public expenditure still strongly reflect the traditional concerns of a resource-based society. The Government's decision to use funds raised from privatisation to increase the national R&D effort is an example of how fiscal consolidation, the renewal of the structure of public expenditure and the needs of S&T policy can be combined.

Education policy provides the foundation for technology policy. New challenges such as skill shortages, interest in science studies, information society skills, entrepreneurship, lifelong learning and high mobility of high-skilled labour are examples of links between technology policy and education policy. Today, the development of the national system of innovation is the guiding principle in the formulation of education policy in Finland.

The co-operation between technology policies and labour market policies creates conditions for job creation, continuous learning and knowledge management, rewarding compensation, trust and sufficient labour mobility. These policies can also address the problems of social cohesion and exclusion. The S&T Policy Council has had a joint meeting with the Economic Policy Council, which is also chaired by the Prime Minister. The Economic Policy Council is concentrating on economic policy and labour market policies. This dialogue has increased understanding of the need for structural changes in the labour market practices and new management principles.

Industrial policy and S&T policy are designed together. The recent Industrial Policy Strategy adopted by the government emphasises very strongly the need to develop the national system of innovation and to renew the regulatory framework in order to promote the innovative performance of the economy. The above-mentioned measures to strengthen the venture capital business are another example of this interaction.

Environmental policy encourages strongly the development of new technology. Scientific evidence is considered crucial for sound argumentation. The cluster programme has opened new channels for environmental administration and research to collaborate with technology-related administration and industry. A joint view of the various links between innovation and environmentally sustainable development is emerging.

Cluster programmes have also played an important role in the promotion of the interaction between health and welfare policies and innovation policies. Today innovation is seen as a major opportunity for the development of health and welfare services and new businesses. Extensive programmes have been launched to explore jointly the opportunities for improved productivity and quality of services and for new business. In these programmes health and welfare services work together with researchers and industry to find new solutions.

The emerging Information Society is changing conventional ways of doing business and managing lives by improving significantly the efficiency of knowledge production, dissemination and utilisation. New information technologies, digital convergence, the rapid growth of the Internet and increased wired and wireless data speeds are main drivers of the emerging Information Society. A strong interaction between technology policies and the policies facilitating the development of the Information Society is currently created through the Information Society Strategies in different sectors.

8. REGIONAL INNOVATION PROCESSES

Perhaps the most remarkable results have so far been achieved through the integration of regional policy and S&T policy. Regional innovation processes have been developed during the last two decades. The first step in this process was the decentralisation of the university system in the 1960s and 1970s. Universities have since become important players in their regions.

Technology centres have developed in co-operation with the universities. Today there are a total of 16 technology centres in different parts of the country. They all play an important role in providing a platform for commercialisation and also a site for high-tech industrial research facilities in the vicinity of the universities. Some of the most recent centres are linked to the new polytechnics but have usually some relation to a more advanced centre in some university campus. About 1000 firms or other organisations are located in the centres with about 10 000 employees. The Federation of the Technology Centres co-ordinates the activities and strengthens the specialisation of the centres.

The Centres of Expertise Programme (CeEs) is an instrument to develop innovative performance in the regions. The CeE programme is being co-ordinated jointly by the Ministry of Interior, the Ministry of Education, the Ministry of Trade and Industry and the Ministry of Labour. In the individual CeEs, the programme is being implemented in co-operation with firms, cities, technology centres, universities, research institutes and other government bodies. The Ministry of Interior is financing the management of the programmes annually with about 1-2 million FIM per CeE.

The purpose of the CeE programme is to bring research, education and business expertise in a region into a close interaction for generating new business, in particular, high-tech start-ups and jobs. This improved interaction should lead to a generation of more advanced projects, which can become eligible for finance from national or EU research funds. The programme should also enhance regional specialisation. The first round of the programme was completed at the end of 1998.

An evaluation of the programme was conducted in 1996. The programme has contributed to the generation of about 8000 jobs. The annual value of the R&D projects conducted in the CeEs is 250 million FIM. In the first round there were eight regional CeEs and three thematic CeEs. The thematic centres concentrated on forestry, agriculture and tourism. The new phase of the programme started in 1999. From 30 proposals, 21 were selected for the next round. Most of the regional CeEs from the first round continue to operate, although all the thematic centres were closed down. Their activities are being transferred to the national cluster programmes. New areas such as design, new media services, quality and culture have been introduced into the programme.

Other important players in regional innovation systems are the Government Employment and Economic Development Centres. All the government regional activities have been concentrated into these centres. They are managed jointly by the Ministry of Trade and Industry, the Ministry of Labour and the Ministry of Agriculture and Forestry. The Centres have also a small technology unit reporting to TEKES. The Regional Councils are in charge of regional planning. Their orientation towards developing regional innovation systems has steadily been increasing. Also, EU Structural Funds are, to an increasing extent, being used to strengthen regional innovation performance.

Some of the regions have been able to develop an effective innovation system, which contributes to regional economic development and job creation. This is already clearly visible in the national demographic statistics. There is a strong concentration of people in those cities where innovation performance is strong. The S&T Policy Council has been following the development with a view to increasing the synergy between national and regional development. The current trend is that national development and regional development are proceeding in the same direction and are mutually reinforcing.

9. FUTURE CHALLENGES

Although developments in Finland have been very promising there is a constant need to find new ways to respond to an increasingly rapidly changing global environment. The innovation system includes gaps, some of its elements need to be renewed and some completely new elements will have to be implemented. The S&T Policy Council discussed these future challenges in March 1999 and concluded that the following issues should be the main development lines for the next few years:

- Basic university funding will be increased selectively. The increased funding has to enhance structural renewal and specialisation of the universities. Universities will have to release 5% of their current funding and reallocate it to their most advanced units. An equal amount of new money will then be allocated to the same units.

- Education of university graduates in the fields of IT and biotechnology will be increased. This will also require some increases in funding for the research in these fields in order to improve the quality of education.
- Commercialisation of research will be enhanced. This requires new incentive instruments, improved education especially in venture capital business and innovation management, and new funding instruments such as the current Capital Loan Scheme.
- Cluster programmes will be expanded. New programmes will be launched and successful existing programmes will be extended. Specific effort will be made to increase networking and knowledge flows over traditional boundaries in order to enhance the emergence of new clusters and the extension of existing ones.
- Finally, new programmes and other instruments will be implemented to foster the development of a user-oriented information society. These efforts will involve projects to enhance the growth of new Internet-based services and other applications. Cultural products, entertainment and design are among the new applications.

All these were seen as important for Finland's development towards a more advanced knowledge-based society in the new millennium.

10. CONCLUDING REMARKS

In retrospect, Finland has changed very much during the last two decades. Finland has been able to move from a resource-based economy towards a knowledge-based economy. This indigenous conversion has taken place in a balanced way. In addition to the rapid growth of knowledge intensive industries the resource-based industries have also been able to benefit from developments. Products and processes have become more knowledge-intensive throughout the economy. A virtuous development circle, when continued, will bring Finland to a completely new position in the international division of labour as a knowledge-intensive society. Recently Finland's development has also been recognised in many international contexts, such as by the OECD and in the World Competitiveness Yearbook by the IMD.

This conversion has been accompanied by developments in science and technology policy thinking. Finland has been able to abandon the linear model of innovation and adopt the systemic approach in the science, technology and innovation policy formulation. When combined with increasing

R&D resources, both in the public and private sectors, this has created an innovation system, which has facilitated development. There is no unique measure to describe the contribution of innovation policy to economic and social development, but there is a great deal of partial evidence, which suggests that the contribution has been significant. The development of the quality of the Finnish science, as documented by the Academy and increasing evaluation evidence, indicates that also this conversion has taken place in a balanced way: scientific excellence, social relevance and innovation performance have been able to develop in a mutually reinforcing manner. There is also a virtuous circle of development within the national system of innovation.

On the other hand, much remains to be done. The innovation system needs to be developed continuously, partly, due a rapidly changing innovation environment, which may unexpectedly raise new concerns or problems and, partly, because the innovation system is far from perfect or complete.

Finland's development will also, to an increasing extent, depend on developments in other countries. European or global development is not a zero-sum game. Improved innovative performance will benefit all the parties involved. Therefore, efforts to enhance innovation and to develop innovation systems should also cover issues, which influence innovation performance in other countries as well as internationally at the European or global levels. The same challenge is faced at the regional level. The studies, which explore the strengths and weaknesses of regional innovation systems, should be continued and expanded to regions which are not yet fully involved.

Although there are a number of concerns, Finland is on the right track. By far the greatest challenge is, therefore, to be able to continue the development along the lines adopted. A virtuous circle is extremely difficult to build but very easy to destroy.

CHAPTER 11

THE SPANISH SYSTEM OF RESEARCH

EMILIO MUÑOZ

1. OVERVIEW OF THE SPANISH SYSTEM OF RESEARCH AND INNOVATION

The Spanish research system, which embraces aspects of both research and innovation, is characterised by its lack both of a tradition in recognising the political and economic relevance of science and technology and of efficient mechanisms for the management of a science and innovation system.

Over time, certain individuals have assumed governance of the research system and have attempted to introduce organisational and institutional reforms to correct its operational deficiencies.

These "windows of opportunity" for introducing science and technology onto the political agenda have had positive, but partial, outcomes, benefiting specific areas or sectors of the research system. However, they have failed to obtain a positive outcome on the wider level.

1.1 – Historical context

The evolution of the research system in Spain has been influenced by the lack of a Spanish tradition. There has been no golden era of science and technology in Spain. The period of enlightenment that occurred at the end of the 19th century and during the first third of the 20th century was cut short by the Civil War (1936-1939). The ensuing autocratic regime was responsible for the isolation of Spain from world trends. As a consequence, while the United States and other Western European countries were making strong efforts to develop their research systems, by increasing both public and private expenditure and the size of the skilled labour force, the Spanish research system remained very small.

1.2 – Main Institutions and the pace of reform

Over the course of the past century, the institutions involved in the implementation of science and technology policies in Spain have experienced a number of changes. After the Civil War, the institution responsible for promoting research and training for research in Spain, the *Junta de Ampliación de Estudios* (JAE) was dissolved. In its place, the *Consejo Superior de Investigaciones Científicas* (CSIC) was created in 1939, its operation modelled on the academies of the socialist countries. The CSIC was established with two goals: to co-ordinate and foster scientific development in Spain and to launch and develop research centres. The CSIC underwent several crises in its evolution, which have been examined by several authors (see the special issue of the Spanish journal ARBOR, 1990; Nieto, 1982; Sánchez Ron, 1992 and Santesmases and Muñoz, 1993, 1997). In summary, it can be concluded that:

- The CSIC failed in its role as an agency to co-ordinate and promote scientific activity in Spain. It did not play a significant role in the design of the Spanish science policy.
- However, it was instrumental in the professionalisation of research careers in the Spanish public research system.
- It was quite successful in the launch and development of research centres and sophisticated scientific infrastructures in a wide range of scientific and technological domains, from chemistry to biology, and from national resources to technological applications in specific sectors (Automation and Robotics, Ceramics, and Building).
- The CSIC was decisive in creating and maintaining high-level libraries related to science and technology.
- The CSIC experienced difficulties throughout its history in managing the divides between science and technology, between basic and applied science, evolving into an academic-style institution.

While the CSIC was unable to fulfil its central role in the Spanish research system, two different trajectories of development were established during the period of autarchy. One was aimed at the creation of agencies and funds for the co-ordination and promotion of science and technology, both in general and targeted at specific areas. The other was directed at the establishment

TABLE 11.1 – List of the OPIs (Public Research Organisations) existing in Spain

OPI	Ministry of dependance (1987)	Ministry of dependance (1997)	Year of foundation
Instituto Geológico y Minero (IGME)	Industry and Energy	Environment	1910
Instituto Español de Oceanografia (IEO)	Agriculture, Fisheries and Food	same	1914
Consejo superior de Investigaciones Cientificas (CSIC)	Education and Science	Education and Culture	1939
Instituto Nacional de Técnica Aeroespacial (INTA)	Defence	same	1942
Instituto Geográfico Nacional (IGN)	Public Works and City Planning	Eliminated* 1996	1944
Instituto Nacional de Investigaciones Agrarias (INIA)	Agriculture, Fisheries and Food	same	1933 Refounded 1971
Centro de Investigaciones Medioambientales y Tecnológicas (CIEMAT)	Industry and Energy	same	1957
Centro de Experimentación de Obras Públicas (CEDEX)	Public Works and City Planning	Economic Development	1957
Instituto de Salud Carlos III	Health and Consumption	same	1986

Note: *Competences were distributed among other administrative units.

or reform of government laboratories, under the administrative responsibility of sectoral Ministries (Table 11.1) and with different objectives and goals for their research and development tasks.

All these reorganisations fitted well with the various efforts to reform and increase the relevance of the Spanish system of research and innovation.

The 1960s

The first attempt to improve the relevance of the Spanish research system in the 20th century took place in the 1960s, with the arrival in government of so-called "technocrats" who attempted to remedy Spain's backwardness by emulating the model of the French National Plans, with reliance on support from the United States. These attempts at reform were driven by the influence of international bodies and trends (the OECD played a significant role in this process). It is worth mentioning that the success in improving the economic wealth of Spain, elevating the country to the tenth position in terms of GDP among the industrialised countries was based on two main sectors (building and tourism), together with the adaptation of traditional sectors of the Spanish economy, such as textiles, ceramics, agriculture, fisheries, to international trends. All these areas were characterised by their limited connections with R&D activities.

The Plans for Strategic Development incorporated R&D as a strategic sector of activity, although these activities had very little influence on economic progress. The First National Plan for Development (1964-1968) was designed to meet the most urgent needs of the Spanish economy. It was suffering from a lack of competitive capacity as a consequence of autarchy. The Second Plan (1968-1971) was aimed at increasing the resources devoted to science and technology, following the suggestions of the OECD. However, the economic crises of the 1970s led to budgetary restrictions, which again reduced the effects on science and technology. Within these Plans for Development, attempts were made to correct some of the endemic problems of the Spanish research system: the lack of professional careers for researchers and deficiencies in the economic and human resources devoted to R&D activities.

The *Comisión Asesora de Investigación Científica y Técnica* (CAICYT) was established in 1958 to cope with those aspects of the implementation of science and technology policies that the CSIC was unable to fulfil. It was created under the "technocratic" influence and was probably inspired by a blend of models (Anglo-Saxon, French and German). CAICYT was established as independent from the CSIC, but was initially closely linked to it. Paradoxically, the capacities of CAICYT to intervene in the scientific and technological development of Spain were limited since CAICYT lacked any dedicated budget to carry out its functions. This changed in 1964 when the *Fondo Nacional para el Desarrollo de la Investigación Científica* (National Fund for Development of Scientific Research) was created with an endowment of 100 million PTA. The annual amount of this endowment, apart from some minor fluctuations, showed an increasing trend until 1971.

Consequently, Spain attained a wealthy economic status while spending only 0.3 % of GDP on R&D and with fewer than 10 000 full-time equivalent research personnel in a population of 38 million inhabitants. It is worth noting also that during the 1960s Spain lacked a good protocol and an appropriate institutional setting for the collection of statistics on R&D.

The new agencies: the promotion of technology and health research

At the end of the 1970s and the beginning of the 1980s, two new agencies were established: the *Centro para el Desarrollo Tecnológico Industrial*

(CDTI, Centre for Industrial Technological Development) and the *Fondo de investigaciones Sanitarias de la Seguridad Social* (FISS, Fund for Research on Health from the Social Security).

CDTI was established in 1978 as an administrative organisation within the Spanish government system and became a *sociedad con capital público* (public enterprise) in 1983. Its objectives were the funding of projects related to the final steps in the process of technological development, fostering a favourable environment for the promotion of technology and innovation. When restructured as a public enterprise, CDTI gained greater flexibility to carry out its functions and to manage its resources, which rose from 779 million PTA in 1978 to 1900 million PTA in 1983. In its second period, its endowment was increased to 4000 million PTA with the following breakdown: electronics and information technologies (40%); mechanical processes (25%); biotechnology and medicine (13.5%); energy (12.4%) and agro-food (8.8%).

The fund for promotion of research in the health and biomedical areas (FISS) was established in 1980 as a sectoral fund, allocated to the Ministry of Health and Consumption. This followed previous experience of collaboration with the pharmaceutical industry, which contributed a part of its profits to the Public Health Insurance System (Social Security System) as an endowment to fund research and training in the health and biomedical domains. The funds allocated to this sector of research have increased steadily from hundreds of millions of PTA to reach figures around 6000 million in the 1990s. The research funded is performed in hospital and university-related centres. While fundraising is distributed throughout Spain, a large proportion is concentrated in Madrid and Cataluña. This situation is shared with other agencies and programmes, although in the case of the FISS, the role of Cataluña is more prominent than in others.

Science and technology in the 1980s

During the transition to democracy following Franco's death, the weak profile of R&D activities further declined. One positive input was the creation of the *Ministerio de Universidades e Investigación* driven by the Social Democrat wing of the *Unión de Centro Democrático* (UCD), the leading party in the two first democratic elections. However, this only lasted for two years.

Until 1979, CAICYT was an organ of the Ministry of the Government Presidency. It then became part of the newly created Ministry of Research and Universities. However, the final period of UCD government saw the disappearance of the Ministry of Research and Universities and the transfer of its responsibilities to a Secretary of State for Universities and Research – a second rank at ministerial level – under the control of the Minister of Education and Science. This is where responsibility for CAICYT remained, until its disappearance following the enactment of the "Law for Science".

The general elections of 1982 were won by the Socialist Party (PSOE), whose electoral programme had incorporated science and technology issues. Within a general context of modernisation, policy entrepreneurs were able to initiate a series of actions. The first decision was to carry out a diagnostic exercise on the Spanish research system. This analysis was the subject of a publication (Muñoz and Ornia, 1986), and among its interesting revelations, the following points are worth recalling:

- Lack of strategic relevance of R&D activities for technological and economic development.
- Small size of the research and innovation system, in terms of both economic expenditure and human resources.
- Limited co-ordination within the public system of research, with a main role played by the Ministry of Education and Science, but with the sectoral ministries allocating resources to different programmes and the Government Laboratories for which they were responsible. No specific allocation for R&D existed in the General State Budget until 1986-1987.
- The agencies involved in the funding of research projects and earmarked programmes through competitive mechanisms (CAICYT, *Comisión Asesora de Investigación Científica y Técnica, Fondo de Investigación de la Seguridad Social y de Investigaciones Sanitarias*) accounted for around 5% of the State Budget, a small percentage for performing real strategic policy work.
- The business sector contributed less than 50% of the effort in research and innovation action, with the lion's share of this effort being contributed by public enterprises.
- Uneven distribution between regions: Madrid accounted for 45% of the public system of research, followed by Cataluña (19%), while Madrid, Cataluña and the Basque Country together accounted for 80% of the research performed by the business sector.
- Human resources also showed a strong concentration in Madrid and Cataluña (37.5% and 21.3%, respectively) followed by Andalucía (11.2%) and the Valencian Community (6.8%).

Following a series of activities,[1] the political effort of the PSOE was consolidated into the enactment of the Law of Promotion and General Co-ordination of the Scientific and Technical Research. This "Law for Science", as it is colloquially referred to, aimed at:

- the introduction of science and technology onto the political agenda (decision and debate);
- fostering co-ordination within the public system by establishing an inter-ministerial Committee chaired by a Minister, designated by the Prime Minister, and with representatives of the main sectoral ministries involved in the performance of research activities;
- promotion of interaction between the social and economic actors and the governments of the autonomous regions with the policies on research and innovation, through the creation of two bodies: the Advisory Council for Science and Technology and the General Council for Science and Technology;
- driving potential action through the establishment of a National Plan for Research and Development, focusing on priorities and funding research through competitive mechanisms, according to appropriate assessment procedures.

La Comisión Interministerial de Ciencia y Tecnología (CICYT, Interministerial Committee for Science and Technology) was set up as a consequence of the Law for Science. It took over CAICYT's responsibility for the co-ordination and planning of science and technology policies and activities, but had greater institutional support, and relied on the General Secretariat for the National Plan for R&D to manage the resources of the research programmes based on priorities and strategic objectives.

Reform in the late 1990s

The 1996 elections were won by the Conservative Party (*Partido Popular*). The new Government brought about change in several aspects of the political arena. R&D policy was suspended for a period of reflection, and this stagnation has been accompanied by a positive trend in the country's economic situation. The Spanish economy has been growing at the fastest rate among European countries, while R&D expenditure has been characterised by slight increases in 1997 and 1998 in relation to economic growth. Moreover, this increase has courted controversy due to the incorporation

1. For a review, see L. Sanz Menéndez (1997).

in the State Budget for R&D of European Structural Funds and some compromises in international programmes related to defence objectives and weapons.

The General Secretariat of CICYT was replaced in 1998 by the Office for Science and Technology (*Oficina de Fomento y Coordinación Científica y Técnica*) as a result of the last reorganisation, and is attached to the Ministry of Presidency (Ministry attached to the Prime Minister, chaired by the Vice-Prime Minister and responsible for the co-ordination of activities of the Ministerial Cabinet). However, this office appears to have undergone more reductions in human resources and infrastructure than the former Secretariat for the R&D National Plan. It has been set up to promote co-ordination and strategic planning, with the approval of a new National Plan for R&D in 1999 as its first important action.

After the elections in year 2000, the leading party (*Partido Popular*), under the impulse of the Prime Minister has set a new administrative organization with the creation of a Ministry of Science and Technology that holds most of the previous competences of the Ministry of Industry and Energy and part of the competences of the Ministry of Education and Culture, in particular in relation to the management of the Fourth National Plan.

2. EVOLUTION OF THE SPANISH RESEARCH SYSTEM SINCE THE 1980s

There has been an increase in the size of R&D expenditure, although it has yet to reach the goal of 1.2% of GERD that was foreseen as one of the goals of the First National Plan for R&D (1988-1991), and which has been kept as an objective in the subsequent National Plans (1992-1995; 1996-2000).

The relative figures have shown ups and downs (see Figure 11.1), although the total amounts have shown an increasing trend, except for a plateau in 1992-1994 (Figure 11.2).

It is worth mentioning that the important increase in R&D investment in terms of current PTA is subject to debate, since it appears mainly to result from the integration into R&D expenditure of Structural Funds and the resources involved in some important international co-operative programmes of the defence sector, funds not hitherto considered under R&D. On the other hand, there is a slight increase in 1988 (constant currency). This lends support to the contention that the research activities are not exerting direct

FIGURE 11.1 – Evolution of R&D expenditure (% GERD) in Spain

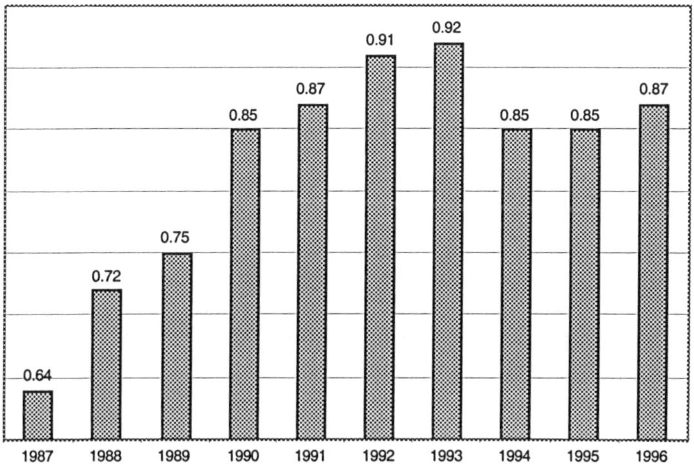

FIGURE 11.2 – Evolution of R&D expenditure 1987-1996 in Spain

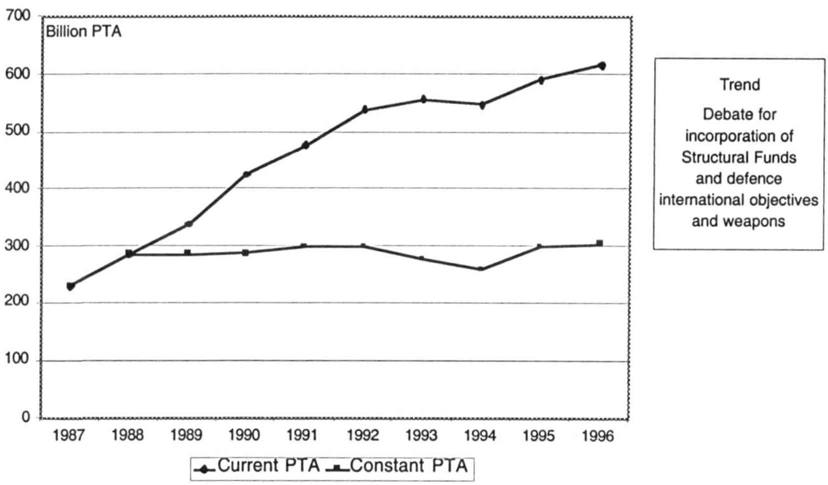

Note: In 1999, 1 PTA = 0.0059 EURO.

influence on the economic wealth of Spain but that the opposite seems to be true. Government push and economic wealth are the factors which influence R&D activities in Spain.

Since 1987, following the Law for Science, there has been a budgetary allocation for the public research system. In 1997, the distribution among

TABLE 11.2 – Distribution in percentage by ministerial departments of the R&D function of the General State Budget (1997)

	%
Ministry of Industry and Energy (MINER)	38.9
Ministry of Defence	20.5
Ministry of Education and Culture (MEC)	19.9
National Fund	8.8
Ministry of Health	5.3
Ministry of Agriculture, Fisheries and Food	3.5
Other	3.1
Total	100.0

FIGURE 11.3 – Evolution of the share of personnel by sector in full-time equivalents employed in R&D activities in Spain

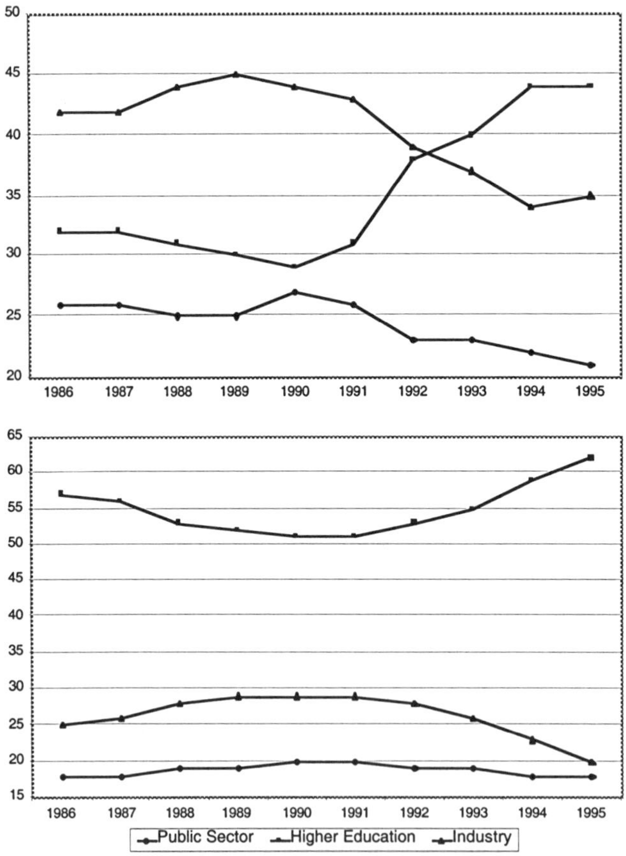

Note: Top figure is in thousand persons and bottom figure in percentages.

ministerial departments is illustrated in Table 11.2, which shows the relatively low percentage allocated to the National Fund, as compared to Ministries such as those of Industry and Energy, Education and Culture, Defence, Health or Agriculture. These figures cast doubts on the efficiency of co-ordination of the National Plan of R&D.

The business sector has been reducing its share over these years as illustrated by the evolution of the research personnel (Figure 11.3).

2.1 – National policies and priorities

The main changes in Spanish science and technology policies experienced since the early 1960s have been driven by attempts to settle and gear priorities within the science and technology domain. CAICYT was the organisation for this type of advance, but its logic of action was hampered by the lack of tradition in the field of designing strategies – the cultural influence of a dictatorial regime deepened this deficit. However, CAICYT was successful in introducing competitive mechanisms for the allocation of funds to projects from the research and the industrial worlds. This bottom-up approach – investigator-driven or industrial manager-driven – led to the shaping of prominent areas of research in institutions devoted to basic and applied research activities and to the emergence of the most innovative areas within the business sector.

Research activities were funded through projects presented by researchers from within the public system of research. Activities in the firm-driven approach were funded through a new form of contract termed *Planes concertados de investigación* (concerted research projects), by which firms developed research and development activities either in-house or in collaboration with public research institutes or university departments, and with 50% of the budget provided by public funds (*Fondo Nacional para la Investigación Científica y Técnica*, see above). This scheme helped to establish links between firms and the public sector, but the collaboration remained incidental.

Tables 11.3 and 11.4 illustrate the resulting maps. A first consideration arising from the data is the apparent separation between the two sets of actions. The activities of the chemical and pharmaceutical sectors by firms clearly surpass the basic research sub-sector potential. The same seems to occur with engineering-related activities (the impact of electronic systems, machine tools and electrical mechanics did not match the effort on basic and applied research in related fields).

TABLE 11.3 – Funding by CAICYT (1985) of basic and applied research activities

Area of activity	Million PTA (1985)	%
Advance of knowledge	*1 369*	*100*
Molecular and cell biology	320	23
Biology of organisms and systems	235	17
Chemistry	233	17
Mathematics, astronomy and physics	231	17
Social sciences and humanities	205	15
Earth and space sciences	109	8
Physical chemistry	42	3
Applied research	*1 117*	*100*
Medical sciences	302	27
Information and communication technologies	252	23
Materials science and technology	122	11
Agriculture	111	10
Chemical technology	95	8
Mechanical and textile technology	75	7
Food technology	60	5
Livestock	59	5
Environmental technology	21	2
Energy technology	21	2

Source: Adapted from Muñoz (1998).

TABLE 11.4 – Funding by CAICYT of *Planes concertados de investigación* according to sectors of economic activity (1969-1985)

	Number of projects	%
Farming and animal husbandry	38	8.2
Agro-foods	23	5.0
Chemicals	53	11.4
Pharmaceuticals	78	16.4
Plastic materials and products	25	5.4
Machine tools	66	14.2
Electrical mechanics	21	4.5
Transport	14	3.2
Electronic systems	139	30.0
Total	464	100.0

Source: Adapted from Muñoz (1998).

In 1981, a short time before the general elections of 1982, CAICYT initiated strategic planning activity, with the launch of figures concerning the *Programas especiales de I+D* (Special R&D Programmes) and the *Planes o programas movilizadores* (Mobilising Programmes).

These programmes were conceived as a series of linked and intertwined research (and development) projects aimed at attaining scientific-technological and socio-economic goals. The programmes were initially focused on the following domains: aquaculture, agro-energetics, biotechnology, microelectronics and transport. These were considered as strategic, taking into account both global trends and actual Spanish capabilities. Despite the strategic rationale, the scientific and industrial communities treated these selections with apprehension. The decision had been taken top-down and with very little participation of scientists and industrialists in the definition of the programmes. This negative attitude on the part of the leading scientific and innovation actors was intensified by the weakness of the party (*Unión de Centro Democrático*) in government at that time.

However, after the victory of the Spanish Socialist Party in the general election of 1982, the science and technology policymakers decided to use this experience as a preliminary exercise in the planning of R&D activities. The programmes and their associated management committees were reshuffled and four programmes on aquaculture, agro-energetics, biotechnology and microelectronics were launched with an endowment of 600 million PTA (1986). The relative success of the experiment served to gain credibility among the scientific community, which was usually reluctant to embrace "oriented" science strategies. It also provided the basis for the future National Plan for R&D. During this period, the Ministries responsible for the most innovative sectors of business, such as agronomy, information technologies, energy and pharmaceuticals, launched sectoral R&D plans (see below).

2.2 – The National Plans for R&D

The National Plan for R&D, established under the Law for Science, is the principal mechanism (or main instrument) for funding national research priorities.

The First National Plan

The First National Plan covered the period 1988-1991. Its priorities were determined by combining top-down decisions with the results of discussions of expert working groups in preselected scientific and technological areas, based on a series of experiences and explicit analytical criteria.[1] In particular, the previous experience of CAICYT and the following criteria were applied to select the priority areas:

- existence of previous national plans established by Ministries or agencies, as in the case of the Electronic and Information Technologies Plan and the Energy National Plan (both launched by the Ministry of Industry and Energy) and the National Plan for Agriculture (Ministry of Agriculture, Fisheries and Food);
- existence of a well-developed scientific and technical community;
- existence of economic sectors characterised by innovation activities and links with the scientific and technical research communities;
- detection of flaws in any one of the scientific and industrial communities mentioned above;
- priorities already existing in international programmes, which may be of strategic relevance for the interests of Spain;
- identification of strategic areas of socio-economic relevance for Spain and her citizens that may be approached through scientific and technical developments;
- reorientation and reorganisation of existing programmes.

The exercise resulted in the selection of 23 programmes, grouped into five large areas: Horizontal Programmes, Agro-food and Natural Resources, Production and Communication Technologies, Quality of Life and Special Programmes (see Box 11.1).

Each programme was drafted by an *ad hoc* working group, set up according to a common structure and to take into consideration the following issues: the definition and formulation of socio-economical objectives; justification of programmes on the grounds of demand and of future prospects, assuming a risk analysis approach; analysis of existing resources, interaction with other industrial sectors and other national and international programmes; well-defined objectives, in terms of scientific, technological and industrial development, and potential social and economic benefits; lines of action; priorities and time-schedule; funding; instruments for management, co-ordination, monitoring and evaluation of the programme; exploitation and technology transfer of eventual results.

---

1. A document was published by the Secretary of State for Universities and Research, Ministry of Education and Science, in 1988, describing the theoretical and economical frameworks and details of the First National Plan on Scientific Research and Technological Development.

BOX 11.1 – The 23 programmes of the First National Plan

Horizontal Programmes	Training of personnel for research
	Interconnection of information resources
Agro-food and Natural Resources	Research on the Antarctic Sea
	Agricultural research
	Research on development of livestock
	Geological resources
	Marine resources and aquaculture
	Food technology
Production and Communication Technologies	Automation and robotics
	Photonics
	Space research
	Microelectronics
	New materials
	Information and communication technology
Quality of Life	Biotechnology
	Cultural and social studies on Latin-America
	Immunology
	Pharmaceutical research and development
	Research on sport
	Historical heritage
	Social issues and social welfare
	Toxicology
Special Programmes	High energy physics (relation to CERN)

Each programme was assigned a committee, an intermediate body between the programme managers and CICYT, composed of representatives from the funding agencies and the sectoral Ministries interested in the area covered by the programme.

The first conclusions to be drawn from the First National Plan were:

- A wave of enthusiasm in the scientific and technological community and among the most innovative industrial firms.
- It was too ambitious in the numbers of both programmes and their goals.
- It failed to carry out internal and external evaluation and monitoring exercises, as originally foreseen. Changes in personnel at different organisational levels in the most relevant ministries, caused a loss of the original philosophy and impetus, and made it impossible for the Plan to progress in an appropriate way.

Subsequent National Plans for R&D (II and III)

The First National Plan became an instrument for funding academic research activities, but lacked any process of evaluation by any of the advisory committees or by the official national evaluation agency (ANEP).[1] The rearrangement of the strategic actions and the selection of priorities resulted from debate within CICYT and from the weight and influence of the sectoral Ministries represented in the Inter-ministerial Committee.

BOX 11.2 – Content of the Second and Third National Plans

Second National Plan for R&D (1992-1995)	**Third National Plan for R&D (1996-1999)**
Quality of Life and Natural Resources	*Life Sciences and Agro-food*
Agricultural sciences	Agricultural R&D
Food technologies	Food technologies
Environmental and natural resources	Environmental R&D
Biotechnology	Biotechnology
Health and pharmaceutical research	Health
Climate R&D	Climate R&D
Marine science and technology	Marine science and technology
Research on the Antarctic Sea	Environment and energy
	Water resources
	Research on the Antarctic Sea
Production and Communication Technologies	*Production and Communication Technologies*
Advanced technologies for production	Advanced technologies for production
Information and communication Technologies	Information and communication Technologies
Materials	Materials
Space research	Space research
	Telematics: applications and services
Fine chemistry (programme raised and managed from Cataluña)	Technologies of chemical processes
Socio-cultural, Horizontal and Special areas	*Horizontal and Special areas*
Social, economic and cultural studies	Social, economic and cultural studies
High energy physics	High energy physics
Information (Network) for R&D	Research of the Links of the Science-Technology Industry System (PACTI)
Research training programme	

1. This agency was created by a Royal Decree in 1987. The ANEP is concerned with the evaluation and monitoring of projects, programmes and their results as well as for forecasting and foresight exercises. For an analysis of the performance of the agency see Sanz Menéndez (1995).

Thus, the most evident change in the Second National Plan for R&D was the regrouping of topics and fields in a reduced number of programmes (17 instead of 23) and their concentration into areas (three instead of six). The Third National Plan for R&D was focused on applied and production-oriented research, with new programmes (20 in total) grouped into three areas. The most relevant feature of the new orientation of the Third National Plan was the incorporation of the PACTI programme (see list below).

The three National Plans each allocated similar amounts of money (around 22 000 million PTA at current values), with similar shares between scientific areas: quality of life, 35-45%, production and communication technologies 40-50%, socio-economic 5-8%; and activity lines: research projects 40-50%, concerted projects with firms 15-20%, infrastructure 10-15%, research training around 15%, and other actions about 10%.

2.3 – Management and mode of operation of priority programmes

With few exceptions, the priority programmes, providing funding for basic and applied research activities, have been managed from the General Secretariat of the National Plan. However, following the enactment of the Law for Science, concerted projects involving firms (reminiscent of the formula of *Planes concertados de investigación)* were the responsibility of CDTI. This decision was based on political negotiation between the Ministry of Education and Science (the chair of CICYT) and the Ministry of Industry and Energy (responsible for the technological and innovation policies). While closer links between firms and the Government was initially favoured, it was subsequently seen as an obstacle to fostering links between academia and industry and to promoting co-ordination among the different actors involved in the processes of research, development and innovation.

The most recent report, corresponding to the end of the Second National Plan (year 1995), enables examination of the breakdown between the different programmes and the distinct actions within each programme as indicators of their relative importance. Table 11.5 illustrates this distribution, expressed as percentages, both vertically between programmes and horizontally for actions in each programme to facilitate comparison.

The data show a division into three groups. The first comprises three programmes: Environment and Natural Resources, Information and Communication Technologies and Materials with percentages higher than 10% of the resources allocated. The second is composed of five other programmes,

TABLE 11.5 – Breakdown of resources (in percentage) allocated to the different programmes of the Second National R&D Plan (year 1995)

Programmes	Training	Projects and infra-structure	Concer-ted projects	Additional expen-ditures	Total % per line	Total %
Agricultural Research	19.0	63.7	13.2	4.1	100	7.5
Food Technology	7.2	58.9	26.5	7.4	100	5.3
Environment and Natural Resources	11.2	78.0	8.4	2.3	100	11.4
Biotechnology	16.1	66.9	13.5	3.5	100	8.9
Health and Pharmaceutical Research	13.6	67.8	14.3	4.3	100	9.2
Climate Research	18.4	81.6	-	-	100	0.9
Marine S&T	5.2	94.8	-	-	100	1.1
Advanced Technologies and Production	8.1	43.3	46.3	2.2	100	7.4
Information and Commu-nication Technologies	3.1	73.8	16.1	7.0	100	15.8
Materials	4.8	69.5	24.3	1.4	100	16.9
Space Research	13.0	64.1	22.9	-	100	2.8
Social, Economical and Cultural Studies	21.4	78.6	-	-	100	1.3
Information for R&D	-	100.0	-	-	100	1.4
High Energy Physics	15.9	39.5	-	44.5	100	4.6
Research on the Antarctic Sea	0.8	51.7	-	47.5	100	2.2
Fine Chemistry	-	100.0	-	-	100	0.2
Training personnel for research	100.0	-	-	-	100	2.9
Total	12.4	64.6	16.6	6.4	100	99.8

Source: Elaborations from *Memoria Actividades del Plan Nacional de I+D, 1995.*
Note: The 1995 data correlates well with the average for the period of the Second National Plan (1992-1995).

Agricultural Research, Food Technology, Biotechnology, Health and Pharmaceutical Research, Advanced Technologies for Production, sharing percentages between 5-10%. Third, the remaining nine programmes show a great diversity among them both in resources and in the internal distribution between actions. It is worth noting that only nine programmes had resources for concerted projects (collaboration between research centres and business) and that the relative importance profile changes with respect to the total resources allocated: Advanced Technologies for Production lead, followed by Food Technology and Materials, and then by Environment and Natural Resources.

It can be concluded, at first glance, that the broad view in 1995 with regard to applied, oriented research activities does not appear too different to that observed for 1985 (see Table 11.3), in spite of some difficulties in matching the data, due to changes in nomenclature. Similar problems affect comparison between Tables 11.4 and 11.5 for industry-related projects (concerted projects).

In the story of the priorities of R&D policies in Spain, the research activities remained stable beyond any structural and political changes and transformation.

2.4 – Co-ordination of R&D activities: the main goal of the National Plan

The co-ordination of all R&D activities in Spain was the aim of the National Plans for R&D. This main goal is illustrated in Figure 11.4, which is taken from the first plan. However, ten years on, it has achieved limited success and co-ordination remains a very sensitive issue.

In fact, only three sectoral programmes were included in the National Plan in 1995 (see *Memoria de actividades del Plan Nacional de I+D en 1995*). These were:

- The *Programa de Promoción General del Conocimiento* (Programme for Knowledge Promotion i.e. funding of grass-roots, curiosity-driven research), administered by the Ministry of Education and Science, the chair of CICYT between 1986 and 1996;
- The *Programa Sectorial de I+D Agrario y Alimentario* (Programme for R&D in Agriculture and Agro-food) administered by the Ministry of Agriculture, Fisheries and Food);
- The *Fondo de Investigación Sanitaria – FIS* (The Fund for Health Research) administered by the Ministry of Health and Consumption, previously referred as FISS (Fondo de Investigación Sanitaria y de la Seguridad Social), which changed its name in 1986 after becoming independent in the funding from social security resources.

As Table 11.2 shows, the National Plan fund amounts to only 9% of the state budget, whereas those three sectoral programmes managed by the sectoral Ministries (Ministry of Education and Culture, Health and Agriculture, Food and Fisheries) account for about 30% of the state budget (1997).

On the other hand, an important share of the state budget (38%) belongs to the Ministry of Industry and Energy. The only R&D promotion activities carried out by the Ministry of Industry and Energy which seem to be co-ordinated by the National Plan are precisely those financed by the Plan itself, namely those integrated under the heading of "Co-ordinated and Co-operative Projects", i.e. projects carried out through collaboration between industries and public research centres (universities and government laboratories).

3. REGIONAL RESEARCH AND INNOVATION POLICIES AND THEIR RELATION TO NATIONAL POLICY

Debate on the relationships between national and regional policies and domains of action has been prevalent throughout the processes of re-organisation of the Spanish research and innovation system, and of the Higher Education system. The Constitutional Act (1976) granted both the state and the regions the rights to develop their own science and technology (innovation) policies, while the State had the right to co-ordinate policies. Consequently, the discourse of co-ordination was supporting the concepts of promotion and planning in the National R&D Plan as it is recorded as a reminder in Figure 11.4.

The Interministerial Science and Technology Committee (CICYT) was established as the official body for the planning, co-ordination and monitoring of the National Plan for Scientific Research and Development. This was considered to be the principal means of harmonising the national scientific and technological effort, through the promotion of research. It was intended to determine priorities and allocate resources between the different R&D actors – company research centres and universities – and in addition to co-ordinating the R&D activities of the different Autonomous Regions (17 in Spain) and the state. The Spanish Constitution assigns responsibility to the national Government for promotion and co-ordination of scientific and technological research in general terms. The 17 Autonomous Communities, through their *Estatutos de Autonomía*, share the responsibilities for fostering scientific research and technological development.

3.1 – Characteristics of regional R&D activities

Regional diversity

One of the main characteristics of the Spanish R&D system in the Autonomous Regions is its heterogeneity. The distribution of R&D expenditure

FIGURE 11.4 – Co-ordination and planning as instruments for promotion

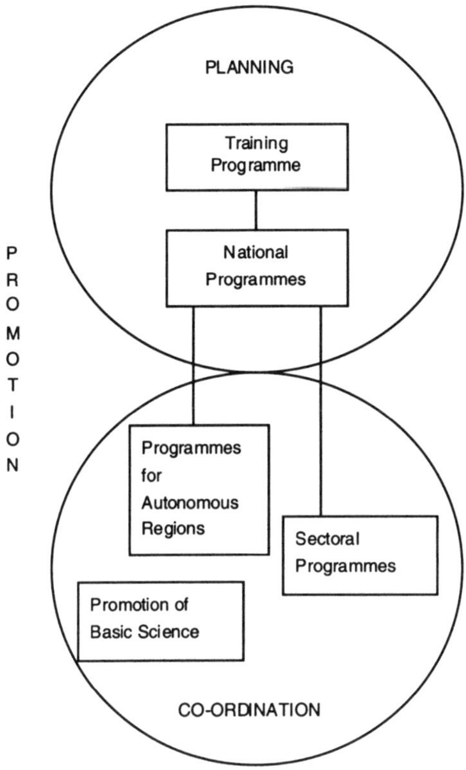

Source: Adapted from CICYT (1988): El Plan Nacional de Investigacion Cientifica y Desarrollo Technologico. 1988-1991. Madrid: CICYT, p. 83.

among the regions and between different sectors confirms this statement (Tables 11.6 and 11.7).

Five regions (Madrid, Cataluña, Andalucia, Comunida Valenciana and the Basque Country) account for 85% of the total expenditure. Madrid, Cataluña and the Basque Country are non-Objective 1 regions and sustain a high technological effort within the business sector (around 50% of the total with all three regions having percentages higher than the Spanish average of 49%). Andalucia and Comunida Valenciana belong to the Objective 1 regions and show a predominance of public sector and university participation (78% and 80%, respectively).

TABLE 11.6 – Share of R&D expenditure by Autonomous Regions

Autonomous regions	1987	1993
Andalucia	7.4	9.4
Aragón	2.4	2.6
Asturias	1.6	1.6
Baleares	0.5	0.3
Canarias	1.0	2.0
Cantabria	0.7	0.8
Castilla – La Mancha	0.6	0.8
Castilla León	2.7	5.0
Cataluña	18.7	19.5
Comunida Valenciana	3.7	5.8
Extramadura	0.7	0.7
Galicia	2.1	2.8
Madrid	45.3	36.6
Murcia	1.2	1.5
Navarra	1.3	1.6
País Vasco	8.5	7.9
Rioja	0.1	0.3
Not regionalised	1.5	0.8

Source: INE, *Memoria de Actividades del Plan Nacional de I+D en 1995.*

The role of Structural Funds

The contributions of European Union Structural Funds to R&D and technological development are organised through the Community Support Framework and involve several different funds: the European Fund for Regional Development (FEDER), the European Fund for Agricultural Orientation (FEOGA) and the European Social Fund (FSE). FEDER makes the most important contribution to innovation and technological issues.

From 1986 to 1994, the Structural Funds related to R&D activities were devoted to the creation of physical scientific infrastructures. In the period from 1994 to 1999, the CDTI was charged with the approval of the European Commission, of managing the subventions, as a global allocation, to foster industrial technological development in the Objective 1 regions. The main goals of this action were:

TABLE 11.7 – Percentage distribution of R&D expenditure by sectors of execution and Autonomous Regions (1993)

Autonomous regions	Public sector	Higher education	Business
Andalucia	18.1	60.1	21.8
Aragón	28.0	36.4	35.6
Asturias	17.6	54.6	27.8
Baleares	29.3	60.9	9.8
Canarias	31.6	66.4	2.0
Cantabria	16.1	55.7	28.2
Castilla – La Mancha	17.1	27.9	55.0
Castilla León	4.8	54.0	41.1
Cataluña	10.1	26.0	63.9
Comunida Valenciana	10.7	59.6	29.7
Extramadura	35.2	57.7	7.1
Galicia	20.1	51.1	28.8
Madrid	33.4	13.1	53.5
Murcia	22.0	55.4	22.6
Navarra	2.2	47.1	50.7
País Vasco	2.5	19.9	77.6
Rioja	31.8	16.3	51.9
España	20.2	31.6	48.2

Source: INE, *Memoria de Actividades del Plan Nacional de I+D en 1995.*

– To increase the number of innovative firms in order to create a modern industrial infrastructure.
– To improve the technological base of the most innovative firms.
– To foster technological services to support a favourable climate for innovation (such as diffusion, information and technology transfer).

The distribution of technological projects in Objective 1 regions amounted to a total of 73 000 million PTA for 250 projects, as follows:

Andalucía	31	Castilla-León	23
Asturias	36	Extremadura	5
Canarias	1	Galicia	17
Cantabria	9	Murcia	15
Castilla-La Mancha	28	Com. Valenciana	85

TABLE 11.8 – Funds allocated to the Spanish Autonomous Regions by the different actions of the R&D National Plan

Autonomous regions	Objective 1 region	% of funds from National Plan
Andalucia	Yes	13.6
Aragón	Yes	3.1
Asturias	Yes	1.7
Baleares	No	0.5
Canarias	Yes	1.0
Cantabria	Yes	1.1
Castilla – La Mancha	Yes	0.3
Castilla León	Yes	3.6
Cataluña	No	25.4
Comunida Valenciana	Yes	9.7
Extramadura	Yes	0.2
Galicia	Yes	2.4
Madrid	No	28.0
Murcia	Yes	2.4
Navarra	No	2.2
País Vasco	No	4.7
Rioja	No	0.2

Source: *Memoria de Actividades del Plan Nacional de I+D en 1997*, and Informe COTEC, 1999.
Note: Total funds allocated in 1997: 16403 million PTA.

3.2 – The National Plan and the co-ordination of R&D activities between regions and the state

Prior to the Law for Science and the first National R&D Plan, the strategies of the governments of the Autonomous Regions differed according to their historical traditions and to the political parties governing each region.

Cataluña was governed by the nationalist coalition *Convergencia i Unió* who appealed to the Constitutional Court against articles of the Law for Science, that were impeding the transfer of research organisations and units (specially from CSIC) from the state to the Autonomous Communities.

The Basque Country was governed by a coalition of the Basque Nationalist Party and the Spanish Socialist Workers Party, who controlled the Department of Industry. The Basque Government gave critical support to the National Plan and continued to promote an industrially-driven policy within the region.

Andalucia was governed by the Socialist Party and strongly supported the National Plan. It developed a Regional Plan (*Plan Andaluz de Investigación* 1990-1992), which, in its first version, was inspired by the National Plan through essentially research-oriented objectives.

Galicia, Aragón, Valencia and Madrid were all governed, at that time, by the Socialist Party and were highly supportive of the National Plan. Progressively, the majority parties in government in these autonomous communities changed and began to develop their own strategies, which were essentially research-driven and were related to the research needs of the local universities.

In summary, the issue of shared responsibilities has been the cause of significant conflict in Spain, in spite of the limited significance of R&D policies. It was finally resolved by the Constitutional Court, which stated very clearly that R&D responsibilities are shared by the State and Regional Governments. The Law for Science established a *Consejo General de la Ciencia y la Tecnología* (General Council for Science and Technology) to co-ordinate the R&D initiatives of the State and the different Autonomous Communities, through exchange of information, written approval of the National Plan's priorities and incorporation of themes proposed by the regions into the national priorities as Regional Programmes (though national in scope). As for the co-ordination of activities between the State and the Autonomous Regions, there is little apparent improvement, as can be seen from the *Memoria de Actividades del Plan* (Report on the Plan's Activities). Apart from a breakdown by autonomous communities of all activities funded under the National Plan, a practice established from 1990 onwards, in the 1995 Annual Report there are only two pages (out of a total of over 186) referring to Regional Programmes incorporated into the National Plan. Of only two such programmes, one, on pure chemistry, was proposed by Cataluña, and shows an uneven distribution with Cataluña receiving between 75% and 90% of the funds allocated to that programme that can be identified as a specific programme for that region rather than a national programme managed from Cataluña. On the other hand, the General Council has been virtually inoperative, as reflected by the lack of reports on its activities and the relevance of its commitments.

The general trend in the distribution of resources from the National Plan for R&D is a move towards redistribution between regions (see Table 11.8

as an example with the data from 1997). In general three types of situation can be observed:

- Regions increasing their share: Andalucía, Aragón, Canarias, Cantabria, Com. Valenciana, Murcia (all Objective 1 regions) and Cataluña and Navarra (non-Objective 1).
- Regions remaining stable: Asturias, Galicia (Objective 1), Baleares and La Rioja (non-Objective 1).
- Regions decreasing their share: Basque Country and Madrid (non-Objective 1 region), Castilla-León, Castilla-La Mancha, Extremadura (Objective 1 regions).

3.3 – Evolution of the regional R&D systems

Some of the 17 Autonomous Regions have established their own bodies or even specific Science Laws for the planning, management and implementation of R&D activities. There is a trend towards transferring research organisations and units from the state to the autonomous communities, while keeping the roles of planning, co-ordination and international relations centralised.

From the information gathered, it can be concluded that there are several regional research and innovation systems in Spain, integrated into a national system, still in search of its real identity (the Matriuska or Russian doll metaphor). The most relevant types in this complex arena are the following:

- The system of the Basque Country (and of Navarra), which is greatly autochthonous in nature, is supported by a strong Regional Government, usually directed by the Basque Nationalist Party, and based on the region's industrial strengths such as metal mechanics, machine-tools and agro-food.
- The systems of Cataluña and Madrid, which are well balanced, relying on national support and with increasing intervention by the Regional Governments.
- The systems of Andalucía and Comunidad Valenciana are based on relevant public and university sectors, with increasing industrial and technological involvement.
- The "catch-up" regions (such as Asturias, Castilla-La Mancha, Castilla-León and Canarias) aiming to draw benefits from any valuable initiative. The situation of Castilla-La Mancha is particularly worth noting since this is the only Objective 1 region of Spain where the business expenditure in R&D surpasses 50% (68.4% in 1997).

4. ORGANISATION OF PUBLIC SECTOR RESEARCH: EVOLUTIVE TRENDS

Public sector research in Spain comprises the Government and the Higher Education subsectors. Together they shared around 53 % of total R&D expenditures in 1995 and 1996 (21% for the Government and 32% for Higher Education). In 1997, this percentage was reduced to a figure around 50%, with a slight reduction in the government contribution.

4.1 – The government sub-sector

In spite of the low profile of science and technology in Spain after the Civil War, the majority of the public research centres that exist today were created or reorganised during the 1940s (Table 11.1). However, the Spanish Government subsector is not homogeneous, as it brings together sectoral centres or institutes performing research or development, closely linked to the interests of the economic or industrial sector, and the *Consejo Superior de Investigaciones Científicas,* a multifaceted organisation devoted to the production (and to some extent) application of knowledge in a wide set of disciplines and interdisciplinary fields.

In the years following their establishment, government laboratories operated on tiny annual budgets, which were dedicated mainly to salaries. The impetus given to science and technology in the late 1960s by the Plans for Development represented an important increase in their income.

During the preparation of the Law of Science, the diagnostic analysis identified more than 20 centres ostensibly involved in public research and technical development activities (Muñoz and Ornia, 1986).

The screening process that followed the preparation of the Law reduced these to ten (those listed in Table 11.1 plus the Astrophysical Institute of the Canary Islands) with the status of *Organismos Públicos de Investigación* (OPIs).

The path followed since 1986 has shown the difficulties experienced in reaching this common status, and the author is particularly sceptical about the possibilities. Constant efforts are being made to promote co-ordination between them. A recent example is a Decree (18 April 1997) establishing a “Committee for the Functional Co-ordination of the Autonomous Organisms of Research and Experimentation”, which included the nine centres, whose main structural and functional characteristics are outlined in Table 11.9.

TABLE 11.9 – Main characteristics of the government sub-sector laboratories in Spain (OPI)

Gov. labs	Employees	Function	Fields	Self-funding	Homologous in Europe
CSIC	7 000	Basic and applied research	All multi-disciplinary	High	CNRS (Fr) MPG (Ge)
CIEMAT	2 000	Applied research, development some basic	Nuclear and alternative energies, biology	High	CEA (Fr) KfK (Ge)
ITGE	600	Applied research, development	Geology, mining, water resources	Low	BRGM (Fr) BIRG (Ge)
INTA	1 500	Applied res., development, quality control	Engines, satellites, space	Medium	ONERA (Fr) DLR (Ge)
IEO	500 – 600	Applied res., development, international representation	Fisheries, sea, aquaculture	Medium – Low	IFREMER (Fr) BfF (Ge)
INIA	400 central (1)	Basic and applied research, development	Agricultural sciences, agronomy, plant biology	Medium – Low	INRA (Fr) FAL (Ge)
CEDEX	400	Development	Public works, civil engin-eering	Low	LCPC (Fr) BASt (Ge)
ISCII	1 000	Basic and applied research	Biomedecine, health	Medium – Low	INSERM (Fr) GSF (Ge)
CEHIPAR (2)	Few hundred	Applied research, development	Ship dyna-mics, fluid dynamics	?	BEC (Fr), HSV gmbh, VWS (Ge)

Source: Own elaboration; taken from Informe COTEC, 1999. *Tecnología e Innovación en España.*
Notes: (1) Plus regional centres under Regional Authorities. (2) Canal Experiencias Hidrodinámicas.

4.2 – The Consejo Superior de Investigaciones Científicas (CSIC)

The CSIC or Spanish Research Council is the most significant public organisation involved in research. As mentioned above, the CSIC became instrumental in the establishment of well-equipped research institutes and in developing a professional career for researchers within the Administration. Both objectives were not easy to introduce and to maintain within the constraints of the conventional Administration rules.

One operational problem of the CSIC that followed its evolutionary trajectory was the fragmentation of the in-house research activities. At the beginning of the 1980s, there were around 1200 permanent researchers distributed among almost 200 institutes, in a range of fields of activity from humanities to engineering, passing through physics, chemistry, biology, agrobiology and geology.

Another characteristic of CSIC that became more problematic during the transition to democracy was the high degree of geographic concentration of the research centres: 40% of the institutes were located in Madrid, and the rest were scattered over the Spanish territory in 11 of the 17 Autonomous Regions. Outside of Madrid, Cataluña and Andalucía share the largest number of institutes, with percentages slightly exceeding 10%. In contrast, the Basque Country is one of the regions where there are no CSIC centres. After a process of evaluation in the mid-1980s and early 1990s, the number of institutes was reduced to around 90 (1995 data). This reduction was accompanied by a certain amount of redistribution, with regard to the regional dimension and to collaboration with other institutions. Of the 91 CSIC institutes recorded in 1995, 18 were joint centres sponsored by CSIC together with universities and Regional Governments.

After a period of stagnation during the 1970s, the number of CSIC personnel increased during the mid-1980s and the early 1990s with a further period of semi-stagnation following the economic crisis of the 1990s. In 1995, there were 6400 permanent staff members in CSIC, distributed as follows: 1830 scientific staff or permanent researchers (all holding doctoral degrees); 1801 research assistants, holding different educational or professional degrees; 1825 trainees, doctoral and postdoctoral fellows; and 106 postdoctoral contract researchers, associated with specific research projects.

CSIC is a public research organisation, which resembles the French CNRS more than the Anglo-Saxon Councils, although important differences do exist between both organisations. The CSIC personnel is not committed to teaching activities, although a significant number participate as teachers in postgraduate courses or as supervisors of doctoral theses (a university professor is always required as a guarantor of each thesis). This collaboration is particularly significant in the case of universities, which are partners with CSIC in joint research centres or institutes. Unlike the CNRS, CSIC cannot fund extramural research in universities, nor can it act as guarantor of the

TABLE 11.10 a & b – Breakdown of CSIC research funding (1995)

1995 total budget of CSIC (PTA 55.3 billion)	%
Ministry of Education and Science	62%
Own resources	38%
Total	100%

1995 ventilation of CSIC own resources (PTA 21 billion)	%
Research contracts with firms and private entities	10%
Funds from National Plan, regional funding agencies and EU Framework Programme	64%
Resources from previous fiscal years and patrimonial earnings	23%
Other sources	3%
Total	100%

Source: *Memoria*, CSIC, 1995.

research quality of the universities. Both institutions are required to search for funding from national or regional agencies or from the private sector. In general, a joint venture between CSIC and universities is considered as an asset for getting funds.

Although the CSIC has a long tradition of collaboration with universities, this is not without problems because the CSIC personnel consider themselves to have lower status than the university professors, even during the periods of CSIC pre-eminence over universities. Since the universities took the lead over CSIC in the 1970s, CSIC employees have been struggling to make their status and that of the organisation comparable to those of the universities and their staff.

The funding of CSIC has been very irregular and based in many cases on the political support lent by certain individuals rather than on a general political strategy. The CSIC never played the role of a central public agency for funding research as was originally intended. On the other hand, after the re-emergence of research activities within the universities during the 1960s, the CSIC was thrown into crises that have affected its funding trajectory. The evolution of CSIC funding and the sources for it are illustrated in Tables 11.10a and 11.10b.

Funding from external sources has become essential to CSIC survival in order to cope with the restrictions imposed by the State General Budget during recent years. The incorporation of residual funds from previous budgets represents the recognition of the commercial characteristics of the organisation, able to budget on the basis of more than one year.

4.3 – The Impact of Military research

Spain is not a military power. Even during the Franco regime the military did not control or influence the research domain, except for certain strategic areas of industrial relevance and technological importance, such as energy, particularly nuclear energy, and materials assessments related to aeronautical or space research and shipbuilding. The Ministry of Defence held the authority over two public research centres: the *Instituto de Técnica Aeroespacial* (INTA, Institute for Aerospace Technology) and the *Canal de Experiencias Hidrodinámicas del Pardo* (Canal for Hydrodynamics Experimentation, located in El Pardo, near Madrid).

During the transition to democracy, military influence declined until the Socialist Party came to power in 1982, when an effort was made to modernise the Spanish military with two main goals: to win the support of the military officers and to help in the industrial development of Spain. The Ministry of Defence participated in the framework implemented under the Law for Science: INTA obtained the status of a Public Research Organisation (OPI), under the terms of the Law, whereas the *Canal de Experiencias Hidrodinámicas* retained an ambiguous position. Under the Law for Science, CICYT was charged with the task of co-ordinating the programme of purchase of military equipment together with the Ministry of Defence. Through programmes of development of weapons systems, the Ministry of Defence was increasing its share of the State General Budget, although these expenses had not been included in the R&D statistics.

The current Spanish Government has decided to include spending on military projects aimed to develop weapons systems through international co-operation schemes in the overall R&D budget. This includes money for developing military technologies, including the production of aircraft, frigates and tanks for the Ministry of Defence, at a cost of 1.42 billion US dollars.[1]

4.4 – Higher Education subsector

The low research profile of the Spanish universities that characterised academia during the first half of the 20th century began to change in the 1960s when the strategic political initiatives of the Development Plans began to provide resources for infrastructure and for the promotion of research at

1. See report in *Nature*, vol. 395, 8 October 1998, p. 535.

Spanish universities. The balance between teaching and research at universities changed significantly, mainly due to the spectacular rise in the number of university students: between 1971-1972 and 1984-1985, their number increased from 350 000 to 750 000. Faculties more than doubled during that same period, allowing the lecturer–student ratio to remain practically constant.

During the democratic transition, the reform of the universities became a priority on the political agenda, though the fights between actors' interests and the weaknesses of the majority political party, UCD, led to an impasse. The situation changed with the arrival in government of the Spanish Socialist Workers Party (PSOE) after their general election victory in 1982. The University Reform Law, enacted in 1983, granted self-governance to Spanish universities (right to autonomy and to be governed by specific statutes enacted by each university) and the creation of public universities continued to be promoted.

In 1985-1986 there were 31 public universities and four private universities. By 1994, this figure had risen to 45 public universities and six private universities. The number of students increased to practically double the university population in 1987. The number of lecturers also increased, although not as quickly, with the lecturer–student ratio rising from 1:20 in 1987-1988 to 1:30 at present.

The evolution of universities that followed the implementation of the University Reform Law was marked by the increasing role of Regional Governments in their governance and funding. The Law established a Universities Council with two committees: one comprising the Rectors, the other composed of representatives of Regional Governments. They have to agree on matters such as: the creation of new universities, the establishment of curricula and the number of students, although conflicts between the two committees have arisen in many instances. It is, however, noteworthy that the evolution of research within universities has followed an independent path, moderated by science and technology policies. Also along this path, the National R&D Plan has exerted a highly positive influence in terms of funding.

Another important outcome of the independence granted to universities by the University Reform Law is the progressive transfer of political responsibility for public universities to Regional Governments. This process began in the so-called historical regions (Cataluña, the Basque Country, Galicia and Andalucía) and some fringe regions such as the Community of Valencia and the Canary Islands. This transfer of authority is in the process of expansion to all other regions.

It is also important to note that the transfer of responsibility to regions includes the transfer of public budgets, which refer primarily to personnel and the maintenance of the installations. On the other hand, regions have used their constitutional right to promote research by providing their own resources, which, of course, vary from one region to another, in line with their economic characteristics and demography. These policies have had very diverse results in universities as can be seen from various indicators, including: the proportion of permanent teaching staff; the capacity to raise funds for researchers and the ability to link academic research and socio-economic interests. An exercise of evaluation of universities has been undertaken since the mid-1990s and in spite of some difficulties in the acceptance and use of the results of this exercise it can be turned into a valuable instrument for political decision and management on the Higher Education sub-sector.

4.5 – *Links between public research and industry*

Creating linkages between research and industry has been one of the main policy objectives in the process of modernisation of Spanish science and technology taking place during the 1980s. It has also been one of the goals of the Regional Governments. Several mechanisms have been used to promote these links and, in particular, the following can be mentioned:

- Financial incentives for university staff by means of research contracts with firms. The University Reform Law allows them to increase their total remuneration by including part of the income from these contracts, up to a total maximum level equivalent to double the value of their salaries.
- Programmes to foster the exchange of researchers between trained staff in the universities – and other centres within the Public System of Research (PSR) as well – and companies.
- Programmes to promote the transfer of the results of research from universities (and other PSR centres) to businesses. This programme responds to the acronym PETRI from its name in Spanish (*Programa de Transferencia de Resultados de Investigación*).
- Creation in the universities (and other public research centres, in particular the CSIC) of Offices for Technology Transfer and for the Exploitation of Research (OTRIs).

Spanish universities have responded positively in terms of using these mechanisms for collaboration with industry but the end result has been quite limited.

Innovative Spanish companies have also attempted to collaborate, and multiply contacts, with universities, although in certain cases this has been done purely for image purposes. In any case, the increase in co-operation has meant an increase in patents. Spain ranks fifth among the EU member states with regard to the number of patents registered in the US. Sectors in which patents have been registered include the four which are most active in R&D: the chemical industry, the electrical equipment, the electronic equipment and the automobile sectors. In addition, Spanish (and European) patents have been registered in three sectors of strategic interest: biotechnology-pharmaceuticals, aeronautics and the agro-food sector.

On the other hand, industry has been involved to a very limited extent in setting priorities and in forecasting. The Spanish system of innovation seems to be the object of pushes from the supply side rather than of pulls from the demand side. Several evaluative exercises of programmes of the National R&D Plan through surveys addressed to the project leaders point to a divergence of interests between the sub-sectors of the public system of research and the business sector (Espinosa de los Monteros et al., 1994, 1995a and b, 1996a and b).

5. INNOVATION POLICIES: MAIN ACTORS

5.1 – Innovation policies

After the "technocratic" impetus of the 1960s to incorporate Spain into modern trends, the responsibility for innovation policies has rested, and still remains, under the competence of the Ministry of Industry and Energy. During the first years of the Socialist Government, the collaboration between the Ministry of Industry and the Ministry of Education was decisive for the preparation and launching of the Law for Science and the first National R&D Plan. The collaboration was based on personal ties between the political actors and on the image of a common project of "modernisation" for Spain. However, as the actors changed and the idea of a common project weakened, the situation began to deteriorate and the tight coupling between research and technology was lost, although the formal agreements were kept.

The Ministry of Industry and Energy was the funding agency for innovation and industrial research, either directly or through the Centre for Industrial Technological Development (CDTI), which is the main agency for the promotion of industrial innovation in relation with the resources of the

National Plan. The actions of the Ministry are implemented by the General Directorate for Technological Policy, the name of which has been changed by different Governments, by means of instruments linked to the Plan for Technology and Industrial Action (PATI) or the more recent ATYCA (Technology and Quality Action). A reorganisation in 1997put innovation and technological policies under the General Directorate for Industry, whereas a General Directorate for the Information Society has been newly established. The creation of the Ministry of Science and Technology with the concomitant disappearance of the Ministry of Industry and Energy has introduced a new reorganisation which is moving in the same direction.

5.2 – Innovation and the business sector

The R&D expenditure of the business sector, including public and private enterprises and the research groups associated with firms, amounted to 127 000 million PTA in 1987 (representing 0.35% of GDP) and to 266000 million PTA in 1993, accounting for 0.43% of GDP. The expenditure in current and constant PTA has declined during the period 1991-1994. As shown above, the loss of R&D effort in the enterprises has been accompanied by a decrease in the personnel devoted to these activities (Figure 11.3).

There were 1140 firms registered in 1987 as active in R&D and 1874 in 1993. Most of the firms are small and medium-sized enterprises: more than 80% had less than 500 workers (see Table 11.11) and the trend has been to

TABLE 11.11 – Evolution of R&D and innovative firms by size and resources

1987	Number of firms	%	% of R&D expenditure	% of R&D personnel
Less than 100	532	46.6	16.1	21.2
100 – 499	393	34.5	23.4	27.9
500 – 999	91	8.0	10.4	10.1
1000 and over	124	10.9	50.1	40.8
1993	**Number of firms**	**%**	**% of R&D expenditure**	**% of R&D personnel**
Less than 100	1 089	58.1	18.6	24.5
100 – 499	558	29.8	23.6	26.4
500 – 999	115	6.1	14.8	16.9
1000 and over	112	6.0	43.0	32.2

Source: Adapted from *Memoria Actividades del Plan Nacional*, 1995.

increase this percentage. These SMEs were spending 40% of the total R&D expenditure and accounted for around 50% of the R&D personnel (about 80% of the firms employ less than one researcher in full-time equivalent while only 3.5% have more than 25 researchers FTE). This fact highlights that there are no large innovative business sector or high-tech companies in Spain, most of the innovation effort being performed by firms in the traditional sectors, such as: machine-tools (Basque Country), ceramics (Comunida Valenciana), textiles and pharmaceuticals (Cataluña and Madrid), and the automotive sector (Cataluña, Comunida Valenciana, Aragón, Madrid).

Recent data show a reversal in the trend as the expenditure of the business sector has increased in current terms during 1995 and 1996, although the level in constant PTA still remained lower than that of 1991.

The innovation survey has shown a decrease in the percentage of innovative firms from 1994 (10.7%) to 1996 (9.6%), but this was paralleled by a very important increase in the percentage of innovative firms performing R&D activities (24.9% in 1994; 32.9% in 1996).

The most intensive R&D performing sectors are: pharmaceuticals, electronic components including communication technologies, aerospace, and other transport materials, automobiles and chemicals.

The proportion of GDP devoted by the business sector to innovation amounts to 1.2% whereas in the European Union this figure reaches 2.5%, while the proportion devoted by the same sector to R&D activities adds up to 1.2% in the EU and only 0.4% in Spain. This is a new indicator of "laggard" character of the innovation sector in Spain.

6. INTERNATIONALISATION AND THE RELEVANCE OF EU PROGRAMMES

Spain has used, during and after the period of Franco's dictatorship, science and technology co-operation as an instrument of foreign policy to overcome international isolation. This has driven the Spanish scientific community to be very active in those aspects of science policy and this path has been reinforced after the democratic transition and, particularly, after the integration of Spain into the European Community in 1986. Spanish actors in the R&D system have been actively involved in supporting and implementing a common European science and technology policy. The European Framework Programme is seen as far more influential than its size would suggest. It boxes above its weight, touching upon and influencing every national research programme in Europe.

The debate on the influence of the European Framework R&D programme centres on the question of whether it should be used for supporting research strengths (the countries who are most influential) or for compensating for the weaknesses (those who are most influenced). There is no simple answer but perhaps one can be found in a synergy between the Framework Programme and the Structural Funds whose relevance in Objective 1 regions is obvious.

On the other hand, the trend in the number of participants in the Second and Third Research Framework Programmes show that France and Spain have experienced the highest growth (Figure 11.5). On the other hand, the distribution of funding by type of participant, indicates an important decrease in the share of large firms and a concomitant proportional increase for the other three main sub-sectors: higher education, research centres and small and medium-sized enterprises, a situation that fits with the characteristics of the Spanish System of Research.

FIGURE 11.5 – Number of participants in the Second and Third Research Framework Programmes

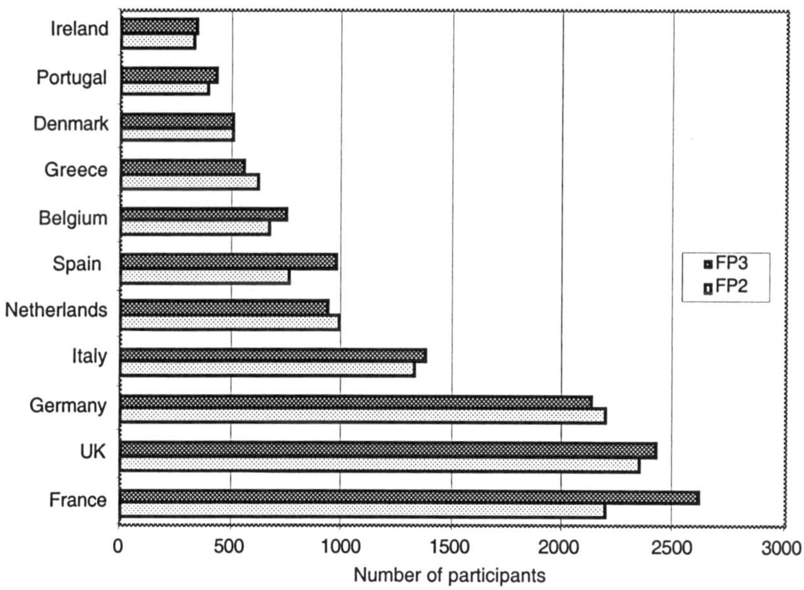

Source: Euro Technology, April 1997.

Recent data gathered regarding the funds obtained from the Fourth Framework Programme by Spanish participants show figures ranging between 11% of the International Co-operation (INCO) Programme with an average of around 6%. The characteristic feature of European Union funding of R&D activities in Spain is that of complementarity to national (and regional) funds.

7. A PROSPECTIVE VIEW

The new impetus for placing research and technological development among the priorities of the political agenda will be monitored with great interest. It has raised new hopes among actors in the Spanish research system, but these are intertwined with fears and doubts about its future.

The general discourse underlying the effort of extra spending on research and development is to promote competition between companies and stimulate innovation to correct the traditional backwardness of Spain in the development of technologies and in industrial innovation. The deficit in the technological balance is constant throughout the economic history of Spain.

The present discourse for Spanish research system matches well with the one pervading the global world in making R&D expenses more accountable to social and economical needs. However, this discourse may be entangled with the attitudes and position of the two main communities of actors of the research system: on one hand, the researchers from the public sector who have been little used to think of the applications of scientific knowledge; on the other hand, the industrialists who have been mainly concerned with short-term actions and are little aware of the strategic value of science and technology.

The priorities established and geared through four-year programmes will be changed as regards the former National R&D Plans. Energy research, biotechnology, archaeology, waste treatment, biomedicine and human health, and information technology emerge as the new priorities, with a mix of scientific strengths and strategic demands as parameters. The reduction in the number of priorities is an interesting fact that should be monitored with special focus with regard to those two parameters.

The current reorganisation at government level may foster the co-ordination within the State Government and the General State Budget, but the co-ordination between the national objectives and the regional ones will continue to experience difficulties in view of the existing heterogeneity among regional policies, their goals and the relative differences in the relevance of actors and interests.

International programmes and activities will continue to play a significant role in the design and implementation of research and innovation of the Spanish research system.

It is plausible to think that a change of strategies by the Spanish scientists producing knowledge will take place in order to overcome their lack of influence – recognition of excellence – in the Science World context.

ACKNOWLEDGEMENT

This chapter has been prepared following the framework proposed by the editors. It has been possible thanks to the collaboration of various colleagues at IESA (Juan Espinosa de los Monteros, María J. Santesmases and Víctor Díaz) and to the continuous dialogue on these issues with Luis Sanz Menéndez (IESA-CSIC) and Fernando Aldana and Jose Luis Puerta (ministry of the Government Presidency). The author's work has been supported by grants from CICYT (SEC-93-0688) and the European Commission (TSER Project No. SOE1-CT96-1036). I am indebted to Maria Angeles Soriano and Maria Carmen Montalvillo for their valuable secretarial support.

REFERENCES

ARBOR, 1990, *El CSIC: una visión retrospectiva*, Madrid.

Comisión Interministerial de Ciencia y Tecnología (CICYT), 1988, *Plan Nacional de Investigación Científica y Desarrollo Tecnológico*, Ministerio de Educación y Ciencia, Centro de Publicaciones Ministerio de Educación y Ciencia, Madrid.

Comisión Interministerial de Ciencia y Tecnología (1997), *Memoria de Actividades del Plan Nacional de I+D en 1995*, Madrid.

Espinosa de los Monteros, J., F. Martínez, M.A. Toribio and E. Muñoz, 1994, *El Programa Nacional de Nuevos Materiales en el período 1988-1992. Su evaluación mediante una metodología dual*, Documento de Trabajo 94-10, Madrid: IESA-CSIC.

Espinosa de los Monteros, J., F. Martínez, M.A. Toribio, E. Muñoz and V. Larraga, 1995a, *El Programa Nacional de la Salud durante el período 1989-1993. Una evaluación mediante metodología dual*, Documento de Trabajo 95-09, Madrid: IESA-CSIC.

Espinosa de los Monteros, J., F. Martínez, M.A. Toribio, E. Muñoz and V. Larraga, 1995b, *El Programa Nacional de Investigación y Desarrollo Farmacéutico durante el período 1988-1993*, Documento de Trabajo 95-08, Madrid: IESA-CSIC.

Espinosa de los Monteros, J., O. Mirabal, E. Muñoz and M.A.Toribio, 1996a, *Recursos Humanos y Política Científica. El caso del Programa Nacional de Nuevos Materiales*, Documento de Trabajo 96-01, Madrid: IESA-CSIC.

Espinosa de los Monteros, J., V. Larraga and E. Muñoz, 1996b, Lessons from an Evaluation of Spanish Public-Sector Biomedical Research, *Research Evaluation*, Vol. 6, 1, pp. 43-51.

Fundación COTEC, 1999, *Informe COTEC 1999. Tecnología e Innovación en España, Madrid.*

Muñoz, E., 1998, La investigación en la España de hoy: mapa de acciones y constricciones y su reflejo en el paisaje de 1997, *ASCLEPIO, Revista de Historia de la Medicina y de la Ciencia*, Vol. 50, pp. 7-29.

Muñoz, E. and F. Ornia, 1986, *Ciencia y Tecnología: Una Oportunidad para España*, Ministerio de Educación y Ciencia, Madrid: Aguilar S.A. de Ediciones.

Nieto, A., 1982, Las dificultades de investigar en España: El CSIC, in A. Nieto et al., *Apuntes para una política científica. Dos años de investigación en el CSIC: 1980-1982*, Madrid: Consejo Superior de Investigaciones Científicas, pp. 41-63.

Sánchez Ron, J.M., 1992, Política científica e ideología: Albareda y los primeros años del Consejo Superior de Investigaciones Científicas, *Boletín de la Institución Libre de Enseñanza*, No. 14, pp. 53-74.

Santesmases, M.J. and E. Muñoz, 1993, Las primeras décadas del Consejo Superior de Investigaciones Científicas: una introducción a la política científica del régimen franquista, *Boletín de la Institución Libre de Enseñanza*, No. 16, pp. 73-94.

Santesmases, M.J. and E. Muñoz, 1997, *Establecimiento de la bioquímica y la biología molecular en España*, Madrid: Fundación Ramón Areces.

Sanz Menéndez, L., 1995, Research Actors and the State: Research Evaluation and Evaluation of Science and Technologies Policies in Spain, *Research Evaluation*, Vol. 5, 1, pp. 79-88.

Sanz Menéndez, L., 1997, *Estado, ciencia y tecnología en España: 1939-1997*, Madrid: Alianza Editorial.

CHAPTER 12

SCIENCE, TECHNOLOGY AND INNOVATION POLICY IN ITALY

ALBERTO SILVANI AND GIORGIO SIRILLI

1. INTRODUCTION

This chapter deals with the Italian science and technology system, the policies set in place by the Government and some relevant developments which may be expected in the coming years.

In the first section a profile of the major R&D players in the country – the largest public research agencies and the university system – is provided.

The second section gives a quantitative picture of the R&D and innovative activities using the traditional indicators – R&D, patents, scientific publications, high-tech products – as well as other data coming from a special survey on technological innovation.

The third section deals with the public policies put in place over the last few decades, in particular since the institution of the Ministry of Scientific and Technological Research (MURST) in 1989, the way priorities are set, and the evolution of the institutional setting.

The fourth section addresses two relevant issues: the changes introduced in university research and teaching and the shifting of the balance from direct to indirect measures for the support of technological innovation in industry.

The final section analyses the prospects of various issues: the increasing interest of public opinion for S&T, the insufficiency of R&D expenditure in the country, the adequacy of public policies for science and technology, the role of firms, of universities, of public research agencies and of private non-profit organisations, the less-developed area of *Mezzogiorno*, R&D evaluation, the financial system, the role of Europe and of regional governments.

The appendix deals with a specific problem: R&D and innovation in the defence sector.

2. THE NATIONAL R&D AND INNOVATION SYSTEM

In the following section the various components of the Italian R&D and innovation system are described. A quantitative analysis of the resources available and of results achieved is made on the basis of appropriate indicators.

2.1 – A profile of the national R&D system

The Italian R&D system consists basically of three groups of organisation: universities, public research agencies and institutions, enterprises. Private non-profit organisations at present are rather few but rapidly growing in number and relevance (Sirilli, 1996).

The Italian university system consists of 73 universities, of which 61 are state institutions, 12 free or private universities. Some universities have ancient origins and an illustrious history (the Universities of Bologna and Salerno date back to the beginning of the second millennium); others were established only recently.

The major public research agencies are: the National Research Council (CNR), which is engaged in general research, the National Agency for New Technologies, Energy and Environment (ENEA), which is concerned with energy, environment and technological innovation; the National Institute of Nuclear Physics (INFN), which deals with high-energy physics, the Italian Space Agency (ASI), which supports and co-ordinates research and industrial activities in the space field. Various ministries have an interest in research in specific sectors and run appropriate research bodies: for example, the Ministry of Health uses the Higher Institute of Health (ISS) for medical, pharmacological and environmental research, the Ministry of Communications calls on the Higher Institute for Posts and Telecommunications for research in this field, the Ministry of Industry and Trade runs a network of experimental stations in support of specific productions (paper, glass, and so on), the Ministry of Agricultural Policies supports various experimental institutes.

Research and development in the business enterprise sector is conducted in a continuous and structured manner by some 1500 private and public enterprises, while this number increases significantly if innovating companies carrying occasional and non-structured R&D are included.

These three groups of organisation display a limited degree of interaction with each other; this feature is considered a weak point of the Italian national system of innovation (Malerba, 1993) and various steps have been taken over the years to foster co-operation.

If the Italian R&D system is compared with other advanced countries, two differences stand out: the large role played by the public sector both in supporting and in performing R&D, and the large share of state-owned enterprises in high-tech sectors – even though a privatisation process is under way.

Public Research Agencies

Public research agencies are highly specialised scientific institutions. Most of them are single-discipline agencies that carry out R&D and provide services for various government organisations in their particular field. The agencies fund, plan and manage research programmes aimed at the objectives identified by the Government. They also implement the national sectoral plans and finance and manage the research programmes associated with support activities for agriculture, industry, health services and infrastructure (De Marchi et al., 1998).

In 1998 about 10000 researchers and 11000 technicians were employed by the public research agencies. In the same year, the government appropriations for research agencies amounted to 3774 billion Lira, equivalent to $2328 million. A brief description of the six major research agencies is given below.

National Research Council (CNR)

The National Research Council's institutional aims are the promotion and support of the scientific and technological research through both its research establishments and the financing of research carried out by scientific institutes and individual researchers. CNR, which was set up in 1923, is supervised by the Ministry for Universities and Scientific and Technological Research (MURST) since its institution in 1989. CNR plays a key role in R&D in Italy both because of its size and of its interdisciplinary and multidisciplinary structure. It is responsible for carrying out basic research, mission-oriented research geared to the country's needs, and for the transfer of its research results to industry, services and government.

As the result of the development of CNR over the last 70 years, and in particular of the last three decades, the scientific and technological network is made up of about 300 scientific bodies (157 institutes, 115 study centres and 17 research groups) mainly located in 18 research areas.

Besides conducting research activities, CNR carries out studies for and provides advice to the Government, grants research contracts and scholarships for training in research; the resources involved in this promotion role represent some 8% of the overall CNR budget.

In the mid-1970s CNR launched its Mission-Oriented Projects involving all sections of the Italian scientific and technological system (university, industry, public research institutions); these are five-year research projects aimed at achieving important national economic and social objectives. The results of these projects – which embrace a variety of sectors ranging from advanced technology to town-planning and environment, from human health to agricultural and food resources, the economy and public services – are to be seen in more than 1400 "products" that are transferable and can be used in the production of goods and services, and in more than 6300 scientific articles published in international and national scientific journals. The share of the CNR budget devoted to Mission-Oriented Projects has been about 10% of the total in the last few years, but more recently it has decreased because of budget cuts (CNR, 1998).

CNR has for a long time (from 1923 to 1989) been in charge of the co-ordination and the support of public R&D in Italy. In the 1990s its co-ordination function has been progressively moved to MURST (CNR, 1999).

The budget of CNR totalled 1306 billion Lira in 1998. Apart from the Government, CNR receives funds from the European Commission and other national and international organisations. In 1998, CNR employed more than 7430 scientific and technical staff (of whom 3574 researchers), supported by 1154 staff dealing with managerial and administrative issues.

National Agency for Energy and the Environment (ENEA)

In the early 1950s public research in the field of nuclear energy was carried out by the National Committee for Nuclear Energy which in 1982 was transformed into ENEA. The Agency is a public-law body whose role in the field of national energy policy is to promote the development and expertise of Italian industry while protecting human health and the environment; it operates in conjunction with other energy agencies and is one of the instruments for the implementation of national energy policy. It is responsible for supervising and monitoring nuclear safety and protecting the health of workers and of the population at large. After the national referendum of 1987, which stopped the nuclear fission activities in the country, ENEA's mission has been enlarged to cover also environmental research and the promotion of innovation in industry and in the services. In 1991 a new law established that ENEA's routine activities should be covered by the ordinary financial transfer from government, while targeted research and technology transfer activities should be financed by contracts placed to ENEA by various ministries and private and public "customers". ENEA employs 3536 staff in a number of interdisciplinary research centres distributed around the country. Its budget for 1998 was around 905 billion Lira.

The Agency's programmes are developed in accordance with the guidelines laid down by the Government and Parliament in the national energy plan. It also manages national programmes involving several agencies, as in the case of the National Antarctic Programme.

National Institute of Nuclear Physics

INFN, which was founded in 1951, was recognised as a public-law body with its own budget in 1971. Its objective is to carry out both experimental and theoretical basic research into fundamental interactions, in particular in the field of particle and nuclear physics. This entails an ever greater use of highly advanced technologies, which for the most part the Institute develops itself by means of parallel R&D programmes in technological fields.

INFN's activities are organised on the basis of five-year plans, which are drawn up by the agency itself and submitted to the Government for approval by MURST. The 1994-1999 plan is currently underway.

INFN has four national laboratories (Frascati, Legnaro, Laboratori del Sud and Gran Sasso), 19 sections, seven associated groups and one national centre, and a central administration. Its activities are mostly carried out within international projects and institutions, with CERN playing a key role.

A close co-operation is in place between the Institute and the universities. This has been an integral part of INFN policy since its foundation. The Institute therefore operates via both its own personnel and university staff.

In 1998 INFN employed 1706 staff, of whom 735 were researchers, 707 technicians and 255 administrators. Some 2800 university lecturers and researchers and 300 technicians were associated with INFN programmes. The Institute's total budget for 1998 amounted to 571 billion Lira.

National Institute of Health

The ISS, which is the technical and scientific arm of the Ministry of Health and the National Health Service, was established in 1934. It is currently governed by Acts of 1973 and 1978 that allow it greater organisational flexibility than other state agencies. The Institute comprises 20 laboratories, and eight technical departments. In 1998 the Institute employed 1924 people. The technical and scientific personnel consisted of 669 researchers and 1080 technicians. The Institute's financial resources amounted to 281 billion Lira.

National Institute of Statistics

The National Institute of Statistics (ISTAT) drafts the National Statistical Programme, carries out the censuses and other statistical surveys laid down in the Programme, directs, co-ordinates and provides technical assistance for the statistical work carried out by the agencies and institutions belonging to the National Statistical System. It also lays down the nomenclatures and methodologies for classifying and recording demographic, economic and social phenomena, and carries out research on the results of censuses and surveys and on statistics relating to phenomena of national interest. In performing its functions ISTAT can call on the assistance of public and private agencies and companies under a number of contracts and agreements and through the sharing of the capital of the bodies and companies concerned. In 1998, ISTAT employed a total of 2423 staff, of whom 614 were researchers; the Institute's budget amounted to 290 billion Lira.

Italian Space Agency

Italy has been involved in space activities for the last decades. Early efforts were fragmented, but strategic planning was gradually introduced which led to the five-year space plan of 1979. The establishment of the Italian Space Agency (ASI) in 1988 to co-ordinate all space activities was meant to give further impetus to the rationalisation of the sector. ASI is responsible for planning scientific, technological and applications programmes in the space sector as well as to improve the expertise and competitiveness of the Italian aerospace industry. The agency drafts and implements the national space plan, by directly co-ordinating national and international programmes contracted out to industrial firms and harmonising them within the international context. It is also an instrument for Italy to participate in the European Space Agency (ESA) programmes. Space activities entail not only co-ordination by ASI but also the involvement of industry, universities and CNR laboratories. In 1998 the budget of ASI was 852 billion Lira, of which 194 billion were devoted to R&D.

National Institute of Physics of Matter (INFM)

INFM is the most recently established government agency (1995) as a transformation of the pre-existing Inter-university Consortium of Physics of Matter. It carries out interdisciplinary research (microelectronics, new materials, telecommunications) mostly in co-operation with universities and takes part in international activities, in particular to the Trieste and Grenoble synchrotrons. INFM's budget for 1998 was 153 billion Lira. The agency operates mostly through university personnel.

Universities

The resources devoted by MURST to universities are allocated as a block grant through a three-year plan. University research activities are financed from various sources, mostly state funding.

Universities also receive funds from public research agencies, such as CNR, INFN, ENEA and ASI. Further funds come from industry, from private institutions and from international research projects, such as the EUREKA initiative, the EU programmes, and so on.

In 1998 the R&D budget appropriations of universities amounted to 5713 billion Lira. It may be emphasised that these data are calculated according to the international methodology which envisages that estimations are made of the time devoted by university teachers to various activities (teaching, research, administration), and that salaries represent about two-thirds of the R&D expenditure in universities.

In the academic year 1996-1997 the number of university students in Italy was around 1 673 000, of which 1 596 000 enrolled in Laurea courses and 77 000 in university Diploma courses. The former had decreased over the previous year by 1.4% after a long period of continuous growth, while the latter had continued rising at an annual rate of 15%. The total number of students is divided almost exactly between males and females, while in some disciplines the ratio is very unbalanced (i.e. there is a strong prevalence of males in engineering and of females in humanities). The number of university graduates is about 115 000 a year. In the same academic year, the teaching staff of Italian universities was made of some 60 000 staff, of which 32% were researchers, 27% associate professors and 23% full professors.

Since the beginning of the 1960s school attendance in Italy has increased dramatically. Almost the totality of young people (93% in 1996) chooses to continue school beyond the compulsory cycle, anticipating *de facto* the legislation which has recently been adopted, aimed at extending the compulsory education period from eight to 12 years. The ensuing increased number of diploma holders, together with a generalised propensity to continue school attendance, has favoured the progressive increase of enrolment in universities which in the year 1995-1996 reached two-thirds of diploma holders (65.7%). This trend has not, however, produced an increase of the level of achieved education: the Italian university system is, in fact, still characterised by a noticeable waste of resources, not only of an economic nature, both from young people and from families.

Looking at the education path of Italian students, the drop-outs from the schooling system stand out as a preoccupying phenomenon. Out of 1000

BOX 12.1 – Educational path of an ideal cohort of 1000 middle-school certificate holders (1995-1996)

- Out of 1000 pupils (13 years of age), 75 leave education, leaving 925 entering first year of high school.
- 610 will obtain a high school diploma. 146 (24%) will leave education, 63 (10%) will go into vocational training, while 401 (66%) will be enrolled in universities: 28 in diploma courses, and 373 (61%) into Laurea courses.
- 241 (i.e. 60% of those entering university) will leave university, while 160 (40%) will get a university degree: 15 will get a diploma (success rate: 54%) while 145 will be Laurea holders (success rate: 39%).

middle-school leavers (Box 12.1), only 610 achieve a high school diploma and barely 160 a university degree. In fact about 34% of students enrolled in high school do not achieve the diploma, and about 60% of university students never achieve any academic qualification (ISTAT, 1999b).

The number of young people who enter university is definitely remarkable, even though they have to face various difficulties within the educational system. This may be explained by two factors. The vocational courses offered by Regions, who represent the only alternative to university at the third level of education, are an insufficient substitute for the university education both in quantitative and in qualitative terms: only 10% of diploma holders chooses this type of courses.

Secondly, the labour market displays a poor demand for high school diploma holders, and the unemployment rate is higher for diploma holders than for those who have completed the compulsory education only.

The number of high school diploma holders enrolling in university has constantly increased over time up 73% in the academic year 1993-1994. This percentage has later decreased to about 67% both because of an increase of tuition fees, and because of the spread of a feeling that university degrees do not ensure appropriate job opportunities.

In the period 1986-1995 the R&D expenditure of universities has more than doubled in current terms (from 2004 to 4549 billion Lira) and increased by 36.8% in real terms. This increase has been faster in the 1980s, while in the first part of the 1990s the growth has slowed down in connection with the recessive trend of the economy and with the contraction of the state budget. Provisional data for the years 1996 and 1997 show an increase of the expenditure respectively by 2.4 and 5.2% which in real terms amounts to staying at the 1995 level. University scientific personnel amounted in 1999 to 50400, of which 12900 were full professors, 18000 associate professors and 19500 researchers (for the evaluation of full-time equivalence of R&D in universities, professors are accounted for 50% of their time, while researchers are considered full time).

2.2 – *A quantitative profile of the Italian R&D system*

Italy's total R&D expenditure in 1998 was 22 501 billion Lira, equivalent to $13 880 million and to 1.11% of GDP. Data in Table 12.1 show a continuous decrease in real terms of R&D expenditure from 1992 up to 1995, which was linked to the recessionary trends of the early 1990s, and a recovery from

TABLE 12.1 – R&D expenditure in Italy (1990-1998)

Year	Current prices (billion Lira)	Constant prices (billion Lira)	Variation over previous year (constant prices)	R&D/GDP ratio (%)
1991	17 657	16 396	-	1.24
1992	17 958	15 933	-2.8	1.20
1993	17 613	14 971	-6.0	1.14
1994	17 389	14 286	-4.6	1.06
1995	17 864	13 969	-2.2	1.01
1996	19 155	14 270	2.1	1.02
1997 (a)	21 152	15 355	7.6	1.08
1998 (a)	22 501	15 911	3.6	1.11

Note: (a) provisional data.
Source: ISTAT, 1999a.

TABLE 12.2 – R&D expenditure in Italy by performing sector

Performing sector	1994	1995	1996	1997	1998
Government	3 697	3 774	3 827	4 317	4 707
Research agencies	*3 042*	*3 133*	*3 132*	*3 500*	*3 875*
Other public agencies	*655*	*641*	*695*	*817*	*832*
University	4 487	4 549	5 082	5 465	5 713
Business enterprises	9 206	9 451	10 247	11 370	12 081
Total	17 390	17 864	19 156	21 152	22 501

Unit: Billion Lira.
Source: ISTAT, 1999a.

1996 to 1998. The level of expenditure of 1998 was in any case at the same level as that of the beginning of the decade.

During the 1971-1990 period R&D expenditure in Italy almost tripled in constant terms. During these two decades the ratio between R&D expenditure and Gross Domestic Product (GDP) increased from 0.8 to 1.2% while it has since decreased to 1.1%.

Table 12.2 shows the breakdown of R&D expenditure by performing sector. The largest spender is the business enterprise sector with 54% of the total expenditure. This expenditure is carried out by some 1500 firms, mostly of large size (80% of the total is accounted for by firms with more that 500 employees). The second largest performer is university, with a share of 25%, while the government sector accounts for 21%.

Table 12.3 shows that R&D budget appropriations of government agencies for 1998 are concentrated on a limited number of large organisations: CNR, ENEA, the National Institute for Nuclear Physics and the Higher Institute of Health, which run laboratories all across the country, while the budgets of the remaining 25 agencies is on average rather small.

In the period 1971-1996 the number of personnel employed in R&D activities almost doubled from 80 861 to 142 288 units in full-time equivalent. Table 12.4 shows that in 1996 the figure reached 142 288 units and that the government sector employs a higher share of R&D personnel (56%, of which 33% are in universities) than industry. In public R&D agencies, which typically run large experimental facilities, the researcher–technician ratio is lower than that of universities.

The analysis of R&D expenditure by type of activity, i.e. basic research, applied research and experimental development is the following:

TABLE 12.3 – 1998 R&D budget appropriations of government agencies

National Research Council (CNR)	1 306
National Agency for New Technologies, Energy and the Environment (ENEA)	905
National Institute for Nuclear Physics (INFN)	572
National Institute of Statistics (ISTAT)	290
National Institute of Health (ISS)	281
National Institute of Physics of Matter (INFM)	153
Italian Space Agency (ASI)	194
Other agencies	73
Total	3 774

Unit: Billion Lira.
Source: Scarda, 1999.

TABLE 12.4 – R&D personnel in Italy (1996)

Performing sector	Researchers	Technicians	Other occupations	Total
Government	13 637	13 168	5 420	32 225
Research agencies	*10 158*	*11 145*	*4 078*	*25 381*
Other public agencies	*3 479*	*2 023*	*1 342*	*6 844*
University	35 069	11 550	2 529	49 148
Business enterprises	27 735	22 076	11 104	60 915
Total	76 441	46 794	19 053	142 288

Unit: Full-time equivalent.
Source: ISTAT, 1999a.

TABLE 12.5 – Percentage share of various countries: publications output in the world total

Country	1984-1989	1990-1995
USA	36.5	35.8
UK	9.2	9.2
Japan	7.4	8.7
Germany	6.2	7.4
France	5.2	5.9
Russia	6.8	5.0
Canada	4.7	4.8
Italy	2.7	3.5
Other countries	21.3	19.7
Total	100.0	100.0

Source: Second European Report on S&T Indicators, 1997.

government sector: 44%, 45%, 11% respectively; enterprises: 3%, 44%, 53%. The percentages for the national total are the following: basic research 22%, applied research 44% and experimental development 34%.

Scientific publications in international journals represent an indicator of the scientific output of a country, namely of its public research infrastructure. Table 12.5 shows that the Italian share of the world total has increased between the last part of the 1990s and the first part of the 1990s by some 30%, well above the increases of the other developed countries. At the same time, Italy's output in terms of scientific publications is half of the size of the other large European economies.

TABLE 12.6 – Patent applications at the European Patent Office

Country	1983-1985	1989-1991	1993-1995
USA	28.8	27.9	29.6
UK	7.5	6.3	5.2
Japan	14.7	20.4	18.3
Germany	21.6	20.0	20.2
France	8.7	8.3	8.4
Canada	0.9	0.9	0.9
Italy	2.8	3.5	3.7
Switzerland	4.7	3.6	3.6
Other countries	10.3	9.1	10.1
Total	100.0	100.0	100.0

Unit: Percentages.
Source: Ueb-Cespri Data Base, in Ferrari et al., 1999.

TABLE 12.7 – High-tech products as a percentage of manufacturing imports and exports

Country	Export			Import		
	1980-82	1989-91	1993-95	1980-82	1989-91	1993-95
USA	31.5	41.4	40.7	19.2	26.2	29.6
UK	23.8	33.0	36.8	22.6	28.6	31.3
Japan	21.1	31.9	32.5	23.7	28.1	33.1
Germany	17.9	21.6	22.5	20.3	26.6	26.4
France	15.5	25.0	27.6	19.8	25.6	27.3
Italy	11.4	14.7	14.7	22.6	28.0	27.9
EU (15)	16.9	22.4	24.3	20.0	25.8	27.0

Note: Percentages of values expressed in constant $.
Source: Sie-World Trade Data Base, in Ferrari et al., 1999.

The analysis of the number of patent applications to the European Patent Office (Table 12.6) shows that in the first part of the 1990s Italy increased its percentage share and reduced the distance with the other industrialised countries – this trend confirmed the previous one observed during the 1980s. Italian applications in fact rose from 2.8% of the total in the period 1983-1985, to 3.5% in the period 1989-1991 and to 3.7% in the period 1993-1995. Also in this case, as for scientific publications, Italy's share is lower than that of comparable European countries.

One way to look at the technological potential of a country is to analyse its capacity to produce and export goods characterised by a high content of knowledge which, in turn, is the result of the effort of the whole national system of innovation. The share of high-tech products in Italian exports of manufactured goods is much lower than that of other European countries (Table 12.7). In the late 1980s-early 1990s it was 14.7%, while the figure was 24.3% for the whole European area, 22.5% for Germany, 27.6% for France and 36.8% for the UK. Such difference was much lower at the beginning of the 1980s (11.4% for Italy as compared to 16.9% for the European countries) and therefore it has significantly increased over the decade (Ferrari et al., 1999).

From the import side, the Italian situation does not seem to differ significantly from that of the other European countries. Italian imports have been increasing in line with those of other industrialised countries and do not show any higher external dependency of Italy for such type of goods. Also the share of high-tech products over the total is almost the same for Italy (27.9% in 1993-1995) as for the European area (27.0%). The real weak point of Italy is its production, and export capacity, of high-tech products which is lower than that of the other main industrialised countries.

Innovation is at the heart of economic change. It is therefore crucial to know why it occurs, why and how firms innovate (OECD, 1998). A major exercise for the measurement of innovation in industry was initiated by the OECD in collaboration with Eurostat, a methodology was devised to set the statistical standards (OECD, Eurostat, 1997), and a co-ordinated data collection initiative (the Community Innovation Survey) was launched both in European and non-European countries. Data for Italy have been collected by ISTAT in collaboration with the Institute for Studies on Scientific Research and Documentation of CNR (ISTAT, 1995).

Table 12.8 shows for the manufacturing sector as a whole and for the main firm size classes the number of firms participating to the survey, the percentages of these firms that have introduced product or process innovation in the period 1990-1992, and the percentage of sales and employees accounted for by innovating firms in 1992. The table shows that only one-third of the firms involved in the survey have introduced innovations during the three-year period considered. The innovative phenomenon has involved however a much larger portion of the Italian industrial structure, 61.5% of employees and 70.7% of turnover of the manufacturing industry covered by the survey being concentrated in innovating firms. The table shows that there are significant differences in the percentage of innovating firms across different

TABLE 12.8 – Innovating and non-innovating firms by firm size, 1990-1992

Classes of employees	Total firms	Innovating firms in total firms (%)	Employees of innovating firms on total firms (%)	sales of innovating firms on total firms (%)
20–49	15 109	25.9	27.5	29.1
50–99	4 142	40.8	41.6	43.0
100–199	2 012	48.0	48.7	47.8
200–499	1 041	58.5	59.8	67.3
500–999	292	74.0	74.5	79.1
1000 and over	191	84.3	91.5	95.9
Total	22 787	33.3	61.5	70.7

Source: ISTAT, 1995.

size classes. Only one-quarter of the firms with less than 50 employees have innovated during the period 1990-1992, while 84.3% of firms with over 1000 employees have introduced innovations. This pattern holds also at the level of all European countries participating in CIS taken as a whole (Evangelista et al., 1997).

Significant differences in the percentage of innovating firms also emerge across Italian industries. Industrial sectors showing the highest percentages of innovating firms are aerospace (67.7%), office machinery (64.6%), radio, television and telecommunications (59.8%). Also at industry level a more effective indicator of actual economic relevance of the innovation phenomenon is given by the percentage of sales and employees of innovating firms. In sectors such as aerospace, office machinery and radio, television and telecommunications innovating firms concentrate more than 90% of the employees and sales of these industries, while in most of the industries producing traditional consumer goods the percentages of employees and sales accounted for by innovating firms remain quite low.

The multiform nature of innovative activities and their sectoral specificity have been underlined in a vast amount of literature which has shown the existence of a multiplicity of interdependent sources of innovation. Besides activities generating new technological knowledge, special attention has also been attached to processes of technology adoption and diffusion (both embodied and disembodied), an acknowledged *sine qua non* for technology to express its economic effects to the full.

The relative importance of the various sources should however also be measured using a common yardstick. The Community Innovation Survey attempted to do it according to an input measure such as innovation expenditure, i.e. the breakdown of innovation-related expenditures by items. Table 12.9 shows the breakdown of expenditure incurred to introduce innovations by firm size. The picture which emerges from the table is very clear-cut. Industrial innovative processes consist, first and foremost, of the purchase and use of 'embodied' technologies (innovative machinery and plants), which account for 47% of total expenditure on innovation, and, secondly, of efforts to generate and develop new knowledge inside firms, as measured by the percentage of innovative spending for R&D activities (35.8%). The other components play a relatively minor role: expenses incurred for design and trial production each account for 7% of total expenditure on innovation, while just 2% and 1.5% of the latter are allocated, respectively, to the purchase of patents and licences and to marketing activities related to the introduction of technological innovations.

The relative importance of the various types of innovative sources can be assessed also by looking at the actual involvement of firms in such activities, on top of the corresponding expenditure which can vary considerably depending on their nature. The data on innovating firms in the service sector in Italy show that firms rely on a wide range of activities, often simultaneously. The two more frequently cited activities are acquisition and development of software and investment in new machinery and equipment with percentages of firms performing such activities higher than 70%. Training activities related to the introduction of technological innovation and design have been carried out by 43% and 33% of innovating firms, respectively. R&D, acquisition of know-how, and marketing emerge as much less frequent innovation activities with percentages around 20% (Evangelista and Sirilli, 1998).

The distribution of innovation expenditure shown in Table 12.9 not only reflects the distinctive profile of the Italian manufacturing industry – with its accentuated specialisation in medium- and low-technology sectors. It highlights a more generalised pattern of innovation activities which holds across all European countries (Evangelista et al., 1998).

The importance of the different sources of innovation in business strategies is, however, strongly influenced by firm size, especially as far as R&D expenditure and investment are concerned. Small firms have a high propensity to innovate by acquiring machinery and plants, against the greater propensity of large firms to internally generate new technologies. Table 12.9 shows that for firms with fewer than 50 employees, R&D activities account for 15% of total innovation expenditure against a percentage close to 47% in the case

TABLE 12.9 – Innovation expenditure by firm size, 1990-1992

Classes of employees	R&D	Patents and licences	Design	Tooling-up and trial production	Marketing	Innovative investment
20-49	14.9	1.5	9.4	7.7	1.9	64.6
50-99	16.3	1.3	8.4	8.5	1.7	63.8
100-199	19.8	1.7	12.8	9.0	2.2	54.5
200-499	27.6	2.2	9.1	9.6	2.2	49.3
500-999	26.0	1.6	13.4	8.1	1.3	49.6
1000 and over	46.7	0.8	4.8	5.7	1.2	40.8
Total	35.8	1.2	7.4	6.9	1.5	47.2

Notes: Expressed in percentages, rows add up to 100%.
Source: ISTAT, 1995.

of firms with more than 1000 employees. Data on investment show an opposite pattern. Innovative investments of firms with less than 200 employees account for more than 50% or more of total innovation expenditures. The other components of innovation also do not appear systematically correlated to firm size. All that emerges is the greater importance of design activities in intermediate size categories, with over 10% of total innovation expenditures in firms with 100-200 and 500-1000 employees. This again is a pattern which holds across most European countries.

3. PUBLIC POLICIES AND THE INSTITUTIONAL SETTING

An important trend in the Italian scientific and technological sector is the recognition that the country has to enhance the effectiveness of its public research system. Cumulative historical factors have created a fairly large sector, with a considerable number of institutions which have grown in size and number over the decades but with little co-ordination. The establishment of MURST in 1989 was a step towards a better co-ordination of initiatives in the public sector.

Closer relationships with the demands of industry and society as a whole are desirable. Programmes like the National Research Programmes managed by MURST and the Mission-Oriented Research Projects of CNR have moved over time in this direction. However, it also appears that, in spite of their intentions, they could be further enhanced to improve the collaboration with industry, increase the focus on topics which can concretely address needs of firms and public services, and more generally create stronger interactions between the scientific community and the production system.

A similar concern is about the efficient production of scientific and technological outputs by the Italian public research system, including both universities and the other public research institutions. There are profound gaps in the system, with some institutions being centres of excellence, and many others being far less efficient in their research. This state of affairs, which is typical of all countries, is mostly due to the fact that in Italy for many years the allocation of public resources to research has not always been made on the basis of a discrimination in favour of the quality of the research proposal and of the research team, while they have normally been distributed according to more "egalitarian" criteria which allowed to give something to almost everybody (Braga et al., 1999).

The reorganisation Plan started in 1998, which aims at consolidating smaller and less efficient units into larger and more efficient institutions, can be a good step forward *vis-à-vis* the present state of affairs if a real rationalisation is implemented. However, given the rigidities of the public research system, the changes will most likely be more formal than substantive.

Another issue which has been addressed over the last few years is the creation of institutions for technology transfer. Overall, the success of the few initiatives started by public research agencies and universities has been rather limited for various reasons: the splitting between generation and utilisation of new knowledge with an intermediary agent in the middle is often ineffective in a knowledge-based economy, the institutional rigidities of these organisations did not allow them to be in tune with firms' needs, companies are not often capable of endogenising and elaborating a new technology but need a "market opportunity" which is well beyond the reach of technology transfer offices. At the same time efforts have been made to transfer new technologies to public services in a non-market context; the results are mixed and show that, where there is a mutual interest, the co-operation between public institutions may be particularly successful.

3.1 – Setting priorities

Research in the public sphere, and in particular in academia, has for a long time been granted a high degree of autonomy and has been managed by elective bodies responsible, by definition, to the relevant scientific constituencies (disciplinary, geographical and so on). Representative bodies of the scientific community have been the National University Committee (CUN) established at MURST and the Conference of Italian Rectors (CRUI), which covered the universities, and the CNR National Advisory Committees representing the whole public scientific community. The latter, which were formally advisory bodies, were in fact entrusted with the evaluation and management of the CNR research establishments (about 300), of the Mission-Oriented Projects, and of the financing of research across the country, mostly in universities. Overall, the research objectives in the public sector have basically been self-determined by the scientific community in pursuance of their own internal needs and equilibria.

Government support to research in industry, which has been conditioned by the scant propensity of Italian firms to invest in innovation, has been channelled through projects submitted by individual firms in the absence of a clear government vision, even though sectoral plans have been used for selecting amongst different proposals. Overall the system of R&D priority setting has been lacking coherence and clarity, and the linkage between the S&T policy and the other government policies has been weak. The actions taken by the Government to finance research and innovation in industry, often on a small scale because of the difficulty of concentrating resources on a limited number of high-priority objectives, have had little effect in reducing uncertainty in firms by defining national priorities (at the same time, the question whether government money is used by firms for launching additional projects or rather for financing initiatives that they would have carried out anyway – the additionality issue – is still an open question). Moreover, state-owned and state-participating enterprises in high-tech sectors, which were meant to be a means to implement government strategies, have not been able to raise the level of the national technological specialisation.

3.2 – The institutional setting

In Italy science and technology policy has emerged after World War II in a rather fragmentary way, with different institutions – the university system, the National Research Council and a small number of important industrial R&D laboratories – developing in separate directions.

In 1963 a Minister for the Co-ordination of Scientific and Technological Research was appointed, with little powers and no autonomous budget. Only in 1989 was the Ministry for Universities and Scientific and Technological Research (MURST) established and started to play a more effective role in co-ordinating the national research efforts. Since its establishment, MURST has been in charge of setting national priorities, overseeing the university system as a whole, supervising the largest public research agencies and managing some of the instruments provided by legislation for directing and supporting industrial research, in particular the Applied Research Fund. A National Council for Science and Technology (CNST) has been set up to advise the Minister.

The Law establishing the MURST provided for certain forms of autonomy to be granted to R&D-performing institutions in the public sector. In order to speed up the decision-making process and to improve efficiency, the Parliament has passed, in the mid-1990s, a bill giving the universities a substantial degree of autonomy. At the same time, the public research agencies have been granted more autonomy but the process has been moving ahead more slowly. Specifically, in the past decade MURST took up the following functions:

- the drafting of a plan for the development of the universities, and the submission of a report to the Parliament on the state of higher education in Italy;
- the drafting and submission to the Parliament of a tri-annual report on scientific and technological research;
- the development of annual and multi-annual programmes for scientific and technological research;
- the co-ordination of the Italian participation to international and European higher education and research programmes;
- the allocation of the resources allotted by the public budget to universities and public research institutions;
- the submission of proposals to CIPE – the Inter-Ministerial Committee for Economic Programming – to launch programmes geared to enhance scientific and technological research in the private sector.

The Ministry did undertake some important steps in the direction set forth in the new legislation. However, their implementation has been hampered by various factors, both institutional and procedural.

The institutional factors are related to the reluctance of other Ministries to give up some of their responsibilities or simply to co-ordinate their activities with MURST's, to the lack of additional resources available to MURST, and to the inability of the bureaucratic infrastructure of the Ministry to implement the new institutional and managerial tasks because of the lack of the necessary skills. The procedural factors were basically linked to the complexity of the selection and financing processes which discouraged SMEs to apply for public support via the financing of R&D projects.

During the period 1994-1996 MURST introduced some adjustments in order to overcome the above mentioned problems, in particular *vis-à-vis* the support to industrial research. Although some steps have been taken in this respect, it appears that the gap between public and private research in Italy is still pronounced. As noted above, the Ministry implemented several programmes to finance research in firms, but this objective was only partly achieved for two reasons: firms carrying out the research projects could not appropriate their results (even though they had the pre-emption right to exploit the results of the research) and therefore the projects were often rather far from their commercial strategy, and because the projects often had an academic bias.

The changes introduced over the period 1994-1996 have been inadequate because of the weaknesses of the earlier structural reform and its implementation. In fact, it turned out to be a foolish ambition to expect that MURST could play a supervisory role towards other Ministries placed at its same hierarchical level without attributing it the necessary institutional tools.

The new Government elected in 1996 launched a general reform of public administration in which research and innovation have been assigned a key role. This stimulated a debate after which the Parliament in 1997 asked the Government to examine the whole national S&T system in order to identify the deficiencies which prevent the national S&T potential from being profitably used to meet the research and technology needs of a modern society. The following weak points of the Italian research system have been identified in a report (MURST, 1997):

- an insufficient systematic approach;
- the lack of a systematic assessment;
- insufficient planning and weakness in the formulation of strategic programmes, also due to total lack of systematic foresight activities;
- little diffusion of the results achieved by the S&T system.

The analysis included in the above-mentioned report, which follows the guidelines set out in the Bassanini Law No. 59 of 1997, asked for a clearer specification of the functions and the activities of the various public research institutions, including the general government planning. This process was based on a framework law passed by Parliament which entrusted the Government to pass the ensuing legislation through different decrees. From an operational point of view it turned out that various Ministries pursuing different institutional aims and policies have adopted specific decrees which have not necessarily been coherent between each other and with the framework plan. Specifically, three main points are identified in the report as the milestones of the reform process:

- a new architectural organisation of the governance of the public research system;
- a re-organisation of the mission of public research institutions;
- a revision of the mechanisms for the support of R&D and technological innovation.

As far as the first point is concerned, the process introduced several innovations. In the first place, it is established that, in order to link the S&T system to the more general economic strategies of the country, the triannual National Act for Economic Planning (DPEF), which is passed yearly by the Parliament, should include guidelines and the amount of financial resources devoted to R&D. In order to implement the interministerial decisions and co-operation, all relevant government actions should be decided by CIPE (Interministerial Committee for Economic Planning). CIPE acts as the institution that provides a general assessment of the big options in the allocation of resources. CIPE adopts the multi-annual National Programme for Research and Innovation. In so doing, it verifies the consistency of the Programme with the budgets available to the various Ministries, and with the European union programmes and research strategies. CIPE also allocates the resources of a newly created Fund for Interventions of National Relevance, which will support multi-annual programmes of national priority through additional financial resources.

The new architectural organisation was aimed at fostering the co-ordination of the various departments' policies through interministerial decisions (CIPE) rather than enlarging the competencies of the sectoral Ministry of R&D (MURST). Various decisions have been made by the Government in this respect:

- The chairmanship of CIPE meetings dealing with R&D issues was entrusted with the MURST's minister.
- The setting up of a permanent CIPE committee in charge of the approval of the National multi-annual R&D programme including the Fund for Interventions of National Relevance, and of specific decisions aimed at fostering co-ordination among different and sectoral actions taken by various ministries and agencies.
- The strengthening of the operational capacities of MURST through the setting up of four consultative bodies: the Minister's consultative committee made of nine members appointed by the Minister (CEPR, Research Policy Expert Committee); the S&T Assembly,

mostly elected, aimed at bringing into the Ministry a representation of the scientific community and of the social and economic interests; the Technical Secretariat supporting the Minister's decisions and the interaction with CIPE activities; the Committee for the co-ordination of the activities in the field of R&D evaluation (CIVR).

The reorganisation of the public research system, and primarily of institutions like CNR and ENEA, is being implemented through:

- a reduction of the number of small institutions which will be merged with or acquired by larger ones;
- a greater autonomy and responsibility and a more stringent evaluation of each institution through better information on their activities and potentials;
- a greater flexibility of the instruments and activities of these institutions;
- a reduction of their bureaucracy;
- support to their human resources, which will include actions for the mobility of personnel and the professional qualification of their employees.

Concerning the revision of the mechanisms to support R&D, the main criteria introduced by the Government in revising the existing instruments are the speeding up of the bureaucratic procedures and the integration of different means to support innovation. While the need of speeding up the procedures tackles one of the major problems of the Italian industrial and technology policy instruments, the second one focuses on the idea that an effective support to research cannot be based solely on financial aid, but on a number of other elements, like an efficient public administration and an effective public research system all across the board, direct mechanisms for technology transfer, and a greater mobility of researchers especially between public and private research laboratories.

In summary, the Italian public research system is undergoing a process of reorganisation. The objective is to improve the efficiency of the public research system by encouraging greater managerial and administrative decentralisation and by shortening the distance between the science and technology policymakers and the beneficiaries of their policies. The process of reorganisation also identifies some of the mechanisms to improve the efficiency of public research in Italy – most notably, an effective evaluation system, wide and systematic co-operation amongst the various actors and institutions involved, competitive mechanisms based on the quality of projects, and more effective allocation of the resources devoted to research.

4. DEVELOPMENTS IN SCIENCE AND TECHNOLOGY IN ITALY

In the last few years various developments have taken place in the S&T sphere in connection with decisions taken at the Parliamentary and government level. These decisions, which implied new legislation, aimed at making the S&T system more suited to a modern society and open to the international competition.

4.1 – Universities

Until recently, the Italian university system has been very centralised and uniform both at the management and educational levels. The students had to attend the courses and to pass a number of tests (according to the discipline from 20 to 33 in four, five or six years) to obtain a degree (Laurea, ISCED level 6, equivalent to the third level education). In a given discipline the path to follow was the same in any university all over the country, and in terms of number of tests to pass and of compulsory disciplines to follow, it was largely determined by the central education authority (i.e. MURST). The main reason was to guarantee anyone the same "amount" of knowledge anywhere in order to award a degree whose consistency and completeness is the same independently of the university awarding it. Of course the quality of courses, professors and facilities affected very much the final quality of the graduates, but the state was supposed to offer everyone the same opportunities, at least in principle.

The experience of training for research in Italian universities, structured as a formalised set of courses for postgraduate students, is a rather recent one. Only in 1980 has a proper system of postgraduate education been settled as an independent process leading to a postgraduate qualification in research. Up to 1998 almost 20 000 Ph.D. degrees had been awarded.

In the last few years Italian universities have undergone a number of changes which are having a quite relevant impact. In the future more changes are expected to be introduced and implemented.

First, university courses are being streamlined with the European standard. Instead of four- to six-year courses leading to the Laurea degree, three levels have been introduced: the University degree (three-year course), the Master degree (two more years) and the Doctorate (three more years). This new system is intended to harmonise the whole European system, to address the problem of producing graduates entering the labour market too late, and to reduce the drop-out rate of Italian universities which is considered too high.

Secondly, universities have been progressively granted greater teaching, organisational and management autonomy. This will allow MURST to play a more strategic role rather than directly managing programmes, personnel and resources; at the same time this implies that universities need to develop their ability to take responsibility for themselves.

Third, the hiring system of university teachers has been profoundly changed after a very long and intense debate. University teachers and researchers are now hired through a selection process managed at the level of each university, rather than at the level of MURST.

Fourth, and as a corollary of greater autonomy, an evaluation system of universities has been introduced.

Fifth, the *Corte di Cassazione* has recently ruled that, in order to implement European directives, in Italy the *numerus clausus* should be introduced in all universities and faculties – at present only the faculties of architecture, medicine and a few others have introduced the *numerus clausus*. If and when this ruling becomes reality, the whole university system will undergo a profound change in the sense of stressing much more the quality of teaching of fewer motivated students. At the same time the university teaching will have to be harmonised with the new high school system whose curricula are being reformed, and more coherence is expected to be introduced between students' high school profiles and their ensuing university education.

Sixth, the recently introduced national teaching system provides that post-secondary education will be delivered in a co-ordinated manner by three educational institutions: high schools attached to the Ministry of Education, regional vocational schools and universities, adopting comparable modules and using a credit system. While at present they are offering rather disparate courses of various quality, in the future the specialisation of each institution is expected to correspond to better and more effective teaching.

Finally, the Italian university system is characterised, on the one hand, by mega-universities (for instance the University of Rome "La Sapienza" has 150 000 students) and, on the other, by micro-universities with a sub-optimal critical mass. MURST is currently making available incentives aimed at implementing the existing legislation according to which each university should not exceed 40 000 students. The problem of local universities still remains to be solved in terms of quality of teaching and research; one may wonder if peripheral universities will ever achieve the necessary level of excellence, on top of playing the role of providing scientific and technical support to the local communities and their production specialisation.

Evaluation of the university system

The Italian university system is organised according to three levels: the Ministry is the highest political decision-making level responsible for setting general objectives and making sure that they are achieved, the second is represented by single universities and the third by operational units within each university (faculties, degree courses, departments and so on).

In order to check the general results of the whole system and achievement of the objectives, it is necessary that at each level, in particular at the highest level, a central evaluation body controls the validity of the results of the activities carried out by the units of the system according to a methodological framework and established procedures. This central body is entrusted with verifying that the results are consistent and compatible with the objectives set forth. Therefore a complete evaluation system has been put in place, defining the interaction and the necessary relations among the self-evaluation, internal and external process.

Law no. 537/1993 establishes the MURST National University Evaluation Council (NUEC) at central level and that each university sets up an Internal Evaluation Unit (IEU).

The universities have already faced the problem of internal evaluation and the institution of the Units under the guidance and the thrust of the Conference of Italian University Rectors (CRUI). In the last few years there has been an acceleration in the setting up of the IEUs. Moreover in each university many committees have been established, at faculty and department level, to deal with the evaluation of degree and career-oriented courses and other activities.

The IEUs have the task of checking the management of government resources, the productivity of research and teaching as well as impartiality and good management, by performing comparative cost-efficiency analyses. Law no. 537/1993 has also established that the IEUs have to prepare an Annual Report on universities to be sent to MURST and to the NUEC to evaluate the efficiency and the effectiveness of the results of research and teaching activity, and to check the development and rebalancing programmes, also for the subsequent allocation of the resources.

Overall, the experience of the evaluation activity carried out in the Italian university system during the last few years may be considered important and fruitful, even though there are still implementation problems (Biggeri and

Scarpitti, 1998). There is still a long way to go, also because a significant part of the university staff, both teaching and managerial, is still very sceptical about the effectiveness of the whole evaluation exercise. In order to be fully successful in the long run, various preconditions have to be met: the setting up of an appropriate information and statistic system, a strict co-ordination between the central and peripheral levels, keeping the evaluation bodies autonomous and independent, the need to show that evaluation reports are taken seriously by policymakers in decision-making.

One step forward has been recently taken: along with the establishment of the Minister's Committee for the co-ordination of activities in the field of R&D evaluation (CIVR), NUEC has been transformed into the Committee for evaluation of universities which has been entrusted with carrying out evaluations – NUEC was basically a technical body in charge of collecting and supplying data and of conducting studies to be used by other instances responsible of the evaluation of universities. The new Committee, which replaces NUEC and is independent of CIVR, supervises universities' IEUs, and those universities not sending their yearly evaluation reports are excluded from the MURST financing. This for the first time makes evaluation a key step in the MURST's decision-making process and shows that it is taken seriously.

Evaluation of the university research projects

Until 1993 research in university was financed by MURST through two channels: the first one was managed at the local level (60% of the total) and the other one (40%), which was used for the support of inter-university projects of national relevance, was managed by the Scientific Discipline Advisory Committees of MURST's National University Council (CUN). The allocation of the latter turned out to be inefficient in the sense of poor-average quality and too many small projects because the selection was left to university professors elected by the constituency which was supposed to receive the funds. In 1994 the 60% funds have been included in the block grant of each university (the money is no longer earmarked for R&D and can be used by the university for any purpose).

Concerning the national fund, a new procedure was introduced in 1997 which envisages that project proposals are evaluated by a committee of national guarantors (seven experts) and two or three anonymous referees, both national and foreign. This was the first time ever that the Italian university system was systematically reviewed.

The new system required that a directory of referees was drawn up, project proposals had to be submitted both in Italian and in English, and a telematic procedure was adopted in order to ensure confidential and quick communication between the guarantors and the referees.

In 1998 1645 projects were submitted to the Ministry involving some 10 000 research groups and, on average, 44 researchers per project. Projects were reviewed by some 2000 experts out of the 12 000 included in the directory (4300 Italian and 7700 foreign). Comparing the results achieved in 1998 with those of the previous year, some positive adjustments have been achieved in terms of improving the "quality" of the proposals submitted:

- a 20% decrease in proposals submitted;
- a significant decrease in the number of research groups per programme, as recommended by the Ministry with the aim of building specific research groups rather than thematic networks;
- a 12% decrease of the total financing requested, even though the maximum share of financing from MURST was raised from 60% to 70%.

The 40% co-financing fund has been supported by the increase of available resources, from 100 billion Lira in 1996 (the last year in which the previous procedure was adopted) to 250 billion Lira for the year 2000. This fund, which has basically replaced the CNR research contracts used up to the early 1990s to selectively channel government funds to university research, is, however, still a modest share of the resources explicitly devoted to research out of the total amount of universities expenditure (about 10 000 billion Lira, most of them for wages).

Overall, the new system has unanimously been considered a success for various reasons:

- it has made it possible to concentrate the resources on a limited number of large projects, selected on the basis of quality;
- it has helped improve the ability of Italian academics to compete for funding on the basis of international standards;
- it has shown that good evaluation can be introduced in universities, while in the past there was scepticism and even opposition to the spread of evaluation procedures;
- it has helped to diffuse the telematic tools among university teachers and departments which, in many cases, were not appropriately equipped. This has also enabled the avoidance of using and sending out paper forms in the refereeing process.

It may be expected that in the coming years the evaluation system, which has been considered as a model in other countries, will undergo a continuous process of improvement and adjustment to changing needs.

Student mobility

The development of an international, and specifically European, dimension in Italian universities is one of the objectives to be reached in the coming years.

Each university should enhance its international activities and promote their development. The implementation of the SOCRATES Project by the European Union is perceived as a useful tool for promoting and orienting the mobility of students around Europe. This project offers students a multiple series of opportunities for attending part of their university course in another European university. Italy has actively taken part in this project and in its predecessor, ERASMUS.

The attractiveness of Italian universities for foreign students is for some groups of countries quite remarkable. Most enrolments take place in the medical faculty: 5530 of the total 22 342 foreign students enrolled in Italian universities in the academic year 1996-1997.

More than half of the foreign students enrolled come from European countries (15 350) with a large share from Greece (8040 out of the total 10493 from EU countries). The remaining number mostly come from Asia (2708, notably from Israel, 813 and Iran, 712) and the rest is divided over the remaining continents as follows: Africa, 2800, America 1043, Oceania 45.

In 1996 the number of foreign students qualified in Italian Universities was 1682; 501 of them in medicine. Most graduates come from Europe (990), mostly from Greece (312), followed by Asian countries (327), Africa (216) and America (140) (ISTAT, 1998).

4.2 – Industrial research and innovation

In this section the most recent developments in public policy toward research and innovation are dealt with. In particular, the following issues are analysed: the Applied Research Fund managed by MURST which has been in place for 30 years, various recent measures supporting innovation through fiscal incentives granted to SMEs, and an evaluation of the Government's policies.

The Applied Research Fund

A major policy tool is the Applied Research Fund, which is administered by MURST and managed by the Italian Industrial Credit Institute (IMI), the main medium-term industrial credit institution (IMI, 1999).

The incentives to enterprises take the form of grants, low-interest loans and risk capital for research consortia. This scheme, first introduced in 1968, until the early 1980s provided the only assistance for industrial innovation. Between 1968 and 1998 about 2500 billion Lira were allocated, of which 1640 billion Lira were granted to firms. Grants for 1998 have been 154 billion Lira.

A new granting procedure was introduced in 1994 with various objectives: the reduction of the timing of project examination to a maximum of eight months; the "defence" of expensive project proposals from large companies in front of the Ministry; and the increase of resources to small and medium-sized enterprises through the financing of individual and co-operative research projects.

In 1997 the Applied Research Fund was changed in order to improve the efficiency of its procedures. A co-financing mechanism, requiring the sharing of the cost of the R&D project from the applicant firm, was introduced. This allowed setting in place a multiplying mechanism of public resources and, at the same time, a more direct responsibility of firms. In addition, it introduced several new measures characterised by more automatic and simpler mechanisms. Related to this is the attention paid by MURST to SMEs, as epitomised by the introduction of a simplified application procedure for smaller firms through Law No. 46/1982. The new procedure envisages that a separation has been introduced between the financial auditing and the technical analysis of each project – while formerly IMI was responsible for both, now the technical analysis is carried out by a MURST committee. Moreover, IMI lost the monopoly of the management of the fund and various financial institutions and banks finance projects on a competitive basis.

Other measures to support research and innovation

In 1983, within the framework of the legislation for the promotion of technological innovation (Law 46/1982), the National Research Programmes (NRP) were launched. They aim at developing strategic technologies through research contracts with enterprises and consortia within sectoral plans. At the end of 1992, 18 sectoral programmes have been proposed by MURST and approved by the Inter-ministerial Committee for Industrial Policy. From 1985 to 1992 the Government assigned 139 contracts for a total contribution of 1534 billion Lira. Some changes have recently been introduced in the NRPs: in 1997 a co-financing procedure has been adopted in order to encourage firms receiving government aid to be more involved in research projects and the attendant innovation strategy.

In 1991 a new scheme (Law 317/1991) was introduced with the aim of creating a favourable environment for SMEs for the development of innovation, in particular within "industrial districts" or local systems of production. This scheme includes the possibility of setting up financial intermediaries which can share the enterprise risks of SMEs and to promote the development of consortia amongst firms aimed at both production activities and the delivery of services such as commercialisation and management services.

Science parks could play an important role in restoring the balance in the distribution of innovation activities across the national territory. Following an agreement between MURST, the Ministry of Budget and the Ministry for the *Mezzogiorno*, 13 regional and local science parks mainly located in the *Mezzogiorno* have been selected for public financing by means of 50 technological projects which started in 1996 and are now close to completion. Even though southern science parks are intended to involve the financial participation of local entrepreneurs, up to now the majority of them are mainly publicly financed. In some cases, for example Tecnopolis in Bari, science parks are pre-existing agencies, while in others they are more "virtual" networks in competition with other local actors for the promotion of technology transfer.

Other initiatives aimed at the establishment of science parks in central and northern Italy are being taken by private and public agents at local level, involving research consortia participated in by companies, universities, public research bodies and local Governments. The promotion of science parks is also supported by private initiatives without any national support, and in some cases European funds are used.

Overall, the experience of science parks is still too recent to evaluate if they have been a success or a failure. Nevertheless, their evolution is expected either to end up with their transformation into public agencies for the promotion of innovation attached to Regions, or to become intermediary structures like Relay Centres. The fact that science parks have grown with weak links with the market will in any case hamper their full development as fully-fledged economic entities.

Other initiatives supporting research and innovation have been launched within local development policies (*sviluppo territoriale*). In the final stage of the Extraordinary Intervention for *Mezzogiorno* (late 1980s, early 1990s) Programme Agreements between the Government and public institutions (i.e. CNR and ENEA) and Programme Contracts between the Government and private and public firms (*Contratti di programma*) have been signed with the objective of implementing integrated projects. Both of these instruments have been unsuccessful mainly for implementation reasons.

Another type of intervention has been implemented through agreements between various local actors (such as provinces, municipalities, firms and trade unions) (*Patti territoriali*) aimed at fostering local development. These agreements may imply that the cost of labour may be reduced in comparison with national standards. This experience has been considered at the European level as an interesting model to replicate in Objective 1 Regions, i.e. less-favoured regions (local agreements for development). All these initiatives have been financed with specific additional funds which, even though not explicitly earmarked for innovation, have often been in part used for achieving technological objectives.

New measures for industrial research and the employment of qualified personnel

Since 1997 the Italian Government has introduced some new measures to encourage R&D expenditures by SMEs, and most importantly to encourage them to employ qualified human resources, particularly graduates with university Laurea and doctoral degrees. These are important measures for a number of reasons. First, one of the main problems of Italian SMEs is that they employ unqualified personnel, and they do not typically employ people with university degrees, or doctoral degrees. This is consistent with their specialisation in traditional sectors and the difficulties that Italian SMEs face in acquiring and using new technologies, in moving into technologically more advanced sectors, in participating in R&D projects or investing in R&D themselves. Second, the new measures rely more systematically on indirect incentives, and particularly tax credit. This allows for greater automatisms in the instruments, thereby addressing, at least in part, the problem of bureaucratic sluggishness that has characterised the use of Italian instruments of industrial and technology policy.

There are three laws passed in 1997 on these matters – the Laws No. 196, No. 449 and No. 140.

Law No. 196 focuses on the employment of professionals with Laurea or doctoral degrees to be employed in research activities by SMEs. It allows for a contribution of 30 million Lira (equivalent to $18 000) per year for a maximum of two years for each new employee with a doctoral degree obtained in Italy or abroad, and a contribution of 20 million Lira (equivalent to $12 000) per year for a maximum of two years for each new employee with a Laurea degree. The maximum contribution that can be granted to each firm cannot exceed 60 million Lira (equivalent to $36 000). The new employees have to be employed with full-time contracts lasting for at least two years, and their salary should not be lower than the "average" salary of people with that professional qualification (Mazzola et al., 1998). The funds made available in 1998 amounted to 5.5 billion Lira (equivalent to $3.4 million). In the first year of enforcement of Law No. 196 MURST has received 137 requests for doctoral degree holders and 246 for Laurea graduates.

Law No. 449 has the same objective of Law No. 196. It aims at encouraging the employment of people with Laurea or doctoral degrees by SME but the incentive takes the form of a tax credit of 15 million Lira ($9000) per each new employee, up to a total of 60 million Lira for each beneficiary firm.

In the first year of enforcement of Law No. 449, 368 requests have been addressed by firms to MURST for employees with a doctoral or Laurea degree. Altogether Laws No. 196 and No. 449 can allow Italian SMEs to employ more than 600 highly qualified personnel with a modest financial effort.

Law No. 449 introduced an additional measure. It allows firms to apply to universities or other public research institutions (for instance the National Research Council) for seconding researchers or technical personnel to the firm for a period that cannot exceed four years. The individual keeps his/her employment relationship with the university or research institution, while the firm is asked to provide an additional compensation as an incentive. The university or research institution can deny the request of the application, but they have to take a decision within 30 days from application.

Up to now this mechanism has been a failure even though firms have shown an interest in acquiring Laurea and doctoral degree holders. Some reasons may be mentioned. First, it has proved difficult to identify individual competencies within public research agencies and universities to be made available to firms. Secondly, research agencies and universities have not yet adopted the necessary regulations for governing this temporary mobility and, by and large, they have been very reluctant to address the problem. Thirdly, a modest part of the research carried out in public research institutions may have a direct industrial application. Fourthly, many researchers in public research settings look at the involvement in industrial and business initiatives as a sort of "prostitution" of science and therefore are not available to spend their time working in firms.

The above Law allows firms to use the fiscal incentive to pay for R&D projects carried out on their behalf by public research laboratories; this scheme has the objective of fostering co-operation between industry and public research institutions in a more effective way than the one envisaged by Law No. 46/1982, which established a directory of public laboratories available to provide R&D services, and which is rather unsuccessful.

Law No. 140/1997, which is managed by the Ministry of Industry and Trade, adopted tax incentives which are linked to the presence of a specific R&D and innovation item in the beneficiary SME's balance sheet. The requirement of a specific item is intended to encourage SMEs to make explicit their innovation activities. At the same time this increases the probability that firms receiving this automatic incentive are inspected by the fiscal authorities for the verification of the correct implementation of the Law – not always the trade-off between the limited economic advantage and the nuisance of an inspection is considered acceptable to SMEs.

All these supporting mechanisms are now inserted in a Law (No. 297/1999) which covers all MURST interventions concerning the co-ordination, promotion and implementation of measures aimed at sustaining research and innovation in industry by means of a unique fund which absorbs all the previous ones. The following aspects are defined in the law: beneficiaries, possible types of intervention, procedures, the modalities of co-ordination between MURST and the Ministry of Industry and Trade, other interventions for the support of innovation. The implementation regulations of the Law are expected to be issued by MURST in January 2000.

Measuring the impact of government policies in Italian industry

Technology and innovation play a major role in wealth creation and economic growth. Identifying and measuring the impact of public support to industry is a difficult task.

While the identification of the various measures taken by Governments and their comparison across countries has been done in a rather satisfactory way, the quantification of the funds involved is still a difficult task. This information, however, provides a picture of the financial effort made by Governments in pursuing their innovation and industrial policies.

Additional elements are necessary to evaluate the effect of government intervention in terms of performance at the firm, industry and national levels, looking not only at the financial incentives, but also at the support that firms receive from the public scientific and technological infrastructure. This can be achieved by asking firms whether government policies have been relevant for firms' innovations. A key instrument in this respect is the data gathered through innovation surveys based on the "Oslo Manual".

It may be pointed out that evaluating public policies in a market economy is quite a difficult task: in particular, government measures in the field of technology often have various objectives (competitiveness, wealth creation) and their effect is linked to many other factors such as the firm's strategy and

TABLE 12.10 – Percentage of firms declaring that different types of government support have been relevant in the introduction of innovations, 1990-1992

Classes of employees	Government funds	Indirect financial incentives	R&D services from public organisations	Technological services	Public procurement of R&D	Public procurement of supplies
20-49	34.3	26.2	7.6	9.7	5.3	7.8
50-99	42.1	24.7	8.6	11.1	5.6	7.9
100-199	47.0	26.2	13.0	16.2	6.4	8.9
200-499	47.6	22.0	9.9	15.6	6.2	7.9
500-999	51.9	25.5	14.4	15.3	8.8	8.3
1000 and over	70.8	39.1	28.0	26.7	21.1	18.0
Total	40.0	25.8	9.3	11.8	6.0	8.2

Source: ISTAT, 1995.

general economic conditions. Evaluation requires a complex set of considerations on the different objectives and criteria which are relevant. A first step for this task is the assessment of the direct impact that specific policy measures have on the individual firms involved.

While the "Oslo Manual" sets forth methodological standards and provides guidelines for collecting information on various subjects such as the impact of innovation on the performance of the firm, the diffusion of innovation through the socio-economic system, the use of advanced technologies, it does not provide any specific guidelines for measuring the use and impact of government policies on firm's strategy and performance. Some countries have collected data on this issue asking firms whether government policies have been relevant in the introduction of innovations by the firm. Italy's innovation survey provides rich evidence on this subject.

Table 12.10 shows the data on the relevance of different types of public initiatives in support of innovation, ranging from funding programmes to the provision of services and to public procurement, broken down by firm size (Pianta and Sirilli, 1997).

It is remarkable that a large majority of the 7553 innovating firms surveyed declared that the existing innovation policies have had no relevance for the introduction of innovations: 60% of firms attached no importance to government funds; 74% did not use indirect financial incentives (including easier fiscal and credit terms, and tax cuts for investment); while other policy tools, such as the provision of R&D and technological services or public procurement are not relevant for 90% of innovating firms.

The share of enterprises recognising the importance of government funds for innovation policies steadily increases with firm size.

High shares of the largest firms acknowledged the relevance of the various policy tools: 70.8% in the case of government funding programmes, 39.1% for indirect financial incentives, 28.0% for R&D and technological services, and about 20% for public procurement. In most cases a clear break can be identified between the answers of the medium-sized firms (up to 999 employees) and those of the largest firms, which appear to be the main recipients of Italian innovation programmes.

Smaller firms, on the other hand, are marginally involved in most actions; they attach the greatest importance to government funds (by one-third of respondents, a share growing with firm size) and to indirect financial incentives (by one-quarter of firms). In fact, the indirect financial incentives (easier fiscal and credit terms, and the tax cuts for investment) represent the policy which has the most uniform impact across size classes. The share of largest firms declaring that such a policy has been relevant is only one-third higher than the share of smallest firms, while for government funds it is twice as high. The almost automatic nature of indirect incentives appears to reduce the entry barriers for small and medium-sized firms (relatively to the large ones); even in such more favourable conditions, however, only one firm out of four acknowledged such a policy as important.

An additional insight into the dynamics of innovation policies, incentives and constraints is provided by the section of the Italian innovation survey of the early 1990s devoted to environment-related innovation. Firms have been asked to rank the importance of different environmental problems – atmospheric emissions, water emissions, waste production and recycling, acoustic pollution – in the innovations they have introduced in the 1990-1992 period.

The results are shown in Table 12.11 which shows that 49.2% of firms assigned some importance to the reduction of noise levels in their innovations, 42.8% paid attention to atmospheric emissions, 38.4% acted on waste and recycling, 35.5% were concerned with water emissions. The average score of firms declaring that environmental factors were relevant is between "moderately important" and "very important".

The lack of familiarity of firms with questions on such issues suggests some caution in interpreting the results. In particular, it is likely that issues of workplace safety and health (such as reduction of noise levels) have been

TABLE 12.11 – Role of environment dimensions in innovations declared by firms (1990-1992)

% of firms mentioning environment dimensions on total firms	%
Atmospheric emissions	42.8%
Water emissions	35.5%
Waste production and recycling	38.4%
Acoustic pollution	49.2%

Source: ISTAT, 1995.

included among the environmental aspects. Such shares are in the same range of the 40% of firms considering important government funds for innovation (Table 12.10), but are much higher than the impact of the other innovation policies.

The innovations introduced by Italian firms in the 1990s appear to have been equally influenced by government technology policies and by the environmental policies setting standards or providing incentives for introducing less polluting and dangerous products and processes.

It should be pointed out that in the early 1990s no large-scale environmental policies have been carried out, little funding was available for anti-pollution programmes and that the enforcement ability by public authorities was generally low. However, environmental issues have often been supported by the setting of specific standards and regulations requiring firms to comply. It appears that such environmental constraints might have been more influential on firms' innovation choices than the offer of incentives, technology services and procurement.

These data suggest that firms in their innovative choices pay substantial attention to external constraints on environmental issues, including workplace safety and health, on which public policy, market requirements (such as the ISO quality standards) and a variety of social forces may exercise some pressure. This impact may be even greater than that of most deliberate policies aiming at particular aspects of innovation *per se.* In other words, innovating firms appear to react to a similar degree to the "carrot" of specific innovation policies and to the "stick" of external constraints, standards and pressures of non-economic and non-technological nature imposed on them by policy and society. This evidence may suggest some rethinking of the most effective means for inducing technological change in firms, and shows us how much we can learn from indicators and surveys for improving our S&T policy (Sirilli, 1997).

An assessment of the mechanisms for the support of industrial R&D and innovation

Government policies for the support of research and innovation in Italy may be categorised into two periods with reference to the balance between selective and automatic mechanisms and their internal composition.

In the period 1982-1995 the stress was on financing R&D and innovation projects autonomously submitted by firms, while automatic incentives were limited to the contribution to firms for the purchase of machinery incorporating new technologies.

Autonomous projects were submitted, by both large firms and SMEs, in the absence of sectoral priorities, and the selection criterion was the scientific and technological quality of the project and the "credibility" (also financial) of the proposing firm. At the end of the day, the system favoured those firms, usually a limited number of large ones, which could submit formally credible proposals and which could afford to pay the costs connected to the delay in receiving the money by the Government (delays could exceed more than two years).

The first period can therefore be characterised by the fact that selective mechanisms did not discriminate between sectors (virtually all firms could put forward a proposal) and the financing was decided simply on the basis of the S&T merit of proposals and on the availability of funds, and that automatic mechanisms could not support the innovative effort of firms without a formal research capacity and operating in sectors characterised by low investment levels in machinery.

In the second period, which started in the mid-1990s, the policy aimed at tailoring the various mechanisms to the features and needs of different categories of firms. SMEs, which typically have difficulty in translating their innovation activities into projects to be evaluated on their technical merit, have been offered, on top of the traditional financing of innovative machinery, various automatic incentives: tax incentives for the performance of R&D and innovation activities as reported in the balance sheet, for the hiring on a temporary basis of Laurea and Ph.D. graduates, and for the mobility of researchers from public institutions. Moreover, SMEs can receive public financing for research projects through a new tax credit system which allows them to use the money for granting research contracts to public laboratories (*Mediocredito Centrale*, 1997).

The procedures of the Applied Research Fund for the financing of research projects have been simplified and harmonised with the European standards. Moreover the introduction of a monitoring system, the provision that the public financing is a co-financing and that the results of the project can be appropriated by the firm, are supposed to render public intervention more effective than before.

Finally, in the second period the top-down mechanism has become more effective as a clear public policy has been established through the National Programme which defines by means of CIPE decisions and the Strategic Additional Fund managed by the Ministry of Treasury and Budget, priorities coherent with the national strategic objectives.

First Period (1982-1995)	
Selective mechanisms	*Automatic mechanisms*
Allmost all government intervention based on a bottom-up approach which in practice covered also formal top-down procedures.	Only support to the acquisition of new machinery (investment) embodying new technologies.
Second Period (1996-1999)	
Selective mechanisms	*Automatic mechanisms*
Substancial increase of resources granted to firms through top-down approach (e.g. Strategic Additional Fund administered by the Ministry of Treasury and Budget) and through co-financed projects. Less resources to firms projects granted via bottom-up approach. Simplification of procedures and European harmonisation.	Support to the acquisition of new machinery embodying technologies. Introduction of fiscal incentives for the acquisition of skilled human resources, co-operative projects in collaboration with public institutions, contribution to R&D and innovation expenditures as reflected in the balance sheet.

5. SOME FUTURE PROSPECTS

Science, technology and society change so rapidly that it is very difficult to make forecasts. Still, governments and other actors have to make choices and plan the development of a socially valuable S&T system. The following developments may be envisaged for the future in Italy.

Public opinion is increasingly aware of the importance of the role played by science, technology and human resources. Some worries have emerged about the country's position in the international division of labour and employment problems especially in science and technology-based activities. Greater attention and policy action is also devoted to universities. However, the social returns of science and technology are not yet clear to the general public and politicians. A special effort by the scientific community to identify clear objectives for their research would certainly help in bridging this gap.

It is generally agreed that *Italy spends too little on R&D*, and the Government should invest more in this field. Even though the public debt does not allow increases in expenditure, especially now that the European Stability Pact puts additional pressure on the government budget, innovation is among the priorities on the government agenda and there are clear signs that, as soon as additional resources are made available, they will be channelled to innovation. New resources can be attracted to R&D through partnerships between public and private organisations. Moreover, the research dimension is expected to play a part in broader policies for industrial sectors and regions.

Public policy for science and technology has been changed in the sense of strengthening the government role in orienting the choices of the various actors and in evaluating their results. Universities and public research agencies and institutions are expected to strengthen their capabilities to manage their extended autonomy. While this strategy has been welcomed in various quarters, there is the risk that the gap between strong and weak institutions and organisations will widen. However this is part of the price of introducing more competition into the scientific system. Concerning the public support of industrial R&D and innovation, it may be expected that, at variance with the past, the emphasis will be placed on automatic mechanisms rather than on discretionary ones. However, fiscal incentives may not produce the expected results because SMEs claiming the incentive are more likely than others to receive an inspection by the finance authorities and may not welcome additional controls linked to a tax reduction producing only a marginal economic benefit. Finally, the Italian Government and public institutions have not developed, as other countries have done, a long-term strategy based on technological forecasting and assessment: the limited ability to evaluate the relationship between technological advancement and socio-economic needs hampers the building of an appropriate science and technology policy.

Too few *enterprises* play a leading role in the scientific and technological arena, while the great majority adopt a follower or "niche" strategy and play the role of successful adopters rather than original innovators, especially in products. This is also due to the unbalanced structure of Italian industry, with few large and very many small firms. The more active role played by government in establishing S&T priorities, in enhancing the technological infrastructure of the country, in encouraging technological development through direct financing and procurement will be a prerequisite for a more effective R&D system. One of the challenges of the business sector will be the ability to respond to the opportunities offered by the public sector through the co-financing of R&D and innovation projects. Overall, it may be expected that in the near future the structure of the Italian national R&D system will remain unchanged, with an insufficient investment from the industrial sector, only partially offset by the public sector (the percentage of R&D performed by the business enterprise sector in Italy is 54%, while it is 67% in Germany, 62% in France, 65% in the UK and 74% in the US). Another big issue is the privatisation process taking place in the whole economy, which also involves large high-tech companies and groups owned or with state involvement. At the moment it is not clear what is going to happen to their technological strategy, and to what extent globalisation processes will relocate their research activities. Many large firms in high-tech have been acquired by foreign groups and this trend is likely to continue in the future.

It may be foreseen that the *university* system will continue to undergo a significant change as a consequence of public scrutiny of its contribution to society in general and to the training of students in particular. A very old and respected institution which has been reluctant over the last few decades to modernise, resting on the laurels of its past successes, will have to use the autonomy recently received by the Ministry to become more responsive to the social needs and be less the preserve of the interests, both scientific and professional, of academics. At this stage the institutional and political decisions have been made: it is now up to university professors to consider this change as an opportunity and not a threat, and to respond positively to the demands coming from society. There is the risk that the majority of professors will not be prepared or willing to take the responsibilities implied in the reform and that they will use their power to stop the process – in Italy academics are quite a powerful lobby, both in public institutions and in the political milieu. In spite of the above-mentioned problems, it should be pointed out that during the last few years academia have shown a significant and, in some respect, unexpected capacity to react positively to many changes imposed on them by society and the Government. It may be expected that in the future this process will continue creating a split between those universities, departments and teachers that will be able to reach or maintain an appropriate level of teaching and research, and the less efficient ones that will strive to survive. A possible consequence of this split would be the emergence of a limited number of "centres of excellence" that will absorb the great bulk of scientific resources.

The *public research agencies* are implementing a system which stresses their autonomy and responsibility *vis-à-vis* the Government. They will need to show that they are capable not only of conducting good research, but also of taking an active role in the national system of innovation through initiatives aimed at co-operating more closely with users of their research – both public and private. Two challenges will have to be faced by these agencies in the near future: their organisational restructuring and the management of personnel. The new organisation provides for a more decentralised and focused research network, with fewer research institutes endowed with more autonomy in conducting their work. Moreover, bureaucracy will be reduced significantly, while the power of governing bodies has been strengthened *vis-à-vis* the previous situation in which participation all too often led to the inability to make choices or even to paralysis. It remains to be seen whether this move will succeed, or whether it will fail to introduce real changes in a scientific setting jealous of its freedom. On the challenge concerning personnel, the agencies will have to implement a new government policy shifting from a situation whereby almost all researchers have tenure, to a more dynamic personnel policy with only a limited part of tenured researchers and younger researchers hired on contract. This change, which has been also implemented in other countries, is aimed at making the research infrastructure more flexible, less bureaucratic and more responsive to external demand. However, there is the risk that in a small and insufficiently flexible labour market of researchers and university teachers this move might do more harm than good to the scientific productivity of the system.

The creation and maintenance of *human resources* for science and technology is one of the issues brought to the attention of policy-makers. In Italy there are too few researchers and more steps should be taken to improve their degree of internationalisation; it is unlikely, however, that a substantial increase will take place in the near future. In the public sector the average age of researchers is too high, also because a large number of professors were hired during the expansion of universities in the 1960s and 1970s. It may be expected that the average age will decrease with the spreading of contract positions, rather than tenured ones. At a more general level, the combined effect of the introduction of three levels of university degree (which is expected to reduce the drop-out rate) and the expansion of the third level non-university training through the combined effort of high schools, regional institutions and universities will offer the labour market a more qualified personnel and will provide better opportunities for continuous education.

Private non-profit R&D organisations are very few and small in Italy. It may be envisaged that they will increase in number and size because of the action of charities funding research on specific diseases (such as cancer or muscular dystrophy) and of the emerging role of banking foundations in the promotion of R&D. The financial incentives available to this sector have not been sufficient to allow them to grow fast. It might be possible, however, that banking foundations could channel to R&D and post-secondary education financial resources devoted at present to other objectives.

MURST is in charge of the S&T initiatives in the South of Italy, which in the past were entrusted with the Ministry for extraordinary intervention in the *Mezzogiorno*. These initiatives consist of the expansion of the public S&T investment in the south, the establishment of new infrastructure (such as science and technology parks) with public funding and the management of the R&D part of the European fund for the less-favoured regions. The post-war history of Italy is full of initiatives to bridge the gap between the north and the south of the country, but the disparities still persist. What is at stake now is to make the best use of the S&T assets in the *Mezzogiorno* (good quality of human resources, attractiveness of the environment, availability of additional resources from the national Government and from Europe, flexibility of the labour market) in order not only to bridge the gap with the centre-north, but also to exploit a fundamental potentiality for the whole country.

In the country there is an increasing demand for *evaluation* of results of public spending, and also research and education are under scrutiny. The starting point of R&D evaluation in Italy is at a low level (OECD, 1992). Evaluation is mostly confined to the selection of candidates for financing, it is basically carried out by the members of the scientific community, it is not a part of government management procedures, and, when carried out, it has little effect on decisions. The limited experience in carrying out evaluation, the inherent risks of its improper use, and the difficulty of reporting results in the presence of unclear objectives, represent the major obstacles to establishing an articulated evaluation system (Silvani and Sirilli, 1995, 1998). Even in the presence of the above obstacles, progress on the adoption of evaluation procedures has come with various laws passed in the early 1990s. This new legislation has been prompted by the request for more accountability of public resources devoted to R&D, as well as by other factors, i.e. the limitation of resources, the appearance on the scene of a new generation of internationally-oriented researchers, the growing internationalisation of science and technology, and the rules set by the European Union. The institutional changes introduced up to now and the implementation made in various settings have brought the system to a point of no-return: the problem will basically be the pace at which the system will adjust to the new regime, rather than if evaluation will be accepted and will spread across the public institutions.

One area which is particularly weak in Italy, and which has been neglected by technology and innovation policy, is the *financial system*. This system is specialised in providing low-risk credit based on real warranties. In order to provide the necessary means to innovators it would be appropriate that policy instruments support financial intermediaries that develop adequate tools to finance innovation or that have a larger number of innovation projects in their credit portfolios.

The impact of *Europe* on the Italian S&T system has been beneficial in many respects, including the push to integration and co-operation, and the adoption of common and more advanced procedures. European integration represents at the same time a threat and an opportunity as Italy, in order to be competitive, has no choice but to raise its innovation level – in the past this weakness was often remedied through adjustments of the exchange rate. The participation of Italian scientific and technical structures in European projects has been increasing in quantity and quality over the years, but remains rather limited. In the case of ESA, EUREKA and CERN the Italian participation has been quite satisfactory, while participation in the first four Framework Programmes has a limited extent and impact. At present the quantitative participation of Italian teams is commensurate with the national potential; however, the challenge of the future is the ability to play a leading role in European teams on the basis of enhanced managerial capabilities and of less bureaucratic procedures.

Since 1970, the *Regional governments* have assumed a growing role in economic planning, including industrial and innovation policy (Iammarino et al., 1999). Some of them have been particularly active in creating a favourable environment for local industry and economic development while others, especially in the *Mezzogiorno*, have basically continued adopting the old model of transfer of resources with little competition and control of the results. Apart from regional laws supporting innovation and the development of small and medium-sized enterprises, some measures included the promotion of services to industries and firms like the creation of financial regional companies to support local investment, promotion of consortia, and other types of services. The role of regions has become even more important in recent years with the new EU policies, mainly the Structural Funds. The challenge for regions is to raise the techno-economic and managerial capability to manage development and innovation projects. This may be achieved through a substantial investment for the education and training of public administrators. Moreover, given that there is too much intermediation between the instruments and the beneficiaries of available funds, it would be appropriate to introduce greater automatisms and less bureaucratic procedures in regional policies: regions should probably move away from the creation of agencies, to policies where the incentives take the form of indirect measures like tax credit or of direct financing using simple procedures like the Sabatini Law for the purchase of new equipment. In the meantime, regions could entrust the implementation of the existing measures, the selection criteria, and their *ex ante* and *ex post* evaluation to independent institutions; this solution has been quite successful in a number of cases.

APPENDIX: DEFENCE R&D AND INNOVATION

According to indicators like the total expenditure for military R&D or the ability to manage the development of major weapon systems, the "military innovation system" of Italy appears smaller than that of countries like France or Germany.

A detailed study of Italy's defence R&D (Perani, 1999) shows that until a few years ago public funds for R&D and procurement have been spread over many industries and firms, in order to maintain a basic domestic production capability in all defence sectors. Funding shortages and the need for restructuring the military innovation system after the end of the Cold War have led to the identification of some priorities in a few Italian "areas of excellence".

After a period of reduction of R&D expenditures, since 1995 the Ministry of Defence (MoD) has been increasing the share of its budget devoted to research activities. In fact, the year 1995 can be considered the lowest point of a spending cycle. Italian MoD spending for R&D can be estimated at $304.5 million in 1995. In spite of problems of definition and statistical reliability, Italy's R&D data can be considered quite low compared to those of other European countries: leaving aside the UK and France which are big R&D spenders operating large military research facilities and nuclear laboratories, Germany has devoted more than $1.5 billion to military R&D in 1996, Spain is spending around $600 million per year in military R&D and even Swedish military R&D expenditure has been in recent years higher than Italy's. As to the percentage of Government Budget Appropriations for R&D devoted to military purposes, Italy is largely below the level of 10% which can be considered the threshold separating countries which consider military R&D as an essential element of their overall R&D and innovation effort from those which are not willing to enlarge their technological competencies in the defence field.

Data from the MoD show a constant effort of the MoD over the last few years to maintain a level of funding sufficient both to fulfil its international commitments and to support the restructuring activities which are taking place in several military firms. The MoD is obviously considering R&D spending as a component of total investment; acquisition of new weapons therefore reduces the resources available for R&D.

MoD R&D spending is concentrated in a few large procurement programmes. Several programmes began in the early 1990s and are expected to last well into the next decade. These programmes can be considered a heritage of the previous period of military investments in Italy, when the MoD supported the acquisition of autonomous capabilities of system integration by the Italian defence industry in selected fields. On the other hand, the next group of acquisition programmes, mainly launched since 1998 include mainly licensed productions and international collaborations.

In Italy, most of the scientific and technological activities related to MoD needs are carried out by industrial firms. Four main industrial groups currently control most Italian military production: Finmeccanica, Fincantieri, Fiat and Marconi (controlled by the UK parent firm). Finmeccanica and Fincantieri account for more than two-thirds of total military production in Italy. The former produces combat and transport aircraft, missiles and missile systems, radars, electronic warfare equipment, avionics, optronics, artillery and armoured vehicles, helicopters and submarine defence systems, while the latter produces military surface ships, submarines and naval engines. The two companies are large state-owned groups which are slowly going through a process of privatisation. A main consequence is that, while such industrial groups are currently spending an increasing share of their turnover on R&D (between 5% and 10%), they do not seem in the position to identify their own priorities in research with reference to a long-term strategic planning based on market evaluation. The future prospects for these groups are quite uncertain. On the one hand the setting up of a "common" European arms market will give them chances for defining collaborations or strategic agreements or even mergers with larger and stronger European defence groups. The current process of privatisation is also aimed at making such a process of European integration easier. On the other hand, the process of restructuring and privatisation is taking place in a period in which both MoD procurement spending and other sources of public support are constrained.

The strategies of the two main private defence groups in Italy, Fiat (aircraft engines, military vehicles, ordnance and explosives, naval electronics) and Marconi (strategic and tactical telecommunications, electronics), being part of large multinational groups, appear more open to the challenge of the integration of international markets and more flexible in facing the rapid evolution of the international arms industry. The military branches of Fiat and Marconi mainly produce high-tech, R&D intensive military products for which demand is likely to increase and is diversified in different countries.

In July 1997, MURST issued a document (*Guidelines for reforming the National S&T System*) describing the policy proposed by the current Cabinet to "put in order" the "national system of science and technology" which includes laboratories as well as firms and universities. In a chapter of this document, military R&D is mentioned too, describing the attitude of MURST towards it. The document points out firstly the need to promote collaboration between the Ministries concerned. Secondly, a primary operational task of the MoD should be to integrate national military R&D activities with similar activities carried out at the international level. According to the Ministry of Research document, the MoD will have to consider not only international programmes developed by the Western European Armaments Group or by the *Organisme Conjoint de Coopération en matière d'Armement*, but also all the activities carried out, for instance, within the EU Framework Programme which could be of interest to Italian defence. The last point, but no less important, is the promotion of dual-use activities. As already mentioned in previous official documents, MURST has pointed out once again the need for the MoD to exploit existing dual-use capabilities in the Italian research system.

The Italian defence system displays a relevant level of internationalisation but this seems to be the result of a "defensive" attitude of both the MoD and Italian firms towards a fast process of globalisation, rather than a strategic choice. It seems difficult for Italy to play a major role, for instance, in the process of European arms market integration. This is mainly due to a tendency to "defend" the existing structure of the domestic defence system against rapid market development.

A key need here is the setting of priorities. Considering the Italian activities related to international collaborations in military R&D or production, priorities are quite often lacking. Several international agreements for arms production involving Italian firms (as well as the MoD) have been agreed upon, but no clear information emerges about what have been the priorities of Italy in pursuing such international collaborations.

From a strategic and managerial point of view various kinds of international agreements have been experienced by Italian firms and several different partners have been involved in collaborations and consortia. It is hardly possible to describe a unique pattern of collaboration for the Italian defence system.

REFERENCES

Biggeri, L. and L. Scarpitti, 1998, Evaluation in the Italian University System, *International Conference on Evaluation: Profession, Business or Politics?*, Rome, 29-31 October.

Braga, M.C., D. Cini and A. Gambardella, 1999, *The Co-evolution of National, Regional and European Science, Technology and Innovation Policies*, Country study – Italy, TSER, November.

CNR, 1998, *CNR Report 1998*, Rome.

CNR, 1999, *Il nuovo Consiglio Nazionale delle Ricerche*, Roma: D'Anselmi Editore.

De Marchi, M., B. Potì, E. Reale, M. Rocchi and A.M. Scarda, 1998, *Il sistema scientifico pubblico italiano*, Milano: Franco Angeli.

European Commission, 1997, *Second European Report on S&T Indicators*, Luxembourg: OPOCE.

Evangelista, R., G. Perani, F. Rapiti and D. Archibugi, 1997, Nature and Impact of Innovation in Manufacturing Industry: Some Evidence from the Italian Innovation Survey, *Research Policy*, 26, pp. 521-536.

Evangelista, R., T. Sandven, G. Sirilli and K. Smith, 1998, Measuring Innovation in European Industry, *International Journal of the Economics of Business*, Vol. 5, No. 3, pp. 311-333.

Evangelista, R. and G. Sirilli, 1998, Innovation in the Service Sector. Results from the Italian Statistical Survey, *Technological Forecasting and Social Change*, 58, pp. 251-269.

Ferrari, S., P. Guerrieri, F. Malerba, S. Mariotti and D. Palma, 1999, *L'Italia nella competizione tecnologica internazionale – Secondo Rapporto*, Milano: Franco Angeli.

Gammarino, S., M.R. Prisco and A. Silvani, 1999, Alla ricerca di un modello vincente di innovazione regionale: alcune considerazioni sull'esperienza italiana, *L'industria*, Vol. 20, No. 3, pp. 537-565.

IMI, 1999, *Panorama degli incentivi*, Roma.

ISTAT, 1995, *Indagine sull'innovazione tecnologica. Anni 1990-1992*, Roma.

ISTAT, 1998, *Statistiche dell'istruzione universitaria*, Annuari, Roma.

ISTAT, 1999a, *Statistiche sulla ricerca scientifica*, Consuntivo 1996 e preventivi 1997 e 1998, Roma.

ISTAT, 1999b, Lo stato dell'università. I principali indicatori, *Indicatori statistici*, No. 1.

Malerba, F., 1993, Italy, in R. Nelson (ed.), *National Systems of Innovation: A Comparative Study*, Oxford: Oxford University Press.

Mazzola, M., L. Criscuoli and F. Cobis, 1998, *Alcuni dati consuntivi sul primo anno di funzionamento dei nuovi meccanismi di intervento per la ricerca industriale*, Roma: MURST, November.

Mediocredito Centrale, Osservatorio sulle piccole e medie imprese, 1997, *Indagine sulle imprese manifatturiere*, Milano: Il Sole 24 Ore Libri.

MURST, 1997, Relazione alle Camere del ministro dell'Università e della ricerca scientifica e tecnologica, Roma, 31 July.

OECD, 1992, *Reviews of National Science and Technology Policy – Italy*, Paris: OECD.

OECD, 1998, *Science, Technology and Industry Outlook*, Paris: OECD.

OECD-Eurostat, 1997, *Proposed Guidelines for Collecting and Interpreting Technological Innovation Data "Oslo Manual"*, Paris: OECD.

Perani, G., 1999, *Working Paper 2: Italy*, METDAC (Managing European Technology Defence and Competitiveness Issues), CREDIT (Capacity for Research on European Defence and Industrial Technology) Research Group, TSER Project. Web site: http://www.spsg.org/credit/index.html.

Pianta, M. and G. Sirilli, 1997, Impact of Innovation Policies: Evidence from the Italian Innovation Survey, *Science and Public Policy*, Vol. 24, No. 4, pp. 245-253.

Scarda, A.M., 1999, *Le risorse per la ricerca scientifica e lo sviluppo sperimentale nel 1998*, Rapporto Tecnico No. 19, ISRDS-CNR, Roma.

Silvani, A. and G. Sirilli, 1995, R&D Evaluation in Italy: A Science and Technology Policy View, *Research Evaluation*, Vol. 5, 1.

Silvani, A. and G. Sirilli, 1998, Lo stato dell'arte della valutazione della ricerca in Italia: un difficile equilibrio tra domanda e offerta, *Università Ricerca*, Vol. 9, No. 3.

Sirilli, G., 1996, Scientific Research in Italy, *Euromecum*, Stuttgart: RAABE.

Sirilli, G., 1997, Science and Technology Indicators: The State of the Art and Prospects for the Future, in G. Antonelli and N. De Liso (eds), *Economics of Structural and Technological Change*, London, Routledge.

CHAPTER 13

FRENCH RESEARCH AND INNOVATION POLICY: TWO DECADES OF TRANSFORMATION

PHILIPPE LARÉDO AND PHILIPPE MUSTAR

1. THE STEREOTYPED VIEW OF THE FRENCH SYSTEM OF RESEARCH AND INNOVATION RECONSIDERED

During the 1980s and the 1990s the French research system has undergone profound transformations and has been the object of multiple organisational innovations. These two decades have been, for public authorities, an exploration phase characterised by the trial of numerous alleys, by the emergence of new public actors and of new approaches to the implementation of priorities, and by the adoption of new regulations and of new frameworks. These experimentations and the transformations they have given rise to, question the previous analyses of French science and technology policy and their unanimous verdict, namely that the French model is characterised by the dominance of a centralising, Colbertist state. Chesnais describes the three principle characteristics of this model of the French system:
"(1) the organization and funding of the largest part of fundamental research through a special institution, the CNRS, distinct from the higher education sector entities ..., (2) a dual higher education sector producing at least one type of senior technical person little known elsewhere, namely the *Grandes Écoles* technical experts elite of engineers cum industrial managers ..., and (3) a pervasive element of State involvement in the production not just of general scientific and technical knowledge, but often of technology *per se* in the form of patentable and/or immediately usable products or production processes" (Chesnais, 1993, p. 192).

Chesnais takes up elements of a description popularised by numerous authors, who attempted, during the 1970s and 1980s, to analyse research and innovation in France (Gilpin, 1970; Salomon, 1986; Ergas, 1987). They are all in general agreement, using the same diagnostic approach, one concerned with a few selected features, and leading to a stereotyped view of France, with emphasis on the absolute power and rigidity of the state. The corollary of the dominant state is a marked centralisation of choices and decisions shaped by policy for large civil technology programmes and significance accorded to defence research and programmes. The public authorities concentrate their support on large programmes (space, nuclear, defence), which contribute to national prestige and independence. This analysis is also heavily critical of the separation between the universities and the CNRS (seen as the most important research organisation in Europe). It emphasises a congenital separation between research and the firm (where researchers are not interested in firms and industrialists look down on research). It stigmatises the split situation of industry with a few high-technology sectors made up of a limited population of firms investing heavily in research and development on the one hand and a majority of weakly innovative firms with little faith in research on the other. Finally, it draws attention to the division in the higher education system between the *Grandes Écoles*, which attract the best students and succeed in both distancing them from research and instilling in them a sense of superiority, and the universities, which are engaged little in research and which, with very few resources, produce battalions of trained students with scant knowledge of the professional world.

This chapter would like to question the validity of this monochrome vision of a Colbertist system. It was doubtless relevant 20 years ago, even though historical works have begun to show the flaws in this description. However, during the 1980s and 1990s, the French research system, in both its public and its private components, has undergone profound transformations, which challenge the relevance of this analysis of the national situation. It can be left to the historians to show that the picture is infinitely more complex. More simply, this chapter is seeking to examine, from the viewpoint of the current situation, the major evolutions that have characterised the French research and innovation system since the beginning of the 1980s (Mustar, 1994 and 1998; Paillotin, 1993; Amable, Barré et Boyer; 1997; Dodet et al., 1998; Guillaume, 1999). Clearly, the French situation is witnessing numerous worrying problems and deficiencies; however, the concern here is not to deny they exist, but to examine the changes that have occurred.

The objective of this chapter is, firstly, to show that the stereotyped view has run its course, and that, after a slow but far-reaching transition, the current situation is completely different. After a presentation of the long-term trends which illustrate this change (section 2), this leads to an examination of the decline, indeed, the near disappearance of the role of large programmes (section 3), the rapidly increasing, not to say systematic, intertwining of the CNRS and the universities (section 4), the transformation of mission-oriented research institutions, which, following statutory reform at the beginning of the 1980s, have progressively converged towards the paradigm of "collaborative academic research" (section 5), and finally, after the failure of the so-called "mobilising" programmes, the reorientation of support for industrial research towards SMEs (section 6).

Second, an interpretation of the current situation is offered, focusing attention on the emergence of new public powers. These currently make it difficult to closely identify national policy with public intervention, as has been done all too frequently in the past (section 7). The five main challenges that this paradigm shift poses for the future of French policy for research, technology and innovation (section 8) are then considered.

2. CHALLENGING THE POWER OF THE STATE

Most papers examining the French system portray France as the perfect prototype of technological Colbertism, the archetype of a centralised country with strong state intervention. Its science and technology policy forms part of an industrial policy, in which the state sets objectives (through the government "Plan") and steers directly (through nationalised firms and state control, particularly in the defence sector, but also via state monopolies in infrastructure and communications). This specificity is reinforced by the strength of public research relative to that of the private sector. How has the dominant position of the state evolved since the beginning of the 1980s?

The Orientation and Planning of Research Act (LOP) of July 1982 announced that the national research effort, as measured by the ratio of gross domestic expenditure on research and development (GERD) to gross domestic product (GDP) should reach 2.5% of GDP in 1985; 20 years later, this objective has still not been achieved.

The 1980s were characterised by strong growth in research expenditure in France, which stood at less than 1.8% of GDP in 1979. It reached 2.42% in 1990 (Figure 13.1, upper curve), with the state and the private sector sharing responsibility equally. After this date, this ratio steadily decreased, going back down to 2.26% in 1997 (its 1985 level), due entirely to the steady decline in public expenditure from 1990 onwards.

The breakdown of French support for research between funding by the state (Figure 13.1, middle curve) and the private sector (Figure 13.1, lower curve) provides a more contrasting image. It shows that the share of the private sector has been constantly increasing since the 1960s, while the state's share has experienced much slower growth, and has, in fact, experienced an annual decrease during the 1990s. This decrease is largely attributable to the drop in public expenditure on large civil and military programmes. Currently, the share of R&D funded by the private sector exceeds that funded by the state a situation which has hitherto not been seen in France.

In 1970, 70% of national expenditure on R&D was funded by the state, although only 50% was conducted by state institutions (Figure 13.2). This figure alone allows an assessment of the importance of transfers between the public sector and large firms (one-fifth of total R&D effort). These transfers decreased slowly throughout the 1970s and 1980s. In 1990, they represented 15% of total funding: the state financed 55% and carried out 40%. Although the ratio of research conducted remained stable during the 1990s (60% by the private sector, and 40% by what had developed into quite a significant public sector), public funding "collapsed", dropping below 50%. Although

FIGURE 13.1 – The share of national R&D expenditure in GDP

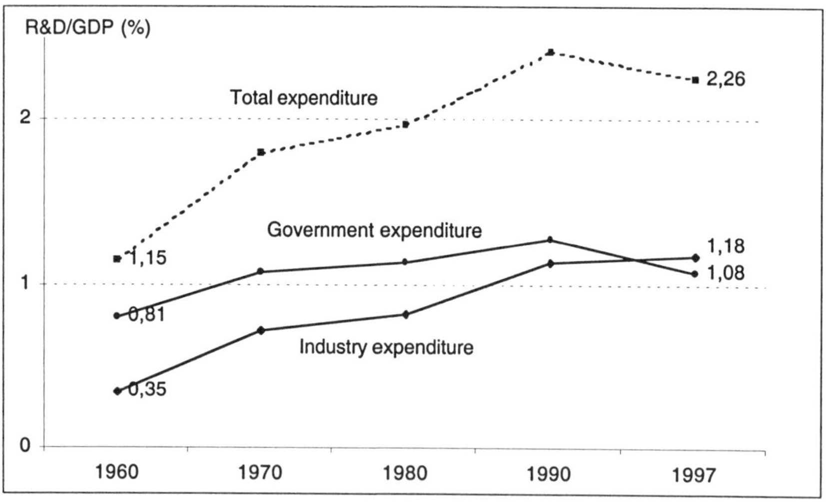

Source: OST, 2000, p. II (data from MENRT treated by OST).

FIGURE 13.2 – French R&D expenditure by performing sector and source of funds

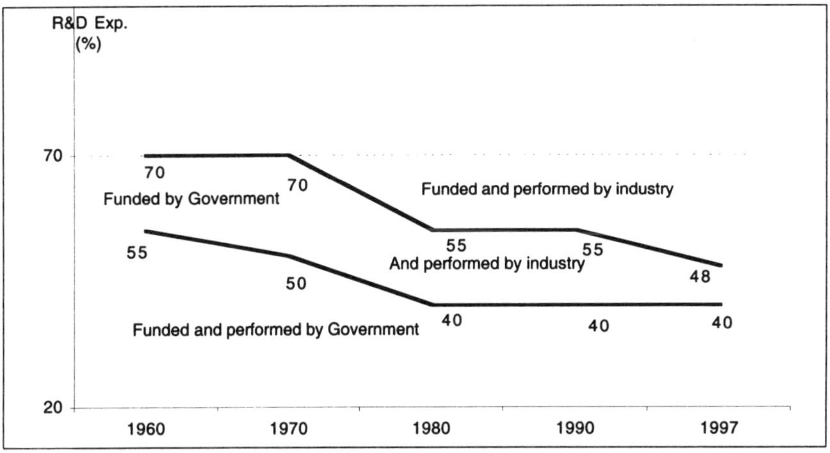

Sources: OST, 2000, p. III (data from MENRT treated by OST).

FIGURE 13.3 – Share of public contracts in total industry R&D expenditure

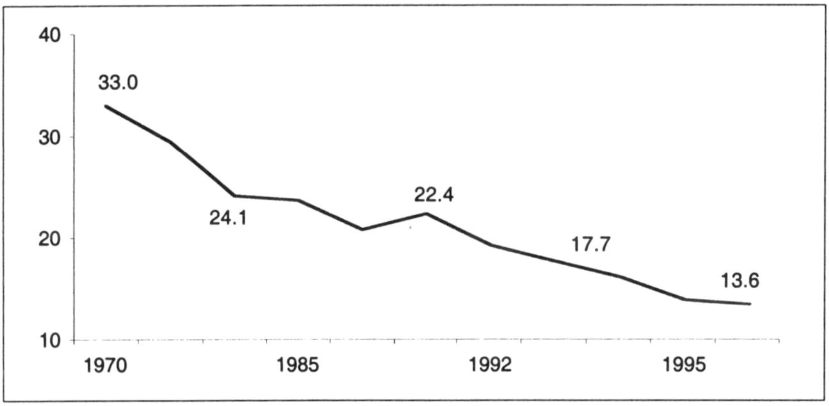

Source: OST 2000, p. IX (data from MENRT-DPD C3, treated by CRERI/OST).

military expenditure still represents more than one-fifth of public R&D expenditure, over 20 years, transfers have decreased by two-thirds to barely 8%, with the private sector carrying out 60% of research and funding 52% of it. By 1996 transfers only represented less than 14% of R&D expenditure within the private sector as compared with 33% in 1970 (Figure 13.3). France can no longer be distinguished from other Western countries with strong military research activity.

The trend described by supporters of Colbertism has been completely overturned: the state is no longer the dominant actor, both in terms of the financing and execution of research. With privatisation, with the sharp decline in funds dedicated to large civil and military programmes, with European integration, "large scale centralised support, technocratic voluntarism, the policy of the 'national champion' and primary protectionism in the form of trade barriers" (Salomon, 1984, p. 119) have vanished. Today only a few avatars remain. The world has changed and France has torn itself away from the Colbertist model, as the following four sections recount. It will be shown that the impetus given to new objectives is less and less a national concern and is moving to the regional and European levels (section 7).

3. LARGE CIVIL AND MILITARY PROGRAMMES, A MAJOR INSTRUMENT FROM A BYGONE ERA

The growth of large civil technology programmes and the importance accorded to defence research and programmes are two marked characteristics of the research policies implemented in France since the end of World War II. In particular, the French public authorities concentrated their efforts on large programmes (such as space, nuclear and defence), which developed national prestige and independence.

One of the specific characteristics of France stems from the fact that the state considered a number of sectors as crucial for the future of the nation and its global position. World War II elevated the importance of science and technology, from the radar to the V2. The Manhattan Project symbolised how a concerted mobilisation of all scientific, technological and industrial actors enabled a decisive advantage to be obtained. Consequently, it was not surprising that one of the first decisions of the new French Government was the unprecedented creation of the Atomic Energy Commission (CEA). While the exact circumstances and reasons merit more detailed consideration, the CEA was not set up with military objectives, but to endow France with a capacity for the production of nuclear energy. Thus, the basic model for the French version of the "large programme" came into existence (large programmes existed in various countries, but they were most widely and systematically implemented in France).

This model was based on a public pilot, which was located within the administration (as in the case of aeronautics), took the form of an agency (as in the case of space) or was suitable for a public research institute (such as in the nuclear domain). It relied on the support of one or more national champions, responsible for the industrial exploitation of the programme and for taking it beyond national borders. This always involved the development of public research capacities, which in turn gave rise to public research institutions. Public services or national operators usually performed a lead-user function for the first full-scale applications, and for demonstrating the relevance of the complex objects developed in this way. All the civil programmes shared a similar rationale: genuine long-term risk-taking, at the frontier of knowledge and technology, where no capacity for the generation of new markets was foreseen. Who could have imagined the development of telecommunications satellites when France launched the first "Diamant" rocket? Who could have judged the market potential of a supersonic aircraft, when France and Great Britain (evidence that the "large programmes" approach was not specifically French) embarked on the Concorde adventure?

These large programmes assumed central importance during the 1960s, with General de Gaulle's return to power. They were concerned with information processing technology (the *Plan Calcul*), civil aeronautics (Concorde and subsequently Airbus), and space (CNES). In addition, there were programmes for marine exploitation, through CNEXO, and, during the 1970s, telecommunications, through the *Direction Générale des Télécommunications,* which later became France Telecom. This model was also applied, with much more significant resources, to the development of new weapons technologies (aeronautics, missiles and communications). Only traditional terrestrial and, in particular, naval armaments had recourse to the Colbertist arsenal.

Each large programme was organised in a "bespoke" manner, according to the pre-existing situation and, in particular, its military connections (a history of the large programmes is sadly lacking, even though some evaluations[1] are available and a number of historical works are beginning to appear). Certain characteristics are sufficient to underline both their budgetary significance and their enduring impacts on the French economy and on French research.

One organisation, the CEA, was responsible for the civil nuclear programme and, during the 1950s, its responsibilities were extended to military nuclear activities.[2] The history of the French nuclear industry is an eventful

---

1. In particular see Tenière Buchot, 1982; Rouvillois and Paillotin, 1985; Lebeau, 1985; Loridant, 1991; and CNER, 1994.
2. A similar situation can be found in the United States, where the weapons laboratories depend on the Department of Energy.

one. During the 1960s gas graphite reactors were abandoned in favour of Westinghouse pressurised water reactors in the civil sector, while the industrial producers experienced a troubled history. However, the oil crisis and the decision by France and its state monopoly, Électricité de France (EDF), to rely completely on nuclear energy, led to the emergence of two world leaders, Framatome in boilers, and COGEMA in the combustion cycle. The termination of the fast breeder programme and debate over the management of waste contributed significantly to the reorientation of action centred solely on the CEA. Thus the programme gradually developed into the funding of one institution of 18 000 persons (of whom 11 000 are on the civil side). In the 1960s, the same model was also used for the exploitation of the oceans (aquaculture, polymetallic nodules and marine thermal energy) centred on CNEXO. However, success was not forthcoming and the CNEXO merged with the ISTPM, which was responsible for fishery research, to create IFREMER, responsible for France's large marine research equipment (including the well-known submarine *Nautile*).

Another organisational format, based on the existence of both civil and military industrial producers,[1] was employed in the case of civil aeronautics. Contrary to the situation in the nuclear sector, activities were under military control, in the form of an administrative directorate, with firms active in both civil and military equipment[2] carrying out the work. The management relied on a military research institution, ONERA, for its research. When the first civil programmes were launched, this organisational format was preserved, with the responsibility of the military leadership extended to civil programmes. The difference between military and civil programmes stems from the nature of their funding: for civil programmes, the state pays for neither research nor equipment, but partially subsidises research and provides reimbursable advances on the sales of apparatus or engines developed. There have been numerous debates, at the international level, concerning these arrangements. In particular, several studies (Giget, 1979) have shown that American public support was of comparable significance to French support, but was linked to the different forms of intervention employed by the

1. These were both public (following nationalisations in 1936 and after the war) and private (Dassault).
2. This military involvement is, to a large extent, linked to the reorganisation of the French system, with the establishment in the 1960s of the DMA, which subsequently became the DGA (*Direction Générale de l'Armement*), the technical directorates of which were responsible for the development of new equipment. The directorate responsible for aeronautical equipment took over steering the civil programmes.

Department of Defence (which brought together several aircraft manufacturers in the same project, and which, through this mechanism, was, for example, largely able to finance the development of the Boeing 747, with the construction of strategic production facilities rented out at low price). The success was finally forthcoming in the framework of international alliances (both European and transatlantic). The increasingly strong output of products (the AIRBUS family and CFM56 engines) and of the producer firms (Aérospatiale and SNECMA) led to the role of the state being eclipsed in both its steering and funding roles.

At the beginning of the 1960s, with the American refusal to sell large computers, the question of national independence in the data processing area arose. Under the aegis of the Ministry of Industry,[1] the *Plan Calcul* (plan for computation), was launched, based on the development of national champions. This plan was accompanied by the creation of a research institute, INRIA, which has remained modestly sized. This particular story is too sinuous for it to be summarised in a few lines. All things considered, since the end of the 1980s, while the core of public research funding has moved, towards community programmes, and recognising that BULL is a modest producer on the international level, only ST Microelectronics, based on the European programme JESSI and stemming from a Franco-Italian merger, has succeeded in making an impact on the international level. When, during the 1970s, telecommunications were becoming a "national imperative", activity was organised in yet another, different way. The national operator France Telecom, which already had its own important research institute, CNET, was entrusted with this mission.[2] Once again, progress was not linear, but was sufficiently quick for ALCATEL to become one of the major global industrial actors in the sector and for firms such as SAGEM to assert themselves on the international level. The privatisation of France Telecom included its research centre, and this is the sole exception, leading to a total disengagement of the state.

1. The Ministry of Industry also piloted another "quasi"-large programme on the exploration and exploitation of oil resources (particularly marine resources). Its originality relies on a specific tax on gas sales which provides both funding to support research (conducted for the most part by companies) and finances a research institution, the IFP, which, like the CEA, has generated numerous powerful industrial subsidiaries and is a major actor in petroleum engineering.
2. To be precise, it was necessary to speak of the "Direction Générale des Télécommunications" as France Telecom had yet to be separated from the whole "PTT" organisation.

TABLE 13.1a – Large programmes in the Civil R&D Budget (BCRD)

	1980	1985	1990	1995	2000
Telecommunications	2.7	3.5	4.0	–	–
Electronuclear	4.0	6.9	6.3	6.0	6.9
Space	1.4	4.0	7.2	8.8	8.8
Civil aeronautics	1.2	2.1	2.9	2.2	1.8
Electronics	0.4	3.1	2.2	2.1	1.5
Total	9.7	19.6	22.6	19.1	19.0
Total BCRD (including telecoms)	18.8	42.1	49.4	51.8	55.5
Share of large programmes	51%	47%	46%	37%	34%
Share of large programmes in BCRD excluding space (2)	44%	37%	31%	21%	19%
Share of large programmes excluding internal expenditure of CEA (3)	–	–	–	10%	7%

Sources: Annexes to the Finance Law "loi de finance", initial endowments from all sources (billion Francs – current value).

Despite the progressive phasing out of certain programmes, the 1980s only witnessed a modest decrease in state involvement across large civil programmes, on account of the increase in the significance of the space programme, which was delegated to an agency, the CNES, and which was largely conducted under the auspices of the European Space Agency (ESA). There are numerous studies on the direct and indirect effects of this situation (Beta, 1988). On the national level, emphasis can be placed on its weight within state funding for the civil sector (above 10% for almost a decade), and on the fact that the successes engendered by Ariane Espace and its launcher have yet to have any repercussions on funding requirements. In addition, the completely independent nature of the market for communication satellites (and the emergence of significant French actors, most often in the context of European co-operations) and the new challenges of satellite positioning, with the first direct interventions by the European Union (with the financial participation of the competing European GPS system, Galileo) are also important. In fact, it remains the only large programme that is still significant, but it is not specific to France. There are similar programmes, organised in the same way, in the majority of the large Western countries.

The 1990s has therefore seen a drastic reduction in the mode of state intervention labelled as "large programmes" (see Table 13.1a and b). Certain

TABLE 13.1b – Internal and external expenditure of organisations associated with large programmes (1996)

	Internal expenditure	External expenditure	Share of firms in total expenditure
CEA (civil and military)	98%	2%	2%
CNES	42%	58%	11%
Civil aeronautics (including ONERA)	52%	48%	48%
Electronics (including INRIA)	31%	69%	69%
Total	63%	37%	15%

Source: MENRT (1999), p. 11.

of these have disappeared completely from the scene, such as the Telecommunication programme with the privatisation of France Telecom and, of CNET[1] within it. Others have become of secondary importance and are limited to marginal reimbursable advances,[2] such as civil aeronautics. The nuclear programme now only covers the internal expenses of the CEA, where, currently, two out of the five operational directorates are concerned with non-nuclear domains (life sciences and advanced technologies), and a third covers a broad disciplinary spectrum of "upstream" research in materials sciences. The electronics *filière* and its heavy commitments (the capital endowments to Bull and the JESSI programme within the EUREKA initiative) almost no longer exist: the set of state actions aimed at new information and communication technologies (from multimedia to information highways via microelectronics and including telecommunications and EUREKA programmes such as MEDEA, NESSI, EURISMUS or PEDEA) represented barely 1.5 billion Francs in the budget for the year 2000.

1. The Ministry of Industry has recently created a national network for telecommunications research "to support the projects of operators and parts manufacturers ... in association with public laboratories". This network is putting out calls for tender (funds committed in 1998: approximately 200 million Francs).
2. It is difficult, taking into account French budgetary practices, to obtain the exact net total of public financial commitments. In 1999, the total resources budgeted reached approximately 1.4 billion Francs. Reimbursable advances are made according to sales. The total of reimbursements given from the general budget has not been identified (they are mixed with various state revenues). However, according to estimates, the successes of AIRBUS, of the CFM56 and of the FALCON 2000 generate reimbursements which are at least equivalent.

BOX 13.1 – Military R&D expenditure

At the end of the 1970s, the share of military R&D of GERD was around 37% (OST, 1994, p. 48). In 1992 the figure was almost the same (36%), before entering into a steep downward spiral: in 1996 it was 23% (MENRT, 1999). This figure has continued to decline rapidly since, with financial commitments (evolution of the RDE, source: annex to the *loi de finance 2000*, p. 51) dropping from 34.5 billion Francs in 1990 to less than 20 billion in 1998 and 1999. Expenditure went from 34.7 billion Francs in 1990 to 25 billion in 1996, 24 billion in 1997 and 21 billion in 1998 (source: ibid.).
The analysis of research reports from within the administration (see table below) shows an even more drastic decrease than that in expenditure, notably in external expenses (which included industrial contracts as well as all public civil research contracts).
So, in 5 years (1992-1996), while the share of national public contracts in the GERD of firms went from 20% to 14%, military R&D contracts decreased even more rapidly, their proportion dropping from 78% to 64% of the total (OST, 2000).

Evolution of recorded military expenditure (billion Francs, source: MENRT, 1999).

Expenditure	1992	1996	1997
internal	11.5	10.4	9.7
external	15.4	11.8	10.8
total	26.9	23.2	20.5

Of course, the situation is different in the case of military programmes since the state is a privileged or unique client (as is currently the case of the Dassault RAFALE aircraft). However, two points must be emphasised. First, the rise of European firms (formerly through limited alliances such as EUROCOPTER but now forming truly European entities such as EADS, which arose from the merger of Aerospatiale-Matra and DASA) is considered by all analysts as inevitable, as is the Europeanisation of orders (following closely on from the EUROFIGHTER, competitor to RAFALE, and in spite of the recent failure of the European frigate project). Next, in contrast to the United States, where the decrease in defence budgets has not affected military research (this represents a secondary share of total military expenditure which has fallen), France has experienced a regular and pronounced decrease of public expenditure on defence R&D since 1992 (see Box 13.1).

The total impulse effect of military programmes is greatly reduced and is centred on specialised firms in the defence sector. The large civil programmes, with the exception of the space programme, which has experienced this same specialisation phenomenon (Bach et al., 1995), no longer exist. The 1990s have thus seen the near disappearance of what has been a central mode of state intervention in research during the post-war period. On the other hand, the state has inherited a set of public research institutions from this policy, which have undergone the same transformations as those institutions born out of sectoral policies (see section 5).

4. THE MIXED SYSTEM OF THE CNRS AND THE UNIVERSITIES

Alongside the large technological programmes, another component of the stereotyped view of French research is the division between the universities and the CNRS. This section shows that this division is, today, being replaced by much more closely interlocked structures. Before World War II, the matter of support for scientists had already been addressed, notably with the transformation of the National Fund for Science (*Caisse Nationale des Sciences*) into the CNRS in 1939 (Picard, 1999). However, the choice had yet to be made between what is today referred to as a project funding agency, in the style of the Anglo-Saxon "research councils", and a public research institution, employing professional researchers. This choice was only made after World War II, and clearly marked out France from other Western countries. Of course, there are other public fundamental research institutions in Europe, particularly the *Max Planck Gesellschaft* in Germany, but none are of the same size as, or have the central role of the CNRS. For the record, with more than 26 000 persons, of which 14 400 are researchers and engineers, the CNRS currently accounts for one-quarter of national expenditure on civil research. The creation of the CNRS diffuses the model of the full-time professional researcher, which can be distinguished from that in the French university sector, where civil servants carry out the dual activities of teaching and research.

The creation of the CNRS was justified by the weak state of university research after the war. Nevertheless, historians have emphasised that the universities have always been engaged in research (in chemistry, for example). They deny that there had ever been such a clear-cut division between universities dedicated solely to teaching and the CNRS, devoted entirely to research (Jacq, 1996). At the end of the 1940s, university research was relatively unstructured and its organisation did not, for example, permit the critical mass necessary for the deployment of "big" physics. To help its structure,

TABLE 13.2 – Average composition of laboratories linked to the CNRS in 1998

Number of staff	Persons	%
Enseignants-chercheurs (teacher-researchers)	14	29%
CNRS researchers	9	18%
Researchers from other public research institutions	2	4%
Ingénieurs de recherche	4	8%
Other technical personnel	10	20%
Doctoral and postdoctoral researchers	10	21%
Total	49	100%

Source: CNRS, strategy directorate, January 2000.
Note: Ingénieurs de recherche are qualified engineers engaged in technical and research support functions.

from the middle of the 1960s, the CNRS put in place a system of partnership between its personnel and university groups, on the basis of a periodic evaluation by its committees. These "associated" teams were consequently able to receive CNRS personnel. This mechanism, which became increasingly extensive, had two lasting effects, which distinguish the landscape of French academic research. On the one hand, it dissociates, within the universities, teaching departments from research structures. On the other hand, a growing proportion of CNRS personnel have been working in these groups. Currently, four out of five CNRS units are joint units (*unités mixtes*) between the CNRS and universities, and these incorporate four-fifths of the organisation's personnel.

The hybridisation between the universities and the CNRS is widespread (one aspect of current debate is the extent of the adoption of this mixed system throughout the institution): Table 13.2 shows that the average CNRS unit comprises more university teacher-researchers than dedicated CNRS researchers. This situation is reinforced by the growth in the research potential of the universities. In the 1970s, the number of CNRS researchers was at a similar level to that of French university researchers (in full-time equivalents). This relationship has progressively reversed as a result of the strong and continued growth in recruitment in higher education. At the end of the 1990s, ten positions were created in universities for each one in the CNRS. Currently the research potential of the universities (even when analysed in full-time equivalents) is more than double that of the CNRS (Table 13.3). The second phenomenon is the increasingly pronounced involvement of the *Grandes Écoles*, most of which are engineering schools, in research. This is another structural feature of higher education, which distinguishes France

TABLE 13.3 – Elements of public research in France

Personnel	Academic laboratories	Public civil laboratories	Defence laboratories	Total
Teacher-researchers	44 000	–	–	44 000
Lecturers	12 000	–	–	12 000
Researchers	11 500	19 000	3 050	33 550
Post-doc (and ATER)	12 200	1 000	–	13 200
Doctoral students	13 650	4 350	–	18 000
Technical/administrative staff	40 600	30 000	16 500	87 100
Total	133 950	54 350	19 550	207 850

Source: OST 2000. Composition in 1996.
Notes: Academic laboratories combine the potential of the universities, of the *Grandes Écoles* and of the CNRS. The civil public laboratories comprise other public research institutions and ministerial research services. Defence laboratories are solely concerned with the interior resources of the Ministry of Defence (excluding teaching establishments and research institutions, such as ONERA).
The 12 000 lecturers correspond to teachers qualified in secondary education employed in universities (PRAG) and who only have teaching duties. The ATER (temporary teaching and research attachés) are a recently created type of temporary university employee (at doctoral or postdoctoral level). They constitute the core of temporary employment of young researchers (9200 out of 12 200).

from other Western countries. Currently, across all disciplines, one in every five Ph.D. theses in France is produced in the research centres of these schools, although they only contain barely 6% of all teacher-researchers.

This transformation has marked the disappearance of one particular feature of the French situation: its frequently described dualism no longer exists. Higher education is playing an increasingly central role in the public research system. Taking this analysis to the extreme, the CNRS could be considered, following the example of the Anglo-Saxon research councils or the NSF, as a research support agency, but more specifically an agency concerned not with projects, but with structures, and which makes its contributions in the form of human potential and large technical resources rather than financially.

5. THE TRANSFORMATION OF MISSION-ORIENTED RESEARCH INSTITUTIONS

In France, as in various other countries, different government departments, confronted with various problems over the course of time, have developed their own research structures, which have progressively developed degrees of autonomy, in the form of "government laboratories" (to use the recognised Anglo-Saxon terminology). The basic mission of these laboratories is to respond to the needs of their different parent ministries. These activities entail an unequal combination of three logics. The first is that of the need for innovation in relation to their area of competence or responsibility. The key words are: the security of goods and personnel, quality and regulation.[1] The second relates to their activities as producers of public services for these ministries. In particular, this is the case for infrastructures managed by the state. The third logic is deployed in dispersed sectors, where individual actors are both numerous and too small to innovate, so the state, through its institutions, acts as an innovator on their behalf.[2]

France has not escaped this type of activity, and since World War II has progressively structured these various mission-oriented public entities into a series of "public research institutions" (see Table 13.4). In 1946, INRA was created, followed by a whole series of institutes associated with agricultural activities, including CEMAGREF for forestry and agricultural mechanisation, CNEVA for veterinary research.[3] ISTPM was established for fishery research and was later merged into IFREMER. IRD was created during World War II for "colonial" research, before focusing on the problems of developing countries. Later, the various institutes specialising in tropical agriculture were brought together in CIRAD. Infrastructural consultancy services were incorporated into LCPC (for civil engineering) and CSTB (for building). Other institutes were also set up over time, such as INSERM for health and medical research (reorganised into its current form in 1962), INRP (for research in education), CEE and CEREQ for employment and qualifications as well as INED for population. BRGM, the state mining organisation, also has a research service. The 1960s saw the practice of government departments setting up their own specialised research institution become widespread. Among the last to be established were the Institute for Transport Research (created in 1971 and becoming the INRETS at the beginning of the 1980s, following

---

1. Today, the importance of this logic can be measured in the increasingly frequent "food crises".
2. Agriculture and the INRA, developer of numerous varieties used in France (from fruit trees to wheat, via chicory) is a striking example. Moreover, it is symbolic to see that the work of Griliches on the social benefits of public research focuses on a case of the development of new agricultural varieties by public laboratories.
3. Currently integrated into the recently established *Agence française de sécurité sanitaire des aliments* (AFSSA) – French Health and Food Safety Agency.

TABLE 13.4 – List of French public research institutions

OPR	Status	Staff	Mission
INRA	EPST	8 533	Agricultural research
INSERM	EPST	5 008	Health and medical research
IRD	EPST	1 635	Development research
CEMAGREF	EPST	612	Research on agricultural machinery, water and forestry
LCPC	EPST	574	Research on roads and bridges
INRETS	EPST	414	Research on transport and transport security
INED	EPST	163	Demographic studies and research
CIRAD	EPIC	1 812	Agricultural research for development
IFREMER	EPIC	1 345	Ocean, marine and fisheries research
BRGM	EPIC	890	Research on mining and geology
CSTB	EPIC	568	Building research
INERIS	EPIC	425	Research and studies on environment and industrial risks
AFSSA	EPA	700	Research on animal and food safety
INRP	EPA	263	Pedagogical research
CEREQ	EPA	163	Research on employment and work qualifications
CEE	EPA	75	Employment studies
Instituts Pasteur	Fondations	914	Health and medical research
And also			
CEA	EPIC	11 354	Research on nuclear energy
ONERA	EPIC	2 183	Aerospace research
INRIA	EPST	766	Research on computer science and artificial intelligence

Notes: EPA (*Établissement Public Administratif*) is the general status for French state-owned agencies or institutions. EPST (*Établissement Public à caratère Scientifique et Technologique*) is the specific status created by the 1982 law for research agencies or institutions.
EPIC (*Établissement Public à caratère Industriel et Commercial*) is the general status for state owned organisations which engage into economic activities. They differ from the two previous ones by the fact that employees are not civil servants and that they depend upon the usual commercial legislation. Staff mentioned for CEA excludes defence activities.

the absorption of the national association for research on road safety) and INERIS for the environment and industrial risks (from the former French Coal Board Research Centre).

As in other countries, there have been concerns over the future development of these institutes since the beginning of the 1970s. A series of *ad hoc* audits have been undertaken, but have had no major effects (Larédo and Mustar, 1995). However, the 1982 Research Act (LOP), which followed the new Socialist Government and the national research conferences (*Assises nationales de la recherche*), radically transformed the Government's research infrastructure, changing the legal framework, the operation of these organisations and their integration in the public system. Two central elements should be borne in mind at this point to understand the substantial changes that the majority of these institutions have undergone over the past two decades.

The majority of the organisations were given a new status, that of "EPST" – public scientific and technological establishment (*Établissement Public à caractère Scientifique et Technologique*). This status resulted in all employees, including researchers, becoming civil servants. Procedures for recruitment and career development adopted the competitive examination system, as in the rest of the French civil service. For reasons which deserve detailed historical studies, the CNRS model of recruitment at a "young" age (following postgraduate study) by discipline-specific committees and on academic career criteria, was imposed across all organisations. It is not surprising that 15 years after the Research Act (LOP) was passed, academic visibility is very strongly reinforced everywhere (as explained by the development of a new speciality – scientometrics). This trend is illustrated by the case of INRA, which, at that time, had numerous groups engaged in the process of innovation, interacting closely with the agricultural world. Two examples are when one group developed a new variety of chicory, Zoom, which transformed national and European markets for chicory, and when INRA researchers participated in the "re-creation" of the Beaufort traditional cheese (INRA-CSI, 1998). Currently, such units, labelled by Joly and Mangematin as "technical centres" serving particular professions are rare within INRA (Joly and Mangematin, 1996). The majority of units are concerned with other types of research activities, which are qualified by these authors as centred on the development of "generic softwares" or on "specialised basic research". They also note the existence of an important number of laboratories undergoing change, thus emphasising the importance of transformations in action. This trend is not about the disengagement from research areas within the domain of activity of the institution, but a marked reorientation and a progressive replacement of professional associations as privileged partners by large firms (either individually or as members of "clubs"). The links between research institutions and the economic world are increasingly centred on contractual relations, which take account of the "professional" specificity of both partners. This leads to the development of co-operative forms, with industrial partners using their own research capabilities to capture and mobilise internally the academic work[1] of researchers in public institutions.

A similar trend can be observed in the relations with the public authorities: units within institutions are increasingly involved in research of collective interest, which is carried out on a contractual basis. This is linked to the combined effects of two changes. On one hand, the 1982 Act brought about a financial division between organisations and their parent ministries with the transfer of budgetary responsibilities to a single ministry in charge of research. Even if the parent ministries had preserved "joint guardianship", the unwieldy nature of the budgetary mechanism is too well known for the eventual impact of such a separation not to have been weighed up. The second stems from incorporation into the civil service and from its associated budgetary constraints, which have become more and more cumbersome in budget allocations. Currently, salaries represent 85% of the annual grant to INRA, leaving units and researchers to obtain the majority of their research funds[2] from outside. At the same time, the parent ministries have not been given significant financial operating resources and have struggled to follow the contractual orientation of the famous "customer–contractor relationship" promoted by the Rothschild report (1971). This left other public actors, the regions (an important lever in support of the "technical centre" role of INRA entities) and European programmes (the principal "proxy" customer for a number of new collective challenges), to play a central role in the promotion of the functions traditionally devoted to these mission-oriented organisations.

The development of academic research, the use of classically configured relationships between public research and companies, and an increasing role of incentives for research oriented towards public missions, are nowadays all part of the spectrum of activities covered by public research institutions. Together with their modes of financing, this spectrum points towards reconsideration of the position of the institutes in the national research system, where they can no longer simply be categorised according to some notion of "mission-oriented research". While recognising their frequently central role in research in particular "domains", their activities are becoming only slightly differentiated from the research conducted by the CNRS and the universities.

1. In the sense of work leading to publication in academic journals.
2. This gathers all running costs but salaries for permanent staff.

Mixed units, combining professional CNRS researchers and teacher-researchers, have also seen their activities evolve in the same way. Following closely on the 1982 Research Act (LOP), an extraordinary development in public–private research relations took place, which is all too often underestimated. The number of contracts between laboratories linked to the CNRS and firms increased tenfold between 1984 and 1996: climbing from 350 to more than 3200, while the sums received by the laboratories currently exceed those from national funding schemes. Taking all disciplines into consideration, these relationships affect half the laboratories of or associated with the CNRS. A number of surveys carried out on the laboratories in particular regions showed the extent of the links with the economic world (links which provide them with almost 30% of their budgets, salaries of permanent staff apart) (AURA, 1998). These relationships rely on renewed modes of co-operation, increasingly set up as equal partnerships between public researchers and industrial researchers, which entail a concentration of relations with firms with their own research capabilities (leading large firms and high-technology firms), and no longer focused on technical assistance to firms or specific professions. The survival of the latter type requires strong incentives from the research institutions or from government departments. "Valorisation" and "transfer" have thus become official duties for researchers (as civil servants), at universities within the CNRS or in the world of research institutions. The multiplication of research exploitation offices or transfer services, an increasingly active patenting policy (with new rules for the sharing of licence income between public research institutions and inventors), the creation of subsidiaries aimed at furthering the diffusion of results (such as INRIA Transfert), emphasis on the creation of firms by public sector researchers (currently encouraged by a change in the status of the researcher, by a policy for increasing the number of incubators and by a national competition for firm start-ups): all these activities provide an indication of a trend towards convergence between all public research institutions and with universities. The increase in importance of the research potential of universities (see above) and the ongoing debate on the systematisation of mixed research units further reinforce this trend towards unification. They explain why it is misleading to continue to differentiate between an "applied" pole and a "fundamental" pole in public research. Increasingly, public sector research is guided by a dominant focus, which can be described as "collaborative academic" research.

In comparison with changes in the British situation, where the government laboratories are moving toward a much greater "service" orientation, French reform is heading towards a convergence of the entire public research sector, not a withdrawal into basic research, cut off from socio-economic issues, but the more systematic construction of collaborative research, which takes account both of the requirements of scientific quality and the problems of partners (taken in the wider sense and no longer defined by the needs of parent ministries). This movement is drawing closer to Gibbons' mode two, decentralising initiatives and favouring the diversity of network approaches, but keeping public sector research a central element of the approach.

6. BROADER BASIS AND INCREASING AUTONOMY OF INDUSTRIAL RESEARCH

6.1 – The major trends

In 1980[1] there were 1300 firms carrying out R&D (according to the Frascati definition). Their expenditure had risen to 31 billion Francs, representing 1.12% of GDP and 2.4% industrial added-value, they employed 126 000 persons, including 34 000 researchers. Analyses drew attention to three principal characteristics of this activity: a very strong concentration on the 100 or so firms employing more than 50 researchers (representing three-quarters of the effort); a strong sectoral concentration (five sectors – aerospace, electronics and data processing, chemistry, pharmacy and automotive – were carrying out 70% of the total effort); and a progressive increase in their own funding (their share went from 55% in 1965 to 70% in 1980), public funding being concentrated on two sectors (aeronautics and electronics accounted for 85% of the sums paid out).

Two decades later (1997 figures),[2] the concentrations observed in 1980 endure: the top 100 firms are still carrying out two-thirds of the total effort, and the same five sectors[3] have a similar significance (64%). This stems, in part, from the weak overall increase in resources dedicated to industrial

1. Source: annex to the Finance Act (*projet de loi de finance*), 1983.
2. Source: annex to the Finance Act (*projet de loi de finance*), 2000.
3. The terminology has evolved: "aerospace" is now used in preference to "aeronautic". An important part has been transfered to the sector of "instruments for control, navigation, measurement and precision". This sub-group (which was marginal in 1980) also combines a part of the electronic equipment sector, which has now been relabelled the communication equipment sector.

research: even if the number of researchers has been multiplied by two (70 700), total employment has clearly progressed less (+30%) as has the share of GDP (1.37%, representing a growth of 22% in 17 years). However, these continuities conceal four major changes.

The first stems from the more important relative growth of R&D expenditure compared both with GDP and with tangible investments.[1] This is all the more significant because firms themselves are funding an increasing share of their research investments, which is the second long-term trend. Between 1980 and 1997 the share of public contracts decreased from close to 25% to below 15% (Figure 13.3), including defence funding, as well as those for the large programmes.

The third change, is evidenced by a fourfold multiplication of the number of firms with an organised R&D effort: 5600 in 1997. This reflects a near doubling of the role of SMEs (firms with less than 500 persons) in the national research effort: in 1983, they employed 17% of researchers, while in 1996 this figure was 28% (OST, 2000). The public policies implemented continuously since the beginning of the 1980s (see the following paragraph) have certainly played a decisive role in this increase.

The final major change is that it is no longer possible to speak of a separation between private and public sector research, as in the 1981 conference. On the contrary, there are an abundance of indicators emphasising the explosion of links between the two.

The tenfold multiplication between 1983 and 1996 of the number of contracts between the CNRS and industry has already been mentioned (increasing from a few hundred to more than 3000, representing around three contracts per CNRS associated laboratory per year). Over the same period the total number of industrial contracts let to public research institutions has increased eightfold (from 500 million Francs in 1983 to 4.1 billion Francs in 1997) and currently represents 4% of firms' research expenditure. The annual survey carried out by two major French journals shows that, on average, the large firms surveyed co-operate with 70 different laboratories, hosting or funding around 40 doctoral researchers and publishing, usually jointly, 30 or so articles in the most prestigious scientific journals.[2] The change in

1. The Annex to the *projet de loi de finances* 2000 (p. 54) exhibits a revealing comparison of relative changes over time in the volume of firms' R&D expenditure, the gross fixed capital of enterprises and GDP.
2. Namely those journals taken into account by the "Science citation index". See Larédo and Mustar, 1999.

these relationships is not only quantitative. This period has also seen the appearance of new forms of relationships, which intimately mix public and private laboratories. These mixed laboratories, jointly established by research institutions and companies, together with CIFRE agreements, explain, to a large extent, the increasing involvement of firms in research training.[1] National programmes, launched at the beginning of the 1980s, have also played a role in this trend, despite their eventful existence (see section 6.3). However, it is without doubt the other public actors, who have appeared since the 1980s, European programmes and regional research policies, who have played a central accompanying role in this development (section 7).

6.2 – The increasing role of SMEs in public support for industrial research

The French "technological gap" is a recurrent aspect of policy debate on industrial research. This started with the productivity missions financed under the Marshall Plan, leading to the creation of technical centres.[2] Some celebrated works (Servan Schreiber, 1967) contributed to this movement, which, at the end of the 1960s, led the DGRST to introduce a procedure for pre-development and later for innovation assistance. Since the end of the 1970s, its resources and its management have been delegated to ANVAR (the National Agency for the Valorisation of Research), progressively becoming

1. The "Industrial agreements on training through research" (CIFRE) organise a partnership between a company and a research laboratory to allow a young researcher to prepare a thesis, dividing his/her time between the two partners. These young researchers are recruited by the companies in return for fixed financial support by the State (today estimated at a third of the salary costs of the young researcher). 200 million Francs (MF from now on) are allocated in the 2000 budget, which corresponds to undertaking 800 new theses a year.
2. Fifty years later, there are still almost 50 of these, covering different traditional industrial sectors. These technical centres, the most important of which are CETIM, for mechanical industries, and IFP for the oil industry, represent 1.9% of domestic expenditure on R&D by firms (BERD). Two-thirds of this effort is funded by supplementary taxes deducted from sales in the sector. Taking into account the new rules of the EU this mode of financing is due to evolve, with the abolition of the taxes and the integration of this funding into public research expenditure.

TABLE 13.5 – Firms benefiting from research tax credits

Size of firms	Turnover	research declared: % of total	credits obtained: % of total
Small	< 50 MF	7%	23%
Medium	From 50 to 500 MF	17%	29%
Large	> 500 MF	76%	48%
Total		100%	100%

Source: Annex to the *Loi de Finances* 1999, pp. 16-17.

its core activity.[1] Since the mid-1980s, ANVAR's activities have been focused on SMEs, contributing to the progressive expansion of the realm of firms conducting R&D. Each year, more than 2000 SMEs are affected by this reimbursable form of aid.[2] Since the end of the 1980s, "young" SMEs (less than three years old) have represented a quarter of the firms supported.

Attention towards SMEs was stepped up in 1982 with the introduction of research tax credits, a fiscal arrangement which reimbursed firms one-half of the increase in their R&D expenditure.[3] Using a broader definition of R&D than that of the Frascati manual and with specific dimensions favouring SMEs (Table 13.5), tax credits have contributed considerably to the development of R&D activities in SMEs. In 1987, 3500 firms submitted applications to the scheme, ten years later, this figure had doubled, exceeding 7000 firms. Half of these (3400) had effectively benefited from a tax

1. Created in 1974 to promote the exploitation of the results of public research, ANVAR has progressively transferred this particular activity to research institutions and their own exploitation service (see section 5), while concentrating its activities towards supporting SMEs. It has also progressively extended the range of its procedures: on top of loans granted for supporting innovation projects, it supports the contracts let by SMEs to contract research societies (SRC) agreed by ANVAR, and it also supports the recruitment of Ph.D.s and qualified research engineers by SMES. In 1998, ANVAR has accompanied the recruitment of more than 1000 researchers, which is some 5% of the total of researchers in French SMEs.
2. It is possible to weigh up the reimbursements made by SMEs by comparing the 1999 budget allocation (795 MF) and the total of support provided to firms through all routes in 1999 (1380 MF).
3. It is charged to the tax on benefits and takes the form of a repayment by the Treasury, when the amount deducted exceeds the taxation. This latter arrangement is particularly important for new technology-based firms.

reduction or a reimbursement[1] equivalent to 3 billion Francs, a sum more than twice the advances provided by ANVAR.

There are other complementary procedures of less financial scope, which are largely oriented towards SMEs. In particular, these include assistance for the diffusion of techniques to SMEs from the Ministry responsible for industry (ATOUT)[2] and EUREKA.[3] Although, to an extent, distinct from it, these procedures must be associated with the invitation to tender for projects in the "key technologies" (*technologies clés*) initiative, created in 1996, closely following the study of 100 key technologies (Ministry of Industry, 1995), and launched annually by the Ministry in charge of industry in association with ANVAR.[4] In total, these various different measures represent less than half of ANVAR's annual commitments.

Support for innovation research and tax credits, which have been implemented on a large scale since the beginning of the 1980s, have since become well established in the landscape of French industrial research. These two procedures are characterised by a "bottom-up" approach (where firms take the initiative and submit proposals), by clear mechanisms of attribution[5] and broad coverage, in terms of both types of firms and sectors of activity. These

---

1. In order to be able to benefit from this measure for the years when they increase their research efforts, firms are obliged to submit an annual declaration.
2. Several procedures (microelectronic components, computer aided design and manufacture, machines and facilities of advanced design and new materials), which were created in the 1980s, modelled on those set up by the BMFT in Germany, have been brought together in a single procedure, the ATOUT procedure, under the management of the regional services of the Ministry. In 1998, this represented almost 170MF of support.
3. EUREKA includes certain "strategic projects", in particular those linked to information and communication technologies, such as MEDEA or the recent ITEA and PIDEA programmes. These latter programmes are financed by what remains of the resources from the large programme on information technology and electronics, and have not been taken into account in the calculations carried out. However, the accent is increasingly on EUREKA's second mode of intervention, "bottom-up" projects put forward by firms. All the discussions and all the analyses emphasise the role of EUREKA for SMEs. Thus out of the 43 new projects which were approved at the June 1999 EUREKA ministerial conference in which there was French participation, there were 42 with SMEs (annex to the *Loi de finances pour 2000*, p. 39).
4. In three years, ANVAR has retained 240 dossiers (308 MF of support awarded) and the Ministry of Industry 81 (497 MF of support awarded).
5. ANVAR's activities are decentralised: decisions are taken directly by its 24 regional delegations. Over a decision period of three months, a threefold examination (technical, commercial and financial) is carried out and the result communicated to the firms.

characteristics can be seen as contrasting with those of large-scale programmes (section 3). Similarly, their continuity conflicts with the erratic life of the efforts to mobilise national potential to meet priority aims.

6.3 – The ups and downs of national "technological" programmes

This stability conflicts with the turbulent existence of "technological programmes". Implemented in the wake of the oil crisis, these programmes largely anticipated the results of work on innovation in the social sciences. Their objective is "to encourage heterogeneous actors (scientific research laboratories, centres for technical research, firms) often competitors (two firms or two laboratories) to jointly identify, on specific geopolitical levels (regional, national, European), competences of future strategic significance and to develop co-operative activities which will allow them to acquire and exploit these competences" (Callon et al., 1997). Activities, such as the "Plan Construction" (launched in 1971), and agencies, such as COMES for solar energy, foreshadowed this trend in France. However, it is the British ALVEY programme which analysts (Georghiou, 1999) consider to be the precursor of this new form of public intervention. After the *Livre blanc sur la recherche* initiated by P. Aigrain (Secretary of State for Research, 1980), which advocated such activities, the 1982 Research Act created seven mobilising programmes and a series of priority programmes. These gave rise to "missions". As a result of their success, 20 or so programmes were launched, although few of these survived the change of government in 1986. The only programmes to last concerned the control of energy (conducted by an agency, AFME which subsequently became ADEME with the expansion of its responsibilities to environmental technologies), land-based transport (through PRD3T and then PREDIT), and, through a lesser measure, agro-food industries (the "Aliment" programmes). Carried along by changes in the Government, and under modified labels, the imperative to develop synergies between actors remained. Currently, it takes the form of "thematic research and technological innovation networks", which carry only marginal weight in state intervention (Box 13.2).

Even though "technological programmes" have become the principal form of European Community intervention (as will be discussed later), it must be noted that they have not succeeded in becoming deeply embedded in the French landscape. Without doubt, this difficulty must be seen as related to the very pronounced change experienced by relations between the state

BOX 13.2 – Thematic research and technological innovation networks

"On a well defined technological domain, each network, in a logic of response to socio-economic demand, must identify technological problems, give rise to appropriate research projects, which are able to lead to new products and services and, in this way, to participate in the creation and growth of innovative enterprises" (p. 66).
"The primary resource of networks is the sharing of resources by the public and private research groups participating in them. The projects that they give rise to can benefit over and above the incentive funding from different ministries and agencies" (p. 66).
PREDIT (2150 MF of public support over five years distributed between four ministries and two agencies) and the national telecommunications research network (RNRT, 1300 MF over five years and distributed between two ministries and one agency) are the two principal "networks" mentioned. New networks concern micro- and nano-technologies, civil and urban engineering, and fuel cells (respectively 55, 10 and 20 MF in the 2000 budget).

Source: Annex to *Loi de finances* 2000.

and large firms. This change is primarily linked to the progressive and continued trend in the privatisation of public companies. It first affected those companies nationalised in 1982 (such as Rhône-Poulenc or Saint-Gobain), then spread to other public industrial companies, including symbolic firms such as Renault. This trend is currently spreading to the large operators of collective services: television and telecommunications, followed by air transport. The increasing internationalisation of these firms, the multinational mergers which affect them (for example Rhône-Poulenc into Aventis) also pose the question of the opportunity for public intervention in research: what is the relevance of the notion of "nationality" of a global firm? What is a national company? *De facto*, national public policies have been disengaged from the support of large companies, and all the issues affecting them – from competition policy to technological development – have been transferred to a new public arena: the European Union.

7. SHARED POLICIES FOR SCIENCE, TECHNOLOGY AND INNOVATION: THE ROLE OF NEW PUBLIC ACTORS

The preceding analysis has brought to the fore a double trend, which has redefined research policy. Firstly, large programmes have all but disappeared and direct support for industrial research is concentrated only on SMEs. Secondly, public sector research, whatever its institutional structure, is converging towards a dominant form of activity: collaborative academic research.

Consequently, is public intervention currently decreasing? The response to this question is negative. Public intervention still plays an important role, for both public laboratories and firms, but its modes of operation are widely diversified. Public intervention can no longer be considered solely in terms of national government policy. Other actors have become major participants: regions, with the decentralisation initiated in 1982, and the European Commission with the creation of the Framework Programme in 1984. In addition to the regions and Europe, a third type of actor is emerging, principally in the health domain: foundations and other "non-governmental organisations". This section presents the particular dynamics of each of these new participants.

7.1 – The increase in regional research and innovation policies

After the failure of a first attempt at regionalisation (1969), which led to General de Gaulle's departure from government, France experienced progressive "deconcentration". It was necessary, however, to wait until a left-wing government came to power and for the decentralisation laws of 1982 in order for the French regions to obtain democratically elected representation, their own significant powers and budgetary resources. The modes of involvement of the regions in supporting research and innovation are still characterised by the choices made at this time.

On the one hand, their activities are focused on investments, and the resources they have at their disposal are the objects of a global transfer from the state to each region. The beginning of the 1980s corresponded to the peak of investment in road infrastructure, its subsequent progressive decrease allowed the regions to widely redeploy their capacities for investment, notably in training and research.

On the other hand, the regions did not have their own competences in the support of economic activities, including research. These competences were shared and, in order to be able to participate, the regions had to develop contractual relationships with the state. CPER, *Contrats de Plan État-Régions*, which are pluri-annual contracts, currently lasting seven years (the first ones lasted five years) support the involvement of the regions in scientific and technological activities.

However, it was not a foregone conclusion that the regions would seize upon research and innovation as one major regional policy. Nevertheless, they became very quickly involved in these areas. The first assessment of these CPER contracts (CSRT, 1988) showed an almost universal involvement by the regions, often under the impetus of the Ministry in charge of research, which used them as a means for financing facilities, but also as the main support for the development of the CRITT (regional centres for innovation and technology transfer). One of the first reports of the *Observatoire des Sciences et des Techniques* (managed by R. Barré) showed the extent of the serious inequalities in the distribution of research and technology activities across France (OST, 1993).

The involvement of the regions is based on the state–region plans which organise a *de facto* regional-national co-ordination of efforts. In the previous plans (1994-1999), the state's share of funding dedicated to research equalled 2.65 billion Francs, an increase of 30% compared to the previous plans.[1] In total the regions spent 1.2 billion Francs in 1995 alone on R&D, in contrast with less than 500 million Francs ten years earlier (Mustar, 1998). The regional councils spend on average 2.5% of their total budget on R&D and innovation but differences between regions are very great (this amount varying between 1% and 4%).

In 15 years, the dynamics of the situation have been totally reversed. The regions, which, in the 1980s, were asked by the state to complement its investments (for example by participating in the funding of large items of equipment or by co-financing transfer centres), are currently putting forward their own projects and their own policies in the context of the preparation of new contracts, often following evaluations of previous activities or strategic analyses of the challenges they are confronting.[2] Through these contracts, they

1. These funds do not include European Structural Funds (all addressing less-favoured regions) nor the share of regions related to the investments included in the pluri-annual contracts. They also do not cover other regionalised government expenditure (such as ANVAR funds).
2. Many regions have developed strategic analyses and are building *observatoires* or other monitoring tools to help them in their action. The regional foresight exercise developed in the Gironde department – Delphi Technopolis – testifies to regions' increasing willingness to develop their own research and technological innovation policies. In *Maine et Loire* (a *departement* within the region Pays de Loire), policies, based upon a permanent observatory, have been established to improve the strength of public research laboratories with the aim of better serving the regional economy. The PACA region, very active in funding research and promoting links with SMEs, has developed, in collaboration with the national S&T observatory, OST, its own regional observatory. Four French regions have been involved in EC RITTS or RTP strategic exercises. Other regions, such as Auvergne, have developed similar approaches to their regional policy.

are able to obtain commitments from the state over a series of years (notably the creation of jobs) and an expansion of the effects of their own activities.[1]

Regional policies have been the object of numerous local evaluations, but an overall perspective is still not available. However, the studies that are available concur in tracing two major trends over time. In a first period, the regions have placed a heavy emphasis on public research facilities. This remains a priority, given that French universities are underequipped, but it has been progressively replaced by more active policies in support of the innovative capacities of SMEs. The latter policies relied first on the interventions and regional structures of ANVAR and of the Ministry in charge of industry. Subsequently, their activity has become progressively more specifically focused on intermediation structures which, in a reversal of the situation, no longer come from the laboratories and from their stock of knowledge to be exploited, but from the problems of regional firms. Currently, it is possible to count around 120 "Technological Resource Centres" (TRC), a name which is symbolic in its abandonment of the notion of "transfer" of its former title CRITT.

A recent evaluation[2] shows the existence of two types of structures: interface TRCs and platform TRCs, which add to the intermediation activity of the former, the development of targeted services. Out of ten public organisations usually mobilised by TRCs to respond to firms' problems, seven are university-based and nine are located in the same region. This underlines the importance of the links woven by the TRCs to successfully carry out their activities in support of regional SMEs. These few data convey the decision, which has been progressively implemented by the regions, to centre their interventions on developing networks of regional innovation actors. Certain regions, such as Alsace, use the concept of *filière* to encourage strong interaction, not only between public research and industrial research, but also between all the economic actors in the same domain. This movement, which witnesses the progressive development of genuine regional policies for research and innovation, is reinforced by the unprecedented success of the experimental European community programme aimed at supporting the regions in the formulation of their own policies.[3] Initially limited to about 20 European regions, this programme has been extended, due to pressure from

---

1. See the *Lettre de l'OST 17* (Autumn 1999), where there is an account of the colloquium "regional research policies: tools and strategies" (Paris, 31 March 1999).
2. National Committee for the Evaluation of Research, 1997.
3. The joint DGXIII and DGXVI RITTS/RTP Programme.

regional executives. This direct connection between the regions and Europe has been facilitated by the Structural Funds, which have accompanied the development of less-favoured regions in Europe. For the next five years, the European authorities have planned to accompany the policies for research and innovation in these regions with a sum at least equivalent to that of the Fifth Framework Programme for R&D.

7.2 – The growing role of European technology programmes

The first EEC[1] research programmes, on renewable energy, date from 1974. After difficult beginnings (Kastrinos, 1997), European research policy has experienced rapid growth: from 4 billion ECU for the First Framework Programme (1984-1987), to 14 billion EURO for the Fifth Framework Programme (which began in 1999). We will not dwell here on the tormented history and the complicated interactions of the negotiations leading to the elaboration of the different framework programmes (Guzzetti, 1995). However, at the very most, two points must be underlined. The first relates to the duration of the programmes and projects they supported, which contrasts markedly with the instability in France mentioned above. The second stems from the national control of each "specific" programme: all its actions had to be validated by a "management committee" comprising two representatives of national administrations.

As with regional policies, the French Government – particularly its Minister of Research at the time, Hubert Curien (who occupied the post for almost a decade) – was very active in the rapid development of the Framework Programme, and of the EUREKA initiative, when the boundaries had been delimited to "pre-competitive" research. It is possible to hypothesise that these two interventions had been conceived as outgrowths of the national policy that were allowed to deepen. Thus it can be acknowledged that, in 1989, it was not possible for the European programme for non-nuclear energy to leave aside the national priority of any particular member country. This led to this programme becoming very much a catalogue, as in many other cases at the time (Callon et al., 1989).

This is manifestly no longer the case today. The Framework Programme has experienced the same trend of reversal as regional policies. Two changes illustrate this: first, its increasing significance, particularly with regard to

1. Apart from CECA and EURATOM, which were based on different characteristics, although they would ultimately be integrated into the Community's actions.

equivalent national interventions; and second, the very evolution of the procedures for the preparation and conduct of its activities.

Both macro- and micro-indicators converge to illustrate this first change and to emphasise the central and increasing role of Community actions. In 1982, the flows of research funds to and from foreign countries (measured in accord with Frascati rules) represented 4% of total research expenditure. In 1998 this figure had tripled, and these flows represented 12% of total research expenditure. Over two-thirds of these flows originate in the public sector. They are mainly concerned with the space programme (the joint flows between CNES and ESA) and with the funding of large international facilities (such as CERN and the "very large telescope"). But increasingly they concern EU research programmes. In the Fourth Framework Programme, France was represented in about half of the projects funded and had a return of 17.5%.[1] This represents over 16 billion Francs or over 3 billion Francs per year between 1994 and 1998 of support to French research teams (public and private) within the context of European collaborative research.

Different studies underline the breadth of programmes. The Third Framework Programme has impacted upon more than 2000 different research actors across 4500 participations in its projects (Gusmao, 1996). Almost all of the large French firms conducting research and the majority of high-technology SMEs have been involved in these programmes. Public sector research (universities, the *Grandes Écoles* and public research institutions) has been present in almost all the projects where there has been French participation. This strong presence emphasises the major role these programmes have played in the coming together of public and industrial research. An evaluation study conducted by the CSI (Larédo, 1995) has also underlined the importance of these projects for public and industrial research laboratories: the two projects in which, typically, each laboratory participated in, mobilised, on average, a quarter of its research capabilities. Thus, in 1998, the total of European contracts held by laboratories linked to the CNRS was higher than the number of all the contracts obtained from French ministries and public agencies.[2]

The second trend is, to a large extent, linked to the requirements for transparency in budget allocation. The generalisation of very formalised selection procedures has limited the importance of the unacknowledged practice of *juste retour* (i.e. obtaining a percentage of the contracts equivalent to the national share of finance contributed) facilitated by the functioning of programme

1. See *Annexe à la loi de finance pour 1998*, p. 38.
2. Source: CNRS programme planning service, January 2000.

management committees. Furthermore, the Fifth Framework Programme has made a break with the previous "committology", shifting the management committees to the level of four general priorities. This shift of an operational entity (the specific programme) towards a general priority (for example "information society") has widely transformed the conditions of national control. Even in the preparation stage, the influence of national administrations has markedly decreased. Clearly, they contribute to future orientations, but in the context of a vast consultation activity, which has allowed more than a thousand contributions to be expressed (several French research institutions have, for example, made their own contributions). The consultations also involve economic partners, who have a strong presence in Brussels, both in their own right, and through European associations representing them. This is how the preparatory processes, which were previously carried out and refereed in Paris (and in each of the capitals of the member states), are partly being shifted towards Brussels (cf. Chapter 6 for further developments).

7.3 – The emergence of new policy actors: non-governmental organisations

This very broad terminology covers many different forms of intervention. For research, gathering them together is an important sign: it typically reflects an acknowledged underinvestment by public authorities. Charities, as they are often labelled, have been particularly active in the medical field and in life sciences.[1]

Little is known in France about the global role of research funding by charities. According to Schaffar (1994), there are no foundations such as the Wellcome Trust in France, which, with a budget equivalent to the UK Medical Research Council, represents a major source of funding for British medical research. Its French equivalent, the *Fondation de la Recherche Médicale* (FRM) has been built on very different ground, its promoters not being a rich philanthropic industrialist but professionals within the medical sphere who have organised their own foundation to attract donations and use the funds thus gathered to support research activities. FRM has a scientific board in charge

1. This is to be clearly distinguished from the *Institut Curie* and especially the *Institut Pasteur*. They are both foundations which survive on donations, on income from patents and licences and on government support, which represent just under 500 million Francs, covering both around 1000 permanent positions and corresponding support.

of identifying areas of interest, and bases its selection of supported teams upon classical peer review approaches. The same applies for the two large state-approved associations, *associations reconnues d'utilité publique* (associations of recognised public utility), created for supporting cancer research, the ARC (recently subject to a vast scandal about the use of the funds gathered) and the League against Cancer (*La Ligue contre le cancer*).

This classical type of research "sponsoring", the main role of which is to gather funds, which are then allocated by researchers themselves through traditional community practices, is today being challenged by the emergence and rapid growth of another type, the "patients associations". These are associations created by patients themselves (or their parents) faced with an "orphan" disease, i.e. a disease about which little is known and which is not found important enough to be the subject of either public health policies or research priorities. In France an exemplary case concerns neuromuscular diseases with the AFM (*Association Française contre les Myopathies*) as major actor. The stated objective of this association is to "cure" these diseases.[1] The project of Professors Dausset and Cohen on genome research, and the corresponding opening of a new avenue in genetics, experienced the association's capability to mobilise the interest of the French public TV channels. This gave rise to a unique situation. Since 1987, through the annual "Telethon" lasting 30 hours, the AFM has been able to mobilise over five million "citizens of neuromuscular diseases" and over 300 million Francs every year. To define and manage its research effort, the AFM has developed a unique model, different from both the public and private models of managing research. This has led Callon and Rabeharisoa (1999), after an in-depth study of their research policy, to speak of a new type of organisation, a "reflexive" organisation. This is translated in the research strategy of the AFM and its capability to focus important efforts in given directions (Schaffar estimates that, from 1988 to 1991, out of the 748 million Francs spent in France on the human genome project, 553 million came from AFM!). This is also manifested by the very pragmatic approach, supporting in one case "exploratory research" through normal peer-reviewed channels, pushing, in another, companies to co-invest with AFM on potential therapy (going as far as investing in start-up companies).

AFM has also developed and supported specific research facilities. A particular case is that of the famous Genethon laboratory, thanks to which, in 1993, the first gene map was produced. This first Genethon was followed

1. The following elements are taken from Schaffar (1994), Patterson and Barral (1994) and Rabeharisoa and Callon (1999).

by a second, for localising and identifying the genes responsible for neuromuscular diseases as well as genes of other genetically inherited diseases. At the same time, a CNRS unit was located within the facility. Very rapidly however, AFM stated that this could not last for long, since its priorities would have to be redirected towards therapeutic aspects (and especially gene therapy) as soon as all genes for neuromuscular diseases had been identified. This has finally driven the French Government to take over the Genethon facility, turning it into a "national sequencing centre".

Such a story (although there are others, even at a less important level such as for cystic fibrosis) casts light on a growing debate related to the role of patient associations within the French public sector research. The question, which has been publicly raised, is whether patient associations "distort" national healthcare priorities to follow the terms employed by a previous INSERM director. The issue may however also be differently stated. Why do five million citizens, clearly knowing the effects of their action (research has been on the forefront of all Telethons), and after the scandal of the misuse of funds by ARC, still choose to support AFM? The question is whether our democratic societies face another challenge whereby spokespersons of given problems are able to cast these problems as "collective" ones (it is no longer a question of a rich individual or of a Machiavellian private company) and are given the means to establish themselves as "quasi"-public authorities? After a "territorial multiplication" (both at regional and European levels), of levels away from the state and its natural arm, government, are we now facing another "problematic multiplication", whereby spokes persons of collectively recognised problems are established as a new type of public authority for solving these problems?

8. NATIONAL POLICY IN TRANSITION: DEVELOPMENTS AND CHALLENGES

The preceding section shows why it is currently not possible to reduce the issues of public intervention to one of national research policy. Rough financial estimates are an indicator of this transformation. Financial incentives provided by the state[1] can be estimated at around 3 billion Francs, regional budgets (which are increasingly dedicated to the provision of incentives) at 1.5 billion Francs, and contracts from the European Union at 3 billion

1. This figure excludes the one remaining large programme, in space, as well as reimbursable advances for civil aeronautics.

Francs.[1] Macro-level quantitative data also underlines the growth in direct research–industry relations: all above-mentioned state incentive actions, which address both public and industrial actors, can actually be compared to some 4 billion Francs worth of contracts let by industry to public research (MENRT, Annex to the *Loi de finance* 2000). These data demonstrate the increasing variety of funding sources available to the research community.

This approach by the laboratories and their operation is all the more important, in view of the institutional choices implemented, since macro-data shed little light on the dynamics of research. The above-mentioned 12 billion Francs (namely national, regional European incentive funding and industrial contracts) are often compared with the civil research budget of 55 billion Francs. However, this comparison does not take account of the necessary expenditure entailed by researchers and *enseignants-chercheurs*[2] within the civil service: apart from the space and aeronautics programmes (for which 10.5 billion Francs have been allocated in the 2000 budget), 90% of the funds are actually assigned to research institutions, where more than 80% is spent on salaries (see point 3).

These developments, even if they are rarely brought to the forefront, have progressively entailed a reorientation of government action and a transformation in its methods of intervention. At the dawn of the third millennium, a new de facto organisational equilibrium might be unfolding around five principal characteristics. This final section attempts to describe these on-going developments which, should they succeed and be fully generalised, would propose a new French model. Otherwise, the process of fragmentation which the previous sections have highlighted, will leave the door open for numerous possible types of amalgamation.

8.1 – A new government organisation

Since 1993,[3] research and higher education have been brought together within the same ministerial entity. This, in turn, has been incorporated in a

1. The OST's estimate (OST, 2000) is 3.2 billion Francs in 1997.
2. In France, senior lecturers and professors are civil servants and are obliged to spend half their time teaching and half their time on research. As a consequence there exists the title of *enseignants-chercheurs*, which distinguishes them from CNRS full-time researchers (who also are civil servants).
3. One cannot take as a point of departure the first brief period of cohabitation (1986-1988) with a Secretariat of State for Research and Universities, straddling by two long periods (together almost a decade) of an autonomous Ministry of Research under the leadership of H. Curien.

vast ministry of education, research and technology, at the time of the return to a Socialist Government (1997). France is not alone in doing this, with Germany also following the same trend, although it should be remembered that German universities are run under the auspices of the *Länder*, and with Japan scheduled to complete a process of administrative reform in 2001, in which the Science and Technology Agency and the Ministry of Education will be merged. This bringing together of ministerial responsibilities is all the more important since France can be distinguished from the other countries mentioned by the significance of its research institutions.

Consequently, it is not surprising to note the growing importance of the debates concerning the structure of public sector research, which now encompasses universities, *Grandes Écoles* and research institutions. The debates focus first on the evolution of institutions as demonstrated by the change in 1998 of the status of the Central Civil Engineering Laboratory (LCPC), the transformation in the same year of ORSTOM into IRD (Institute for Development Research) or the amendments to the status of the CNRS, which are aimed at reinforcing the autonomy of its executive (in relation to ministerial supervision). They also concern the basic entities where research takes place: should it still be accepted that the CNRS has its own research units or should all its units be mixed, in other words jointly run with universities or other research institutions? In the same way, should INRA or IFREMER, whose personnel are, for the most part, located in sites separate from universities, bring about changes to their establishments and also actively engage in partnerships, which favour the mixed units? Debates are still focusing on the size of the relevant research teams, with the advocates of autonomous teams (built around a lead researcher at the head of a team of 10 to 20 persons) opposing the supporters of laboratory–institutes, which bring together several teams (as in the German model of large research institutes). The CNRS, which currently comprises 1300 units (with an average size of 50 persons, including teacher-researchers and doctoral researchers), would see, in the first case, their number climb above 3000, and in the second decrease to around 1000. The debates focus, finally, on the reinforcement of links between industrial and public research, with the latter increasingly used, as will be seen in the next section, to support and as leverage for initiatives in favour of innovation.

8.2 – A durable transformation of industrial policy

The majority of new initiatives taken to promote innovation rely on public sector research. The Innovation Act passed in 1999, as well as the fiscal

BOX 13.3 – Elements of the competition for the set-up of innovative firms and for the creation of incubators

The National Competition for the creation of Technologically Innovative Firms
The competition was launched in 1999, with a budget of 100 MF to encourage the creation of new firms. 1913 applications have been received and 244 selected. 165 award winners have received an average of 200 000F (maximum possible: 300 000F) for projects judged to be "emerging" ("at the idea or prefiguration stage, needing to be developed on the technological, organisational, industrial, commercial, legal or financials levels"). 79 prize-winners have received "creation-development" SUPPORT for projects, "the basis of which are already developed and are situated just upstream from start-up stage". The latter support amounts on average to 1 MF (maximum: 3 MF) and corresponds to a maximum of 35% of the costs of development. The Competition was renewed in 2000.

Structures for incubation and starter capital
The call for tender, initiated in 1999 and renewed in 2000, aims to "encourage the implementation of new structures for incubation, coming in particular, but not only, from higher education establishments or research. Support ... will be given, with preference for new projects set up in partnership" (Ministry website, Action Incubators, February 2000). A company incubator is defined as "a place hosting and assisting holders of innovative firm start-up projects. It offers them support in training, consultancy and financing, and accommodation until they join a technology park and can support the cost of industrial premises". In 1999, 18 incubators have been supported at a cost of 106 MF, while 33 MF have been allocated to INSERM, INRA and the CNRS to set up a 200 MF starter fund in biotechnology with private sector partners.

Sources: Annex to the *Loi de finance 2000*, Ministry website (February 2000).

arrangements in favour of innovating firms, to which attention will return, focuses on the development of "relations between public researchers and companies" and "collaborations between public sector research and companies" (presentation of the Act on the Ministry website, February 2000). Amongst the principal arrangements, several are concerned with a relaxing of the status of the civil service, enabling researchers to engage in industrial projects, to take out patents and to set up firms or to participate in their creation. Similarly, there are three principal initiatives put forward by the Ministry in its annual report to Parliament (Annex to the *Loi de finances* 2000): thematic networks for research and technological innovation (presented above); a competition for the start-up of innovative firms; and a competition for creation of incubators (Box 13.3). These initiatives are indicative of the new approaches adopted: the primary role of networks is the facilitation of sharing of access to public research resources around socio-economic demands; the objective of the competitions (which together are provided with around 300 million Francs per year) is to initiate a trajectory, which, in a way, is "seed money" aimed at crystallising a set of actors, who will then be led towards taking complete charge of the implementation of the projects initiated.

The situation has moved a long way from the large programmes described previously. Over the same period, the other main actor in national R&D policy, the Ministry of Industry, has been merged into a vast Ministry of Economy and Finances. Industrial policy is less and less a matter of direct, targeted interventions (through large programmes or policies supporting some national champion, and particularly "nationalised" companies). It has become the construction of a framework favouring competition and the competitiveness of firms. This process passed through the privatisation process, carried out on most industrial and service firms previously owned by the state, by macro-economic monetary and fiscal policies, and the introduction of "regulatory authorities" responsible for organising competition in traditionally monopolistic sectors, such as telecommunications and energy. Innovation policy has since become increasingly centred on the general environment of companies: the taxation of innovating firms is thus the third part of the Innovation Act passed in 1999; the taxation of stock options is the subject of an acrimonious debate which has lasted for almost three years; and intellectual property issues have become increasingly important. These trends also explain the reinforced significance of the innovation agency (ANVAR) to the extent that its mode of operation, which is founded on procedures that are open to all companies, that are quick, that are based on explicit and localised criteria of attribution (through total decentralisation of its activities), follows perfectly this change of direction in approach.

Thus, and this is the second major characteristic of national policies, a re-orientation in favour of innovation, around, on the one hand, public research, and, on the other, the general environment of firms, often borne by other policies (fiscal, intellectual property etc.), is taking place.

8.3 – The issue of co-ordination between public authorities

Consequently, it can be better understood how, *de facto*, complementarity between public interventions has developed. The state is reorienting itself towards public competences in research. The regions are concentrating their interventions on more localised issues, particularly on SMEs, while Europe is focusing its efforts on the large global challenges. Clearly, these orientations are not exclusive to particular levels of policy actor. Through ANVAR the Government plays an important role in supporting SMEs, just as Europe actively supports French high-technology SMEs (Larédo, 1995). Similarly, information and communication technologies are the object both of a major European programme and also of national support (although this has significantly and regularly decreased over recent years). This complementarity is coupled with a growing co-ordination at the European level, which goes beyond the framework of intervention of the European Union. The examples of space policy with the European Space Agency (of which the programmes currently represent more than two-thirds of the activities of the CNES) or the interventions in aeronautics, particularly related to AIRBUS, are all the more striking since they concern the last French large programmes. Similarly, the recent creation of national centres for gene-sequencing and genotyping are examples of new co-ordination becoming established between national policy and non-governmental organisations or patients' associations.

These new actors, as demonstrated earlier, have become participants in their own right, and no longer off-shoot developments which the national Government delegated certain specific missions to. This trend in co-ordination relies on procedures, which are now broadly established: the pluri-annual contracts between regions and the state (so called CPER) on the one hand, the preparation of the Framework Programme on the other. Of course, both have significantly changed over time, the latter with an increasing role played by the European parliament and organisations representing actors established to operate directly at the European level, and the former witnessing regional actors formulating and putting forward their projects themselves. Of course, questions remain, particularly with regard to the exclusively bilateral approach on which the pluriannual contracts are implemented between the state and each region. Whatever the difficulties, co-ordination between public authorities occupies an increasing space and constitutes the third major challenge. Although broadly accepted in principle, this co-ordination is still the object of endless debate on its breadth, its modalities, and the instruments to support it (Papon, 1998). According to the *Commissariat général du plan* (Duby, 2000) there is no "single best way" to promote it, divisions should operate in an *ad hoc* way for each problem, and are evolutionary over time. This presupposes the development of the capabilities to anticipate future problems and the facilitation, through broad debates, of the construction of shared visions. The above-mentioned report considers this "prospective" work as a new role for national government, following the example of the United Kingdom Foresight programme.[1] The role ascribed to the national level would no longer be focused on the provision of direct financial incentives, but also to provide strategic impetus through the development of anticipation actions, of the promotion of public debate on new technologies, and evaluation of the effects of past actions.

8.4 – A new framework for public involvement: pluriannual contracts

The reorientation of research policy toward the public sector has gone hand in hand with a progressive evolution of the relations between the Government and the "operators" of research: sectoral ministerial programmes and services on the one hand, research institutions on the other. The term "operator" has been proposed by Chabbal (Chabbal, 1987) to describe the evolution of these relationships. It has accompanied reflection on evaluation, which has become a central instrument of strategic action. In line with this trend, which affects the majority of the European countries and the European commission, France has selected a distinct path, the objective of which is to ensure the quality and the credibility of evaluations. The National Evaluation Committee of the Universities (CNE), created in 1984, and the National Committee of Evaluation of research (CNER), created in 1989, to evaluate all the other operators of French civil research, have proposed a "Guarantor model" (Larédo and Mustar, 1995). The assessments made of this approach emphasise four complementary aspects.[2]

First, the decade of activity of these two institutions has rooted evaluation in the practices of the operators, almost all of which have set up their own evaluation structures. Second, this has generated "robust" evaluations, i.e. able to withstand criticism and thus to produce new reliable information. Third, while the CNE has succeeded in adapting its procedures to the rhythm of the policy agenda (around 20 universities to evaluate each year), this has not been the case with the CNER, which, after ten years of activity, was far from having completed the round of French research operators. Finally, the link between evaluation and strategic management was made difficult by an

1. "Prospective" work has played an important role in the formulation of national policies in the 1960s and 1970s. Its disappearance from the political scene in the 1980s was such that certain initiatives taken in the 1990s (a French Delphi survey based, as in Germany, on the Japanese one, the Key Technologies Operation conducted in 1995 and again in 2000) had only a secondary impact.
2. See Larédo (1997) for a review of the analyses of this experience.

organisational issue. Cutting the umbilical cord to the parent Ministry has facilitated the independence of evaluation approaches, the quality and the credibility of the work carried out, but it has also made appropriation of results by the Ministry difficult. Analysis of the impacts of evaluations has shown that the effects were significant when the operators could themselves implement the recommendations made, but that they remained limited (indeed erratic) once recommendations demanded direct government intervention. In short, the culture of evaluation has for the most part stopped at the doors of the Ministries and the central administrations. This has led the authors of the report on research (Majoie, 2000) to make the transformation and generalisation of evaluation a priority for national policy.

This priority makes all the more sense because the framework of relationships between the universities and the Ministry has been profoundly transformed with the introduction of "quadrennial contracts". These were first introduced in 1988. Their objective is to define, over a period of four years, the mutual obligations between each university and the Government. After 12 years and three rounds of contracts for most universities, it is clear that they represent a strategic event in university life, especially with regard to training, particularly doctoral training. The contracts also define the human resources (and in particular new posts of senior lecturers and professors) and financial resources that the state undertakes to provide to the universities. Finally, contracts earmark a certain number of research teams (among those not associated with the large research institutions, such as the CNRS and INSERM) on which it concentrates the Ministry's research support: more than 90% of the support for university research, namely 2 billion Francs in 2000, is allocated in this way. Consequently, it is easy to understand why the negotiation of contracts is a major event in the life of every French university. The same approach has been initiated in the 1990s for research institutions with pluri-annual contracts, but these are limited to general orientations, and remain at best indicative of the state's commitments. Nevertheless the trend is widely underway and it is likely to become deeper and more widespread in the coming years, placing pluri-annual contracts at the core of national research policy and the principal support for its general orientations (that the policy of incentive aims to initiate and render visible).[1]

---

1. It is thus possible to consider the competitions for new firms or incubators (cf. Box 13.3) as a means to initiate procedures which would then be perpetuated in pluri-annual contracts.

This change, however, comes up against the limits of the autonomy of the universities and research institutions. Two salient examples can be taken. The debate over the renewal of the French synchrotron radiation source illustrates one of the major problems of French public finance. For research institutions, the impossibility of writing off the capital costs of their facilities leads to the renewal of all the projects, even those of medium importance, at the central level. This makes it impossible (or at least very difficult) to implement a long term policy for major equipment in each organisation. The other example originates from the limits of the reform of the universities. The status of "public scientific, cultural and professional establishment" (note the symbolic absence of the terms training and education in this title), which stems from the 1984 Act, guarantees autonomy. However, the extent of this autonomy is, at the least, limited, according to the analysis of it that is made in the report of the *Commissariat général du plan* (Duby, 2000): the universities "are not in control of recruitment, of the management and promotion of staff, and of their budget. They have neither analytical accounting practices, nor management of accounts, nor control of their property". Consequently it is not surprising to see the National Evaluation Committee of the universities regularly emphasising in its reports the question of the "government of the universities" which, following their analyses, makes any strategic choice difficult.

8.5 – Multiplication of "crises", public debate and "problem-solving" policies

The fifth major reorientation refers to the changes experienced in the relationship between science and society. In France, this question was particularly salient at the time when the decision exclusively in favour of nuclear energy was made in the 1970s. With the accession to power of the Left, and the compromise made in favour of retaining nuclear energy, the social evaluation of technologies became a Parliamentary "challenge" which led, in 1983, to the creation, by the National Assembly and the Senate, of the Parliamentary Office of Scientific and Technical Options (*Office parlementaire des choix scientifiques et techniques*). Largely inspired by the model of the American Office of Technology Assessment, which was abolished shortly afterwards, the French office adopted a different approach. All of its studies are carried out and authored by Members of Parliament themselves. They have at their disposal logistical resources for this work, and also the option to commission studies from specialists (which they widely used, particularly at the time of the evaluation of the French space programme). However, their preferred way of operating has been to use the American-style practice of "hearings".

The early stages were not easy, because such an approach constituted a radical break in the French landscape, which tended to delegate the whole of the phases of preparation of decisions to the government administration, that is to say to the Executive. It required a succession of "crises" – AIDS and contaminated blood, radioactive fallout from the Chernobyl accident – in order for the Government, which was overwhelmed by the problems related to nuclear waste reprocessing and by the systematic opposition from residents close to the Government's experimental waste burial sites, to ask the Office and Député Bataille to deal with the question in 1990. Following a series of completely open public debates, the Office was led, for the first time, to challenge previous decisions, not because of scientific errors, but because of the need to leave policy choices open. This resulted in a law – the Bataille Act – passed in 1991, which forbids all irreversible waste stockpiling for a period of 15 years, the time span which was judged to be necessary to demonstrate the feasibility of the weakest of the three identified options. This same act involved a research programme around the three options and implemented a follow-up Parliamentary process (Barthe, 2000). It was also following the initiative of the Office that the laws on medical ethics and the code of ethics for therapeutic tests were adopted at the beginning of the 1990s.

This *de facto* reduced the role of the consultative bodies established at government level. Since 1982 a "superior council for research and technology" (*Conseil supérieur de la recherche et de la technologie*, CSRT) has been in existence, chaired by the Minister and modelled on a parliament of researchers. Responsible for the production of an annual report on the state of science and technology, its role has progressively decreased to the point that, when the Socialists returned to power in 1997, the new Minister of Education and Research, C. Allègre, decided to appoint a National Council for Science (*Conseil national de la science*). Two years after its creation, it was difficult to assess the extent of its influence, as it had neither produced any publications nor contributed to public debate.

The increase in the number of "crises" – from "mad cow disease" to the potential dangers of genetically modified organisms, as well as the hole in the ozone layer and global warming, or agricultural pollution of drinking water – has reinforced the importance of this trend in increasing the amount of public debate. After the "industrial" policy wave of the 1980s, public debate has revived the importance of research for public policy, and has given rise to a different allocation of responsibilities. The generic delegation carried out at the time of national elections and translated into sectoral government policies is less and less the rule. Solutions are increasingly specific, as in the example of the law on nuclear waste reprocessing. Similarly, a specific agency has been established to organise and finance research on AIDS. A public interest group was created to organise efforts related to genetic research, before the institution of specific national centres and the policy of *génopoles* initiated by a patients' association was taken up at national level. The Dormont Committee, which was created to provide advice to the Government on BSE, was delegated the responsibility for organising and financing the related research programme, with public research institutions providing the human and financial resources for its activity.

This fifth characteristic, has thus instituted the progressive replacement of the coupling "parent ministry – mission oriented research institution" by another, i.e. "public debate – *ad hoc* organisational setting" within which the legislature has regained a role which, in France, it had long since lost.

8.6 – Challenges and new perspectives

The objective of this chapter has been to bring to light the great changes that have characterised the French research and innovation system over the last two decades: the near disappearance of the large programmes, the end of the policy of national champions, the displacement to the European level of technological programmes aimed at encouraging collaborative research, sharing with the regions the strong priority of the innovative capacities of SMEs, the increasing role of non-governmental organisations in the definition of priorities (particularly in health), and the convergence of public sector research (universities, the CNRS and mission-oriented research institutions) around "collaborative academic research". It was not the purpose of this chapter to describe the complex, painful and controversial processes that ended in such situation, but rather to take hold of these developments which signify the end of Colbertism, founded on the central role of an omnipresent and all-powerful state. This, de facto, entailed a radical reshaping of national research policy. This reshaping is however far to be stabilised as was shown by the limitations faced by, and the controversies on, the five main directions followed.

The first is linked to the realisation by national innovation policies of the existence of other public actors: the European Union and the regions.

The co-ordination of policies – in a multi-annual framework whether it is a question of the European Framework Programme or state-region contracts – has become a central activity and a major challenge, with important controversies over the nature of allocations, specialisations or complementarities to be encouraged.

In this context, national policy increasingly has recourse to two principal levers. On one hand, as in the United States, the growing role of other policies (such as fiscal and industrial property) can be observed. On the other hand, it relies more and more on public sector research to promote its priorities (for example, the creation of high technology firms and the promotion of research networks).

It is not surprising then to see that the shaping of public sector research has become both a priority on the policy agenda and the source of a number of debates. This has led the Government to set up a Ministerial department, following the example of Japan, in charge of all the sector's constituent entities.[1]

This has also led Ministerial departments to develop a different approach to their relations with these entities – universities and research institutions – based on multi-annual contracts. The recent report of the *Commissariat général du plan* emphasises both the path travelled and the outstanding problems: the autonomy of actors (compensation for their capacity to articulate and implement strategies) remains too limited, the evaluation system put in place in the 1980s is poorly adapted to this approach, and the capacity of the Ministry for anticipation remains insufficiently formalised to provide support to debates on the orientations and choices.

However, and this is the final characteristic, public debates which question research are increasingly numerous – from AIDS to genetically modified organisms or mad cow disease. They are restoring the challenges of public research oriented towards public needs to the policy agenda, but in a novel way, both because the traditional coupling of "parent ministry – mission oriented research institution" no longer exists and because each debate requires ad hoc structures adapted to the particular nature of the situation and the credibility of the work. This last element may well be the principal issue of the decade just beginning.

1. The recent government reshuffle (May 2000), while keeping a united administration, has however reintroduced a split in political responsibilities between education and research.

ABBREVIATIONS

ADEME	Environment and Energy Research Agency
ANVAR	National Innovation Agency
ARIST	*Agence Régionale de l'Information Scientifique et Technique*
BCRD	*Budget Civil de Recherche et de Développement* – covers all civil government appropriations for research
BRGM	French Geological Survey
CEA	Atomic Energy Commission
CEE	Employment Studies Centre
CEMAGREF	Rural, Water and Forest Engineering Centre
CEREQ	The Centre for Research on Employment and Qualifications
CIRAD	International Agronomic R&D Co-operation Centre
CNES	National Space Research Agency
CNET	National Centre for Research on Telecommunications – now France Telecom Research
CNEVA	National Veterinary and Food Research Centre – now part of AFSSA – French Agency for Health and Food Safety
CNRS	National Scientific Research Centre
CSTB	Building Science and Technology Centre
CRITT	*Centre Régional d'Innovation et de Transfert de Technologie* – French equivalent of technology resource centres
EPST	*Établissement Public à Caractère Scientifique et Technologique*
INERIS	National Institute on Industrial Risks and Safety
IFREMER	National Institute for Oceanic Research
INRA	National Agronomic Research Institute
INRETS	National Institute for Transport and Safety Research
INRIA	National Institute for Research in Computer Science and Automation
INRP	National Institute for Educational Research
INSERM	National Health and Medical Research Institute
IRD	Research Institute for Sustainable Development (formerly: ORSTOM)
LCPC	Central Laboratory for Roads, Bridges, Public Works
MF	million Francs
ONERA	National Office for Aerospace Research
PREDIT	National Transport Research and Technology Programme

REFERENCES

Amable, B., R. Barré and R. Boyer, 1997, *Les systèmes d'innovation à l'ère de la globalisation*, Paris: Economica.

Arrow, K., 1962, Economic Welfare and the Allocation of Ressources for Invention, in Nelson, R. ed., *The Rate and Direction of Inventive Activity*, Princeton: Princeton University Press.

Arthur, B., 1988, Competing technologies, An Overview, in G. Dosi, C. Freeman, R. Nelson, G. Silverberg and L. Soete (eds), *Technical Change and Economic Theory*, London: Pinter.

Bach, L. et al., 1995, les effets indirects des grands programmes de développement, in M. Callon, P. Larédo and P. Mustar (eds), op. cit.

Barré, R. (sous la dir. de), 1993 and 1999, *Science et technologie – indicateurs*, Observatoire des Sciences et des Techniques (OST), Paris: Economica.

Barthe, Y., 2000, La mise en politique des déchets nucléaires, Thèse de doctorat, École des Mines de Paris.

BETA, 1988, *Étude sur les effets économiques des dépenses spatiales européennes*, rapport pour l'Agence Spatiale Européenne, Strasbourg.

Callon, M. et al., 1989, *Évaluation des programmes publics de recherche, le cas du programme communautaire énergie non-nucléaire*, Namur: Presses Universitaires de Namur.

Callon, M., P. Larédo and P. Mustar (eds), 1997, *The Strategic Management of Research and Technology*, Paris: Economica.

Caracostas, P. and U. Muldur, 1997, *La société, ultime frontière*, Luxembourg: OPOCE.

Chabbal, R., 1988, *Organisation de l'évaluation et de la recherche à la commission des Communautés européennes*, Luxembourg: OPOCE.

Chesnais, F., 1993, The French National System of Innovation, in R. Nelson (ed.), op. cit.

Cohendet, P. and A. Lebeau, 1987, *Choix stratégiques et grands programmes civils*, Paris: Economica.

Comité National d'Évaluation de la Recherche, 1994, Évaluation du programme spatial français, in CNER, *Un autre regard sur la recherche*, Paris: La Documentation française.

Conseil supérieur de la recherche et de la technologie, 1986, *L'évaluation de la politique gouvernementale en faveur de la recherche industrielle*, Paris: CSRT.

CSRT, 1988, *Contrats de plan État-Régions dans le domaine de la recherche, rapport d'évaluation*, Paris: Imprimerie Nationale.

Dodet, M., P. Lazar and P. Papon, 1998, *La République a-t-elle besoin de Savants?* Paris: PUF.

Duby, J.J., 2000, La politique française d'innovation face à la concurrence internationale. Cinq meilleures pratiques étrangères pour la recherche française, in J. Majoie, op. cit.

Ergas, H., 1987, Does Technology Policy Matters?, in B. Guile and H. Brooks (eds), *Technology and Global Industry, Company and Nations in the World Economy*, Washington, DC: National Academy Press.

Gilpin, R., 1970, *La Science et l'État en France*, Paris: Gallimard.

Guillaume, H., 1998, *Rapport de mission sur la technologie et l'innovation*, Paris: La Documentation française.

Gusmao, R., 1996, *Recherche nationale et recherche communautaire, un regard sur l'engagement français*, Thèse de doctorat du Conservatoire National des Arts et Métiers, Paris.

INRA, 1998, *Les chercheurs et l'innovation. Regards sur les pratiques de l'INRA*, Série 'Sciences en Questions', Paris: Institut National pour la Recherche Agronomique.

Jacq, F., 1996, *Pratiques scientifiques, formes d'organisation et représentation politiques de la science dans la France de l'après-guerre. La politique de la science comme énoncé collectif*, Thèse de doctorat, École Nationale Supérieure des Mines de Paris.

Kastrinos, N., 1997, *The European Community Framework Programme as Technology Policy: Towards an Assessment*, Ph.D. Thesis, University of Manchester.

Kuisel, R., 1984, *Le capitalisme et l'État en France. Modernisation et dirigisme au XXe siècle*, Paris: Gallimard.

Larédo, P., 1995, *The Impact of Community Research Programmes in France*, Paris: Presses de l'École des Mines de Paris.

Larédo, P., 1995b, Structural Effects of European Research Programmes, *Scientometrics*, Vol. 34, No. 3.

Larédo, P., 1997, Evaluation in France, a Decade of Experience, in *OECD Proceedings, Policy Evaluation in Innovation and Technology*, Paris: OECD.

Larédo, P., 1998, The Networks Promoted by the Framework Programme and the Questions they Raise about its Formulation and Implementation, *Research Policy*, Vol. 27, 589-598.

Larédo, P. and P. Mustar, 1995, France, the Guarantor Model and the Institutionalization of Evaluation, *Research Evaluation* Vol. 5, No. 1.

Larédo, P. and P. Mustar, 1999, Le palmarès des entreprises qui innovent, *La Recherche*, 326.

Lebeau, A., 1985, *Élements d'une stratégie dans le domaine spatial*, Rapport pour le ministre de l'industrie et de la recherche, Paris: CPE.

Loridant, P., 1991, *Rapport sur les orientations de la politique spatiale française et européenne*, Paris: Office parlementaire d'évaluation des choix scientifiques et technologiques.

Majoie, J., 2000, *Recherche et innovation : la France dans la compétition mondiale*, Rapport au CGP, Paris: La Documentation française.

M.E.N.R.T. (Ministère de l'éducation nationale, de la recherche et de la technologie), 1999, *Recherche et développement dans les administrations*, Les dossiers No. 107, Paris: MENRT.

Ministère de l'Industrie, 1995, *Les 100 Technologies-Clés*, Paris: Ministère de l'Industrie.

Ministère en charge de la recherche, annuel, *Annexe au projet de loi de finance pour la recherche*, Paris: Imprimerie Nationale, Web site: http://www.education.gouv.fr.

Mustar, P., 1994, La politique d'innovation en France: le colbertisme entamé, in F. Sachwald (ed.), *Les défis de la mondialisation. Innovation et concurrence*, Paris: Masson.

Mustar, P., 1998, Les transformations du système de recherche français dans les années quatre-vingt, *Annales des Mines*, February.

Mustar, P. (sous la dir. de), 1998, *Les chiffres clés de la science et de la technologie*, Paris: Economica/OST.

Nelson, R. (ed.), 1993, *National Innovation Systems*, Oxford: Oxford University Press.

OECD, 1966, *Politiques nationales de la science: France*, Paris: OECD.

OECD, 1985, *La politique d'innovation en France*, Paris: OECD/CPE.

Paillotin, G. (sous la dir. de), 1993, *Recherche et innovation: le temps des réseaux*, Préparation du XIe Plan Paris: La Documentation française.

Papon, P., 1998, Research Institutions in France: Between the Republic of Science and the Nation-state in crisis, *Research Policy*, Vol. 27.

Patterson, F. and C. Barral, 1994, L'Association française contre les myopathies: trajectoires d'une association d'usagers et construction associative d'une maladie, *Sciences Sociales et Santé*, XII, 2.

Pestre, D. and F. Jacq, 1996, Une recomposition de la recherche académique et industrielle en France dans l'après guerre 1945-1970, *Sociologie du Travail*, 3.

Picard, J.F., 1999, La création du CNRS, *Revue pour l'histoire du CNRS*, No. 1.

Rabeharisoa, V. and M. Callon, 1999, *Le Pouvoir des malades. L'Association française contre les myopathies et la Recherche*, Paris: Presses de l'École des Mines de Paris.

Salomon, J.J., 1986, *Le gaulois, le cow-boy et le samouraï*, Paris: Economica.

Schaffar, L., 1994, Associations, fondations et recherche publique, in A. Esterle and L. Schaffar (eds), *Organisation de la recherche et Conformisme scientifique*, Paris: PUF-Nouvelle Encyclopédie Diderot.

Secrétariat d'État à la Recherche, 1980, *Construire l'Avenir, livre blanc sur la recherche*, Paris: La Documentation française.
Servan Schreiber, J.J., 1967, Le défi américain, Paris: Denoël.
Ténière-Buchot, P.F., 1982, *Évaluation de trois programmes de développement technologique: électronucléaire, aéronautique, recherche spatiale*, Paris: IDRH-CPE, July.
Weisenburger, E., V. Mangematin and P. Larédo, 1999, *Les profils d'activité des laboratoires de l'enseignement supérieur agricole*, Paris: CSI-DGER.

ACKNOWLEDGEMENT

The authors are indebted to Michel Callon, Remi Barré, François Jacq, Frieder Meyer-Krahmer and Bastiaan de Laat. Their ideas, the discussions that we have had with them and the work that we have carried out together have contributed greatly to this chapter. An earlier version of this chapter was discussed at the seminar of the CNRS programme Les enjeux économiques de l'innovation. *The authors are grateful to David Encaoua, the Director of this programme, for this opportunity.*

CHAPTER 14

GENERAL CONCLUSION: THREE MAJOR TRENDS IN RESEARCH AND INNOVATION POLICIES

PHILIPPE LARÉDO AND PHILIPPE MUSTAR

The previous chapters allow the importance to be measured of changes, which, in less than a decade, have radically transformed the landscape, which had resulted from World War II. From the analyses of specific research and innovation policies presented in this book, the editors draw out a number of changes which lead to the identification of three major transformations. First, the repositioning of the technological interventions by public authorities and the end of large programmes, the priority given to SMEs and the marked changes in the role of defence research in a number of countries are all aspects of the progressive establishment of new relationships between international competitiveness and public policies. Second, the focusing of public activity on universities and the reorganisation of government laboratories have given pre-eminence to public sector research in many countries. These two transformations, which are *a priori* opposed to each other, confront public policies with a third tension between, on the one hand, the globalisation of activities and, on the other, the emphasis on specificity and the effects of proximity. How, if this is the case, can public interventions be developed, which are capable both of promoting the development of local links and of contributing to the organisation of a global framework? This conclusion presents our views on these transformations.

TOWARDS NEW RELATIONS BETWEEN INTERNATIONAL COMPETITIVENESS AND PUBLIC POLICIES

As was the case at the time of the comparative study edited by Nelson, the same preoccupation with international competitiveness is evident in all the national analyses. However, the extent to which the economic context has evolved is assessed. It is not necessary to set out in detail the idea that growth depends on the existence of "strong firms", which are capable of positioning themselves in the global market. Rather, it is taken as read in all of the national analyses carried out. Likewise, it goes without saying that for firms to be large they must be multinational. In the majority of the analyses, the term "global" is essential. Firms disengage themselves from their national context in order to become directly established on the global level: what is good for General Motors is perhaps no longer so for the United States. There is, therefore, a tendency for policies to distinguish increasingly between the research and innovation activities of firms which are often classified by surveys under the label of the said country (as in *Fortune* magazine's classification of the world's top 500 firms), and the industrial research activities which are carried out in that country. There has been an increase in the number of studies which calculate the relative weight of "foreign" firms in the industrial R&D of a country. For instance, it amounted to 11% of all US industry expenditure in 1995 (cf. D. Mowery, Chapter 2). It is also possible to find evidence of this in the active participation of Thomson and Philips in the American digital television programme and of IBM in the European information technology programme.

The increasingly evident globalisation of large firms has been coupled with the growing role of industrial research in the national R&D effort. Consequently, the share of industrial research has increased, quite spectacularly, in the majority of those countries where the state plays a central role. In less than a decade, it has gone from 50% to 63% of GERD in the United States and has increased by 10% in France (from 43% to 52%). This increased capacity probably explains a large part of the contraction, or at least the complete redefinition of public intervention with regard to "large programmes". Countries such as the United Kingdom, which is a precursor in this matter, or, subsequently, France, have completely stopped the majority of the large civil programmes which characterised their public intervention in the 1970s, ranging from civil aeronautics to computers, via telecommunications and energy. The efforts made in American policy for technological catch-up, first in the context of the National Cooperative

Research Act (1984) then more directly through Clinton's Advanced Technology Program and the Technology Reinvestment Program, have been toned down for a number of years. Even the very rapid increase in European programmes in the 1980s, which offset the on-going disengagement of member countries (we will return to this point later), has progressively led to a stagnation (certainly at a high level) and, in particular, through the Fifth Framework Programme, as P. Caracostas and U. Muldur show in their chapter, to a change in orientation, with societal problems occupying an increasingly significant, not to say dominant, place. Even in Japan, large programmes, such as the Fifth Generation Computer Programme, are widely deliberated (S. Callon, 1995), and have given way to interventions which focus on public centres of excellence and networks of co-operation between public and private actors in basic technologies.

The decline of explicit technology policies is widespread. Everywhere, direct public interventions are linked increasingly to socio-economic objectives. This does not mean that governments have suddenly lost interest in the international competitiveness of their industrial fabric. Quite to the contrary, although they are using other instruments and other foci. Instruments of the "large programme" or "national programme" type presuppose a strong national capacity for anticipation and co-ordination. This can currently be found more readily in "country-regions" such as Finland or the Netherlands, than in others, where public and private actors are more numerous than 20 years ago. It is, therefore, not surprising to see that in a number of countries, there is greater emphasis on the conditions for technological competitiveness than on competitiveness itself, which is considered as the responsibility of firms themselves. Industrial property, sources of finance (particularly legislation on risk capital), taxation, the functioning of universities to promote links with the economy and society, are among renewed approaches to the same questions. However, these no longer rely on technology policies as such, but mobilise "other" policies to the benefit of the technological competitiveness of the industrial fabric (cf. D. Mowery on the United States). In the same way, the instruments which have gained importance in the implementation of technology policies, as underlined by A. Silvani and G. Sirilli, are characterised by their automatic attribution and by the simplicity of their implementation (linked to the existence of simple criteria or to the greater proximity of the actors targeted). In numerous countries, these instruments have seen a strong increase in their importance among public interventions more generally. They also have the advantage of better corresponding to the new focal point of industrial policies: SMEs.

If there is one point shared by the countries studied, particularly those in the European Union, it is clearly the priority accorded to SMEs. All the governments agree in recognising their importance in creating employment and their role in the development of new activities. SBIR and other long established "manufacturing extension programs" in the United States, ANVAR and the tax credit for research in France, Technology Resource Centres in the Nordic countries, assistance for researchers and engineers in Germany, and fiscal support for investment in Italy are among the types of interventions which have been adopted in numerous countries to create frameworks of active interventions to promote the development and innovation efforts of SMEs. Since the beginning of the 1990s, they have been complemented, and sometimes surpassed in importance, by the numerous mechanisms aiming to promote the creation of technology based firms. The rapid development of these new firms in the United States is developing the role of risk capital. Numerous countries have sought to promote its emergence, both by indirect and direct measures. However, they soon realised that, in their countries as in the United States, the initial phase of the creation of such firms, which is rooted in university life and in public research, requires a bundle of public interventions. Ranging from nurseries and incubators to seed money, via the provision of incentives for firm creation by academics, researchers, even doctoral and postdoctoral students, the number of programmes and actions to promote the creation of high-technology firms, principally in information technologies and biotechnologies, have multiplied. It is possible to note, as F. Meyer-Krahmer does, a fashion for such activities and the imitative behaviour that have affected numerous countries, to the extent of forgetting the invariably central role of other SMEs in economic activity, particularly the employment they create.

These developments underscore the enormous convergence of countries, which is becoming apparent. However, a decade earlier, the liberal model adopted in the United Kingdom and the United States (which B. Bozeman and M. Crow have termed "market paradigm") was opposed to the centralised "Colbertist" model of France, with a strong degree of public intervention, also shared by Japan and Korea. Ten years later, this contrast is now quite distant, although one aspect of it remains and continues to play a central role, that concerning defence. However, the situation has greatly evolved. Ten years ago, Germany and Japan, with their weak defence activities, could be contrasted with the United States, the United Kingdom and France, where military research represented between one-third (France) and a half (the United States and the United Kingdom) of public research and development effort. Today only the United States, despite the decrease in the relative weight of military research and the very strong growth in health research, is distinguished by having extensive military research. B. Bozeman and J. Dietz emphasise its pervasiveness: "75% of all federal money spent on development, 75% of all federal money going to industry, support to university equal to NSF and NIH, 50% of federal money in federal labs and 56% of all funds to FFRDCs". Furthermore, this does not include DARPA and its role in strategic technologies. In France, to the contrary, there has been a strong and steady decrease in military research expenditure throughout the 1990s, which is now solely focused on specific military technologies and on exclusively military firms, thus losing any "pervasive" role. If there is any opposition of models, it is now between, on the one hand, the United States, where the only co-ordinated and centralised research activity is in the hands of the military, and, on the other, Europe, via the Framework Programme, and Japan, through reform in the public administration of research, where there is renewed co-ordination in place. We will examine this in due course, after having tackled what seems to us the second principal movement detected by the national analyses.

EDUCATION, UNIVERSITIES AND PUBLIC SECTOR RESEARCH

In the majority of the countries studied, higher education and training have become a clearly displayed priority and a central topic of interest. It was a question both of increasing resources (of which France and to a lesser extent the two other Mediterranean countries provide evidence) and of bringing academic research closer to the economy and society (as illustrated by the British LINK programme or the Finnish policy). It has been a matter, finally, of better organisation, by promoting the establishment of centres of excellence (Nordic countries) or of mixed research units (France), the critical mass of which is a result of bringing university researchers and CNRS researchers closer together. The generality of this trend is underlined by the United Kingdom, with its "research assessment exercise", linking government support for university research to the international recognition acquired by "research units", Japan with Monbusho's policy of "centres of excellence" (following the ERATO programme), and Germany, which openly poses the question of the reorganisation of its universities (cf. the conclusion by F. Meyer-Krahmer). The quality of a country's university research is without doubt considered by most developed countries as a key strength of its system of innovation. Even in the United States, D. Mowery emphasises the importance of the structural changes in progress together with the continued growth in the contribution of universities to the overall national effort, and above all the continued drop in their dependence on federal government funding. Numerous national analyses (in the Netherlands, Finland, and France, to name but a few) stress the significant developments experienced over two decades, trends which combine both more pronounced international academic integration and a multiplication, not to say an explosion, of the links between university research and the socio-economic world. This is what we have sought to articulate, by speaking of a new paradigm and by proposing the term "collaborative academic research".

This priority should not overshadow the very different modalities of organising and steering this research. In certain countries, the universities fall under the national responsibility, in others, such as Germany, they are the responsibility of "states" or "regions". Whatever the preferred form of organisation, in most of the countries studied, growing regional intervention can be observed. Indeed, for certain of them, such as Spain, a progressive transfer of responsibility from the central or federal state to the regions is visible. Despite these differences, it is striking to see the importance that all central or federal governments attach to university research. In this respect, the modalities of intervention still differ greatly from one country to another: the strong financial autonomy of the universities in Germany, can be contrasted with the two very different models of the Anglo-Saxon "research councils" on the one hand, and of the French CNRS on the other. In the former case, financial support is awarded to individual projects, selected after frequently discipline-specific calls for tender. In the second, the CNRS provides human and structural support for joint projects by means of mixed units, created on the basis of a "bottom-up" initiative and after a disciplinary evaluation. In this sense, the French CNRS can be distinguished, to a large extent, from its counterparts in Germany (*Max Planck Gesellshaft*), Italy (CNR) and Spain (CSIC) by the fact that the near totality of its researchers work on university campuses in mixed teams. It constitutes, in a way, an original response to the question of the long term management of public research competences. This questions is increasingly clearly posed and to which RIKEN in Japan offers a different response, by increasing the number of visiting positions for academics.

This widespread emphasis on university research has gone hand in hand with the reform of "government laboratories", most of which are dedicated to research in line with a specific mission and which have progressively increased in number in all countries. The 1980s have been characterised by the two contrasting routes taken by France and the United Kingdom in their respective reforms. In both these cases, it was a matter of strengthening their autonomy and of dissociating them from the Ministries, which had originally set them up. In both cases, the intention was to open them up more widely towards society and the economy, and to question their "sectoral monopoly" substituting it for open contractual relations. However, the paths followed have been very different. In the United Kingdom greater importance has been attached to the creation of "agencies" oriented towards the service of their operational clients, while, in France, the emphasis was on the professional nature of their research activities, giving researchers from these organisations the same status as those in the CNRS. Even if this has led, 15 or so years later, to very different profiles, these organisations still constitute an important part of the national public research potential in both countries. Numerous other countries have not eluded this trend, such as Spain (with the status of OPI, cf. the developments by E. Muñoz), or Italy with the reform of several sectoral institutions, while certain others, such as Finland, have chosen a third path, as illustrated by the German *Fraunhofer Gesellschaft* or TNO in the Netherlands, which combines basic public support and strong socio-economic involvement. Japan is also affected by this trend, and is marking its entry into the third millennium with a complete overhaul of the status of both its researchers and organisations. In the United States the debate on the national laboratories has been important, although to a limited extent, since, as B. Bozeman and J. Dietz point out, the status quo has prevailed and the measures taken (CRADA) have had little effect on their trajectory. Thus, these institutions, whose role and activities have been under careful examination for some time, have emerged renewed, stabilised, and in certain countries, strengthened from the troubled times of the 1980s and the beginning of the 1990s. This "renewal" probably stems to a large extent from their capacity to offer "independent expertise" faced with the growth in the political importance of health problems, and also faced with an increasing number of food and environmental crises and the public debates they have generated. Moreover, these organisations appear as a way to restabilise public relations, which are absent in the new organisation of largely deregulated public services which are operated by "global" enterprises, as witnessed in the cases of communications and energy.

To an extent, these movements contradict the analyses carried out by the proponents of the "new knowledge-based economy". The "new mode of knowledge production" (Gibbons, 1994) has not limited the role of the universities by making them a partner amongst the others. By transforming their practices, in developing their organisation, universities appear to have had their centrality strengthened in the majority of national system and policies. To grasp their role, it is necessary to implement a small shift in Arrow's theory of the public reservoir. It is no longer a public reservoir of knowledge, where firms come to fish for the knowledge they need, but a public reservoir of competences mobilised by actors in society, both public and private, to help them resolve the problems they face. The evolution of the "government laboratories" shows that this reservoir is not just limited to "the university sector" but that it also encompasses "sectoral" research institutions and agencies. The majority of these have undergone changes in their operation to constitute what is increasingly appropriately named as "public sector research" (Senker et al., 1999).

FROM NATIONAL POLICY TO THE CO-ORDINATION OF PUBLIC INTERVENTION

The repositioning of technological interventions, the marked evolution of the role of defence research in several countries, the focus on universities and the reorganisation of public sector research reflect the importance of the changes that have been in progress for less than a decade, and that are radically transforming the landscape which resulted from World War II. To us, these transformations seem to embody two tensions, which, at first sight, are opposed to each other. On the one hand, they emphasise the globalisation of activities, which may concern firms or basic research, as witnessed by research on climate or the sequencing of the human genome, and indeed even abortive projects such as ITER. On the other hand, they accentuate the specificity and proximity, which may affect SMEs or universities, of which, in the vast majority of cases, the pool of attraction rarely extends outside the region where they operate. How, then, can public interventions be developed, which are capable both of promoting greater links on the local level, of which work on the Italian industrial districts has underlined the results, and of contributing to the organisation of a global framework, to enable the promotion of the global competitiveness of national firms? The federal countries, the United States and Germany, have *de facto* proposed a path, which has been followed by numerous European countries.

Two key words describe this: decentralisation and subsidiarity. The first accentuates the need to transfer actions to public structures close to the territory of operation. Spain, France and Italy have consequently favoured the birth of new territorial organisations. These have very quickly adopted education and research as the two principal priorities for the activities which they may have been delegated. Other countries, such as the United Kingdom and Japan, have sought similar effects through a decentralisation of the administration. Even small countries such as Finland have placed emphasis on territorial organisation of public support for the innovative capacity of firms. The concept of subsidiarity, in this respect, has been mobilised in Europe to take account of the redistribution of responsibilities between member states and their joint establishments, the European Union and Commission. Interest in the concept of subsidiarity relates, to a large extent, to its ambiguity and the capacity for learning that it has enabled and still enables. It has often been considered that the community level should deal only with matters that have not been dealt with by member nations and their governments or which have been explicitly delegated to it (such as common agricultural policy). Over time, the acceptance of the concept has evolved to reflect the existence of an "optimal level" of treatment for each problem, that is to say, of a public authority, which better enables others to intervene in an efficacious manner. For research and innovation policies, there are two significant, dimensions, which are still only just emerging. On the one hand, there is recognition of the plurality of interventions and of relevant public authorities. Strong and symbolic evidence of this can be seen in the fact that, for the period 1999-2003, the Structural Funds (that is to say the funds destined for the less-favoured regions of the Union) are, for research and innovation policies alone, more significant than those for "direct" technology policy in the Fifth Framework Programme. On the other hand, there is the awareness that responsibilities are not allocated for once and for all. The issue of the "adequate" level, or more exactly of the "good allocation of public responsibilities" requires *ad hoc* answers and has, thus, to be reconsidered for each new problem. There is no easy path, although there are numerous examples to show that advances have been made in this trend. As a result, the Union has progressively taken a *de facto* leadership, which is currently uncontested, in public interventions to promote information and communications technologies. As a counter-example, when considering the way in which public interventions on genome sequencing have been organised, it is possible to assess the way ahead. Similarly, in most places, and in particular in the United States and in Germany, states and regions have strongly invested in support for SMEs, even if they were previously not interested or had no institutional responsibility for it (as in France). These few examples, drawn from various countries, show that it is more and more difficult to group public intervention and national policy together. Consequently, the issue arises of the coherence and co-ordination of public interventions. In the very large majority of the countries, the "governance" of the public systems has thus become a central preoccupation. This has led, in the majority of countries, to profound reorganisations, both in structures and steering instruments.

The only exception is the United States, as underlined by D. Mowery in his conclusion: there has been a "remarkably limited extent of any change in the structures and institutions through which science and technology policy is formulated" while "virtually all the central components of the innovation system that emerged in the post-war US economy are undergoing change". Furthermore, D. Mowery mentions the potential importance for the functioning of the various public components of the "government performance and results act" which is still in its "early stages". In several countries, the beginning of the 1990s is characterised by the development of closer links between the Ministries in charge of research and those managing higher education. This trend was initiated by Germany with the transformation of the BMFT into the BMBF, and was then followed by France.[1] It continued with the creation or the overhaul of the Ministries in charge of higher education and research in countries like Italy and Korea. Finally, a similar amalgamation of the Science and Technology Agency and *Monbusho*, which is in charge of education, will, in 2001, complete the profound administrative transformation being undertaken in Japan. This latter is accompanied by an overhaul of the central advisory structure close to the Prime Minister, of which Y. Sato details both its former role and its place in the shift from a sectoral to a national policy for research and innovation. Similarly, E. Ormala brings out the importance of the Science and Technology Policy Council in the transition that Finland has undergone, in moving from an economy based on natural resources to an economy based on new communications technologies. In both these cases, this steering structure, under the direct chairmanship of the Prime Minister, brings together those responsible in the Government and external personalities, building communication between the political world and the world of research and innovation. This approach, which involves all the Ministries concerned, seems to have been more productive than approaches favouring the existence of an administrative co-ordination structure, generally placed close to the Prime Minister, as this has been the

1. A Ministerial change at the beginning of 2000 has reintroduced two different Ministries, while retaining only one administration.

case notably in Spain and the General Secretariat in charge of implementing the National Plan for Research. Other countries have chosen not to tamper with pre-existing structures, probably because they have already experienced the limitations of administrative transformations linked to short-term policy objectives. Their preference has been to emphasise tools to promote concertation between the actors involved and an appreciation of the effects of public activities. The United Kingdom has thus been an instigator in Europe of the evaluation of research policies and then of foresight approaches. The United Kingdom Foresight Programme currently constitutes backbone of a coherent development of public interventions, leading to what L. Georghiou terms "foresight culture" in analysing the transformations induced. Similarly, since the 1970s, the Japanese have been anticipating and preparing for the future, by renewing their foresight exercise and enlarging its horizons every five years. The Netherlands, with the body of councils and other structures which populate what B. van der Meulen and A. Rip term the "intermediary layer", offer another institutional arrangement, which promotes the "bottom-up aggregation" of interests and anticipations, facilitating co-ordination between actors and preparing the articulation of the policy.

This convergence is not concerned only with the central operation of the Government, which is too often sliced into "sectoral" policies, which simply add together rather complementing each other. It also concerns the different levels of intervention. In their chapter, P. Caracostas and U. Muldur present the development of the preparatory procedures of the European Framework Programme, and bring out the importance of its constituent consultations. They underline, in this development, the challenge of co-ordination between national policies and European policy: it is not sufficient for the principles to be embodied in treaties (as has been the case since 1987) in order to achieve co-ordination. It has been built up progressively, and in the majority of European countries, the preparation of the next Framework Programme has become a significant aspect of national policies. The same trend can be observed in the relations between government and regions. Germany has a long tradition of concertation between the Federal Government and the *Länder* for all shared policies, of which higher education is part. Spain, with the autonomy of the regions, has witnessed a progressive transfer towards the regions of the responsibility for universities. Similarly France has accompanied its decentralisation with the use of pluriannual contracts between the regions and the state to co-ordinate their interventions. Largely centred on transport infrastructures, the *contrats de plan État-région* (which have existed since the beginning of the 1980s and were renewed in 2000 for seven years) have progressively accorded an increasingly important role to higher education and to research.

The co-ordination of public interventions has thus become a central question on the policy agenda of all the countries studied in this book, with the exception of the United States. However, B. Bozeman and D. Mowery both agree in saying that the stability of the organisational arrangements allow intersectoral activities, as in the case for the Presidential initiative promoting "information highways". It is possible to see here the other side of the same reality, since in one case, there is an established institutional framework, operating for more than half a century, while in the other, Europe, where the framework is not yet fully built, its construction entails a complete redesign of the structures and a redefinition of the very concept of national policy.

Such issues challenge the perspective of the system of innovation, founded on the triangular relations between university, government and industry, often known as the "triple helix". This line of inquiry is commensurate with the re-examination of each term used here. We have attempted to show that it was, without doubt, necessary to substitute the notion of university, with a richer and also more polysemic notion of "public sector research" embracing universities and other public research institutions (the famous "government laboratories"). Likewise, the analysis of the interventions promoting industrial development, shows a growing differentiation between, on one hand, a limited number of "global" firms, which are less and less connected to any particular nationality, and the rest of the industrial fabric, which is largely constituted by SMEs, most notably in the local territorial area. Finally, the analysis of a Europe, which is still under construction clearly exposes the limits, at least with regard to research and innovation, of the nation–state, tucked tightly between a "continental" level, which is in a better position to tackle the questions of the worldwide organisation of exchanges and to deal with the problems considered as truly global (such as climate change), and a "regional" level, which is more appropriate for taking charge of the problems of "small" individual actors (such as SMEs) or collective actors (such as the municipalities and the universities). This analysis could be made more complex still, if one takes into account, as we propose in the analysis of the French situation, the new collective actors, such as the patients' associations or other non-profit organisations. This is probably another significant lesson of this book, which is to re-examine the simplifications made, which have difficulty in grasping the richness of the present dynamics of change.

RESEARCH AND INNOVATION POLICIES AS KEY TO THE DOOR

Throughout this work, as in this conclusion, we have collectively endeavoured to remain focused on public policies. It is these that have led us, through their transformations, via the evolution of the questions that they deal with, and via the web of relationships that they weave, to the analysis of the different components of what can conveniently be called a national system of innovation or more simply a system of innovation. We feel, quite obviously, that we have left aside many important questions, such as those of the financing of firms, or the global organisation of markets; but equally we are certain, at least in part, of having brought out the very significant transformations which seem to us to characterise the 1990s and which lead those areas studied to convincingly distance themselves from the model inherited from World War II. Similarly, it is striking to see how much Nelson's critique that analyses "play down the existence of a ... well-structured and thought through general policy" no longer seems to apply, and that the search for coherence and the creation of a co-ordinated approach to interventions have become central in the political agendas. Without doubt, the critique has turned out to be profitable, and has led the advisors to our current policy makers to become more concerned with it. The analysts also have an indirect look-out role, endeavouring to highlight the changes, which are now only faintly sketched out, but which seem to them to have to play a strong role in the coming years. Each of us in our own way, the authors in concluding their chapters, and the editors in this conclusion, we have sought to draw attention to them.

REFERENCES

Callon, S., 1995, *Divided Sun*, Stanford: Stanford University Press.

Gibbons, M. et al., 1994, *The New Production of Knowledge*, London: Sage.

Senker, J. et al., 1999, *European Comparison of Public Research Systems*, Final Report, TSER Programme, Brighton: SPRU.

BOXES, FIGURES AND TABLES

THE UNITED STATES

JAPAN

KOREA

THE EUROPEAN UNION

GERMANY

THE UNITED KINGDOM

THE NETHERLANDS

FINLAND

SPAIN

ITALY

FRANCE

INDEX